20 044 459 55

AF606194

Power
Systems
Harmonics

Power Systems Harmonics

Computer Modelling and Analysis

Enrique Acha
University of Glasgow, UK

Manuel Madrigal
Instituto Tecnológico de Morelia, Mexico

JOHN WILEY & SONS, LTD
Chichester ▪ New York ▪ Weinheim ▪ Brisbane ▪ Singapore ▪ Toronto

Baffins Lane, Chichester,
West Sussex, PO 19 1UD, England

National 01243 779777
International (+44) 1243 779777
e-mail (for orders and customer service enquiries): cs-books@wiley.co.uk

Visit our Home Page on http://www.wiley.co.uk or http://www.wiley.com

Other Wiley Editorial Offices

John Wiley & Sons, Inc., 605 Third Avenue,
New York, NY 10158-0012, USA

WILEY-VCH Verlag GmbH
Pappelallee 3, D-69469 Weinheim, Germany

Jacaranda Wiley Ltd, 33 Park Road, Milton,
Queensland 4064, Australia

John Wiley & Sons (Canada) Ltd, 22 Worcester Road
Rexdale, Ontario, M9W 1L1, Canada

John Wiley & Sons (Asia) Pte Ltd, 2 Clementi Loop #02-01,
Jin Xing Distripark, Singapore 129809

Library of Congress Cataloging-in-Publication Data

Acha, Enrique
Power systems harmonics: computer modelling and analysis / Enrique Acha, Manuel Madrigal.
p. cm. (Wiley series in software design patterns)
Includes bibliographical references and index.
ISBN 0 471 52175 2
1. Power systems harmonics – Computer simulation. 2. Harmonics (Electric waves). I. Madrigal, Manuel. II. Title
TK3226. A24 2001
621. 3199–dc21 2001017863

British Library Cataloguing in Publication Data

A catalogue record for this book is available from the British Library

ISBN 0 471 52175 2

Produced from LaTeX files supplied by the authors.
Printed and bound in Great Britain by Antony Rowe, Chippenham, Wilts
This book is printed on acid-free paper responsibly manufactured from sustainable forestry, in which at least two trees are planted for each one used for paper production.

Contents

II CONVENTIONAL POWER PLANT EQUIPMENT 173

Foreword

Ancient history

Electric power quality is an *ancient subject* – at least by electrical engineering standards. The subject deals with corruption of the power supply in such a way that loads are disrupted, and system characteristics are degraded. This has been a concern since Edison's day. The main issues of power quality have generally been confined to short term phenomena particularly voltage flicker, momentary events, and harmonics. Reliability and long term effects such as outages of more than a few seconds have been studied elsewhere in power engineering. Also, certain gross effects such as those that arise from transmission line construction weaknesses have *not* usually been included in *power quality* engineering. The mainstay of power quality engineering has come to focus on AC waveform distortion, and, in essence, this is the subject of this book. The traditional approaches of distribution engineering have been phasor analysis and the manipulation of sinusoids to suit the needs of the problem at hand. The resurgence in interest in power quality analysis is due mainly to the proliferation of the high power ($P > 1$ kW) semiconductor switch in power flow control. Advances in transform theory and operational calculus have also led to new discoveries in analysis methods and new solutions to old problems. Further, pressures of deregulation, as well as new advances in modeling, have also stimulated interest in the power engineering community on the subject of electric power quality. Operational calculus and matrix techniques for the analysis of power system harmonics are the focus of this book.

Harmonics and matrices

J. B. Fourier discovered many years ago that many periodic functions $f(t)$ can be expressed as the sum of sines and cosines – a Fourier series – each of whose terms are harmonics of integer multiples of the fundamental frequency. There are some periodic functions that do not possess Fourier series, but these are rarely of interest in electrical engineering. The Fourier series is the core of analysis of nonsinusoidal functions (including the aperiodic case which is analyzed by the Fourier *transform*). The basic concept of the Fourier transform is simple enough: utilize a kernel of $\mathrm{e}^{-\mathrm{j}\omega t}$ in an integral transform of $f(t)$ and one finds, effectively, a sum of terms that are integrals of the Taylor series expansion for $\mathrm{e}^{-\mathrm{j}\omega t}$ each times $f(t)$. This central idea has not changed very much over the years except for the inclusion of an integrating factor in the integral transform (i.e., the Laplace transform) and various manipulations of the Fourier–Heaviside transform concept, including methods to obtain the real part of $F(\omega)$ given the imaginary part, and interesting effects of algebraic operations under the Fourier transform. Examples of the latter are the fortuitous conversion of linear differential equations in time to algebraic equations – and the reduction of the convolution integral in the time domain to a simple product in the frequency domain. There have been several theoretical advances such as the Heisenberg uncertainty effect which places a limit on the accuracy of the evaluation of Fourier components in practical instrumentation. From a computational point of view, the main advance in the last 50 years has been in the use of a discrete form of the Fourier transform, and the utilization of coefficients of the discrete Fourier series in the form of operational matrices. In this book, these key elements are focused on power system issues of AC circuits. The concept of operational matrices refers to the manipulation of coefficients that describe function $f(t)$ rather than working with $f(t)$ itself. Operations on $f(t)$ (e.g., squaring, or passing $f(t)$ through a saturation function) are accomplished by studying how the cited coefficients of $f(t)$ change as the function undergoes the given operation. As a quick example, $f(t) = \cos(at)$ when squared becomes $1/2 - 1/2\cos(2at)$. Thus the descriptor $[0, 1, 0, 0, \ldots]$ for $\cos(at)$ becomes $[1/2, 0, 1/2, 0, 0, \ldots]$. The way the descriptor maps under various operations and nonlinearities describes, in effect, how voltage and current signals propagate through an electrical network – even when the network contains nonlinearities and rotating elements. The concept of operational matrices is a concept of a matrix map that is used to replace simple linear and nonlinear maps such as differentiation and squaring. Mapping one vector in, for example, the time domain to another vector in the Fourier domain, to another vector in some other domain may give the user advantages in calculation. As an example, electric power $p(t)$ often gives problems because it is a nonlinear function of voltage and current (namely, the product $v(t)i(t)$). If a domain is constructed in which power is treated as a new state variable with special properties, relatively simple solutions methods might be used for engineering calculations. If power were mapped to a current, for example, perhaps a pseudo-circuit could be constructed to study how power flows. (Incidentally, this has been done!)

A strength of the present book is the recognition and documentation of important patterns in operational matrices and how these patterns may be used to solve complicated problems in power engineering. The subject of the discrete Fourier domain is less of a focus in this text – perhaps because discrete Fourier mathematics have filtered down to the undergraduate engineering curriculum, and most students are familiar with discrete frequency methods. However, the real applications are likely to be in the discrete Fourier domain.

Computational methods

Traditionally, power engineering has been dominated by computational methods. This is the case because power systems are large, they are complex, and they are not easily analyzed using purely theoretical methods. This fact, coupled with the advent of MATLAB™ for engineering calculations, makes a natural marriage. In MATLAB™, all quantities are assumed to be complex matrices and this is ideal for the operational calculus cited above. In the text, actual MATLAB™ code is shown and syntax is illustrated. Modern day students will identify with these explanations. There may be some shortcomings to overdependence on computational methods because some unseasoned students may have difficulty in recognizing erroneous solutions. However, graphical representations and observance of certain patterns of solutions will help comprehension, debugging, and identifying errors.

The power of the methods

Some contemporary electrical engineering students feel that power engineering is too rooted in the past. These students need to examine carefully the power of operational matrices and discrete frequency methods in AC circuit analysis. Especially when rotating elements are included, the complex models used, and the elegant – if not efficient – solutions using these advanced methods, have an appeal even to the beginner. It is interesting to conjecture whether nature actually functions in this way, but the important result is that operational matrices and harmonic expansions are powerful in numerical solutions of AC circuit problems. Effectively, the methods of this book are applied signal processing. Perhaps one of the most important factors is the ease of design when operational matrices are used: interaction of harmonic signals, nonlinear processing of $v(t)$ and $i(t)$, and long term solutions are difficult to consider in conventional domains for the design of harmonic filters, active filters, power conditioners, and other power quality enhancements. When the power of operational methods is utilized, the designs often are much simpler. The disadvantage of some solutions, however, is that familiar time domain quantities may be elusive – or less evident. The engineer needs to evaluate this point on the merits of each case. One way to avoid this difficulty is to approach all power quality problems consistently using operational methods, and base all component models on the same technique. There may be a parallel to a famous quote of Dr. R. Bracewell who commented on a complaint that the Hartley transform, an all-real transform similar to the Fourier transform, was less intuitive and less physically realizable than the Fourier transform. Bracewell simply pointed to the complex operator, j1, and queried as to the intuitiveness and realizability of this bane of all electrical engineering students!

One step beyond

The power of operational methods for harmonic analysis suitably praised above now raises the question of "to where from here?" One issue that needs attention is aperiodic (e.g., transient) analysis. Another is the issue of noise in engineering systems in general. Another is the appearance of chaotic signals. In these cases, operational methods will work, but there are issues of how many terms to retain, how to accurately model the given phenomenon, and how the solution methods parallel (or do not parallel) harmonic analysis. Also, the reader needs to evaluate alternative domains to obtain the best window on the problem at hand. The early French mathematicians referred to transforms (and, after all, operational matrices are a form of transform analysis) as "images". These images are different as viewed through different windows, and the challenge is to select the best transform view for the application.

There are potentially valuable results attainable from the eigenanalysis of operational matrices. This type of analysis is usually found in the differential equations to find the modes (i.e., time constants) of components of the solutions. There are advantages in algebraic equations as well: modes in the solution might be associated with physical processes and the design of system components might be assisted. In power engineering, this means the eigenstructure of the operational transformations may give insight into the design of filters, loss reduction mechanisms, and even the attenuation of interference. Eigenstructure is also the basis of the symmetrical component transformation. In this case, the modal matrix of a three phase impedance matrix of the form

$$\begin{bmatrix} s & m & m \\ m & s & m \\ m & m & s \end{bmatrix}$$

is the symmetrical component transformation. This transformation has been extended to the polyphase case. The all-real form of the modal matrix is Clark's transformation and these advances beyond the basic Fortescue symmetrical component transformation have been applied extensively to AC power circuit analysis (e.g., in my book on "Computer Analysis Methods for Power Systems," second edition, Stars in a Circle Publications, Scottsdale, AZ, 1996). What has not been done is to examine the eigenstructure and modal analysis of operational matrices as carefully set out in this present book.

The topics illustrated in this book are certainly motivational and elegant, and it seems that some reader is likely to take a step beyond in advancing solution methods, not only for power quality, but for all areas of dynamic systems.

G.T. Heydt
Arizona State University

Preface

The face of electrical power systems is changing rapidly with the widespread use of power electronics technology in electrical power networks. This new technology has brought about many technical and economic advantages, but it has also introduced new challenges for power engineers. Power engineers must be trained to manage and operate electrical power networks with a high content of power-electronics-based equipment, where one of the challenges is the control and prediction of harmonic voltages and currents. To understand the problems introduced by harmonics and to take informed action at the operational and planning levels, power engineers require a solid understanding of how harmonics are generated and propagated in electrical power networks. From the theoretical viewpoint, the exercise involves the co-ordinated study of electric circuit theory, Fourier analysis, power electronics and power systems analysis. This book on power systems harmonics takes a step towards meeting this educational need.

The book offers the reader a thorough theoretical grounding on the topic of modern power systems harmonics modelling and simulation. The book is self-contained and organised in three main parts, namely Part I: Fundamental Theory, Part II: Conventional Power Plant Equipment, and Part III: Power Electronics Equipment.

Part I covers periodic signals and circuit analysis operating under non-sinusoidal conditions, a topic of great importance in electrical engineering which is normally overlooked in power systems education and training. This is followed by a thorough and systematic study of harmonic domain techniques using both Fourier and Hartley series. This opens the door for the reader with a traditional power systems background to gain insight on the use of alternative, more efficient harmonic domain techniques than those afforded by the Fourier series.

Part II addresses the harmonic modelling and analysis of conventional power plant equipment. The following plant components are studied in detail: the synchronous generator, multi-conductor transmission lines, the power transformer, laminated iron cores and electric arcs. The treatment of the various plant components follows the basic harmonic domain principles covered in Part I of the book.

Part III addresses the harmonic modelling of power electronics equipment. It includes novel material, which may be used to study complex harmonic phenomena which are of current concern to the modern power systems industry and equipment manufacturers. Emphasis is made on the modelling and analysis of FACTS and custom power equipment, a new application area in power systems which is having a huge impact on high-voltage transmission, low-voltage distribution and even industrial and residential systems world-wide.

The book is aimed at both undergraduate and postgraduate students. Practising engineers will also benefit from the material contained in the book. To encourage a hands-on approach to the use of the theory, plenty of simple numeric examples are provided throughout the book. MATLAB™ functions are used to solve the numeric examples and in the modelling of conventional and power electronics equipment.

The book was prepared at the University of Glasgow while M. Madrigal was on leave from the Instituto Tecnológico de Morelia, Mexico. The manuscript was prepared using the LyX(version 1.1.5) document processor running under Linux (readhat 6.2) installed on a Toshiba Satellite Pro 4300 laptop computer.

The authors would like to thank J. Arrillaga, N. Rajakovic, J.J. Rico, A. Semlyen and R. Yacamini for key contributions to the area of power systems harmonic analysis, many of which have provided a source of learning and inspiration to the authors. Special thanks go to G.T. Heydt for his kind contribution to our book in the way of a most illuminating Foreword.

We would like to thank our colleagues at the Power Engineering Group at the University of Glasgow, UK, for their support and assistance during the various stages in the preparation of the book.

We are also grateful to the staff of John Wiley & Sons, particularly Kathryn Sharples for her continuous encouragement throughout the preparation of the manuscript.

Enrique Acha
Manuel Madrigal

Part I

FUNDAMENTAL THEORY

1

Introduction

1.1 Background

The problem of power systems harmonics is not a new phenomenon by any means. It is as old as the power network itself, with the first studies on the subject traced back to the 1890s, when transformers and rotating machinery were identified as the main source of waveform harmonic distortion.

It is reported [1] that in 1893, Charles Proteus Steinmetz arrived in Lynn, Massachusetts, to work for Thomson-Houston. One of his early assignments was to solve a motor heating problem at Hartford, Connecticut. Steinmetz had the mathematical background to conduct harmonic analysis and made good use of it to identify and solve the problem, which was traced to the combined action of a resonance in the transmission circuit feeding the plant and a generator with a substantial amount of waveform distortion. To analyse this problem, he used a device called a wave-meter to construct electric waveforms and, not without a great amount of work, carried out Fourier analysis. Reportedly, it took him about one hour to produce each coefficient of the Fourier series. Fourier analysis is a very ingenious mathematical tool, first published by the mathematician Joseph Fourier in 1822 [2].

A further harmonic problem associated with generator waveform appeared in the literature. The problem came in the form of excessive circulating currents between solidly grounded generators operated in parallel. Today, this is a well-known problem involving zero-sequence third harmonic voltage in star connected machines. In 1910, an additional harmonic problem arose as a result of harmonic interference in communication circuits sharing a common right-of-way with power lines.

At that time, a reduction in harmonic distortion was achieved through a series of practices such as improved generator and transformer designs, improvements in grounding methods and the use of suitable three-phase transformer connections. Transmission lines were kept as short as possible following generator manufacturers recommendations, who described their products as being limited to transmission lines of a certain length, e.g. 5–10 mi, because of generator's wave shape. Such measures were technically and economically successful thanks to the operational and structural properties of the early power networks. From the utility viewpoint, harmonic distortion was no longer considered to be a problem and for several decades the phenomenon was generally ignored within the power community – harmonic analysis was generally considered as a theoretical complexity rather than a practical engineering tool.

In the 1960s, there were many individual shunt capacitors used for reactive power compensation on a large number of industrial power systems, all of them installed without much concern for harmonic resonance problems. This apathy among power engineers began to change when significant resonance problems were encounted, the problem being traced to the harmonics generated by the recent introduction of static drives. In fact, traction applications had been using thyristors and diodes since the late 1960s, a trend that was exacerbated in the following decade with the development of faster, cheaper and more reliable power electronics valves. The improved technology found great favour in self-commutated applications such as industrial choppers and inverters. By that time the trend was irreversible and the development of the Darlington and Triplington bipolar power transistors in the early 1980s, together with the development of the gate-turn-off thyristor (GTO) and the insulated gate bipolar transistor (IGBT), saw power electronics become the inseparable companion of electrical power systems.

Due to the unique technical advantages afforded by power electronics equipment in voltage and power control, in power quality improvement and in process control, it has permeated all sectors of the electricity supply business, namely generation, transmission, distribution and utilisation. However, an undesirable side effect of power electronics equipment is that it achieves its intended function at the expense of generating harmonic distortion. In this respect, power electronics equipment may be seen as a pervasive load and, coupled with the traditional sources of harmonics such as saturated transformers, synchronous generators and electric arc furnaces, has become a new challenge to power systems engineers who have to operate and supply electricity to a large and varied number of loads characterised by the risk of causing major harmonic distortion in the power system. The problem that Steinmetz faced in 1893 has resurfaced, but in today's context the harmonic problem has grown in complexity. It is fortuitous that at this time we are better prepared to control the harmonic problem, with advanced software and hardware methods, new instrumentation and new equipment.

Paradoxically, power electronics equipment, the root cause of today's harmonics problems, are now being used very effectively to enhance the reliability of supply and to deliver sinusoidal voltage and current waveforms to the end user. In fact, power electronics equipment is being used to enhance power quality in general, of which power systems harmonics is a subset.

1.2 Harmonics in Power Systems

The voltage and current waveforms in AC power circuits are expected to be sinusoidal, with constant amplitude and frequency. However, to a greater or lesser extent, all power plant components possess the undesirable property of introducing distortion into the AC power circuit causing the voltage and current waveforms to deviate from their otherwise intended sinusoidal form.

Under the assumption of a constant waveform distortion, the voltages and currents may be decomposed into a set of sinusoidal waveforms of varying amplitude and phase, with frequencies that are integer multiples of the fundamental frequency. In Fourier analysis, the frequency multiples of the fundamental are termed harmonic frequencies.

In spite of their long history and the potentially damaging effects that harmonic distortions may cause to utility and customer equipment alike, until quite recently this was a topic considered to fall within the realm of groups of specialists drafting standards and recommendations, conducting research and providing consultancy services. However, the increased presence of harmonics and the threat that the phenomenon represents to security of supply, together with the associated energy and economic losses, is begining to raise awareness among a wider community. Their effect on equipment may be attritional, although power factor capacitors are known to have developed terminal failure following ferroresonance exposure. Harmonics will induce overheating in wiring, motors and transformers resulting in premature breakdown in insulating material and in a considerable reduction in the useful life of rotating machinery and transformers.

Equally important, they interfere with highly sensitive processes in industry governed by programmable microprocessors. Variable speed drives used in assembly lines are prime targets for harmonics, and so are robots and numerically controlled machines.

1.3 Power Quality

Power quality (PQ) is an umbrella term that embraces all aspects associated with amplitude, phase and frequency of the voltage and current waveforms existing in a power circuit. Adverse PQ environments may result from transient conditions developing in the power circuit, the installation of one or more non-linear loads of relatively large rating or large numbers of non-linear loads of low rating.

Many years ago, apart from power factor and flicker concerns, PQ problems were confined almost entirely to the sphere of regulated, public utilities. However, over the last decade, the issue of PQ has gained renewed interest due to the increasing use of loads sensitive to service quality, e.g. computers, industrial drives, communications and medical equipment. In its present form, PQ is an even more complex problem than in the past because the new loads are not only sensitive to service quality but also responsible for affecting adversely the quality of power supply. It is widely acknowledged that both customer and utility equipment will suffer undue stress in an unfriendly PQ operating environment. PQ disturbances at the utility level may result in the maloperation of remote control, protective devices and energy metering; overheating of cables, transformers and rotating machinery. PQ disturbances in the customers premises may result in the loss of computer data, due to voltage sags with a duration of only a few milliseconds. Even shorter voltage sags may cause the tripping of industrial drives. This has prompted professional bodies to enact recommended practices in order to limit the severity and number of disturbances to customer equipment.

A large number of PQ disturbances have been reported in the open literature, with some of them being transient in nature and others being related to periodic, steady-state operation. There is no general agreement between the various authors on a unique classification of PQ disturbances but some of the more common disturbances are: voltage and current harmonics, voltage sags (dips), swells, electric noise, impulses, notches and flicker.

1.4 Custom Power

The concepts of flexible AC transmission systems (FACTS) and custom power were both developed in the late 1980s by N.G. Hingorani [5,6]. Since then, they have generated great interest from the major manufacturers of electrical equipment and utilities from around the world. FACTS uses the latest power electronics technology so as to make the high-voltage transmission network electronically controllable. From the operational point of view, it concerns the ability to control, in an adaptive fashion, the path of the power flows throughout the network. The ability to control active and reactive power flows, line impedance and nodal voltage magnitudes and angles at both the sending and receiving ends of key transmission lines with almost no delay increases significantly the transmission capabilities of the network whilst enhancing considerably the security of the system. The most popular FACTS equipment are the static compensator, the thyristor-controlled series compensator, and the unified power flow controller.

Custom power addresses the needs of low-voltage distribution systems. It focuses primarily on the reliability and quality of power flows. However, voltage regulation, voltage balancing, power factor correction and harmonic cancellation may also benefit from this technology. Custom power emanated in response to reports of poor PQ and reliability of supply affecting factories, offices and homes. The vulnerability of modern automated equipment and production lines to PQ disturbances on the utility's network is at the centre of such problems. Such disturbances may be in the form of voltage sags, swells, harmonic distortion, impulses and power interruptions. With custom power solutions in place, the end user should see tight voltage regulation, no power interruptions, low harmonic voltages, and acceptance of rapidly fluctuating and non-linear loads in the vicinity without affecting the terminal voltage.

It must be remarked that although FACTS and custom power share the same technological base, they obey different technical and economic objectives. FACTS equipment is aimed at the transmission level whereas custom power equipment is aimed at the distribution level, in particular at the point of common coupling between the electricity supply company and clients with sensitive loads or independent generators. The most popular custom power equipment are the distribution static compensator, the dynamic voltage restorer, the solid-state transfer switch, and the high voltage direct current (HVDC) light.

1.5 Engineering Tools and Methods

The planning, operation and management of electrical power systems with a significant number of power electronics loads and controllers are complex endeavours that require well-trained engineers, together with state-of-the-art power systems software running in fast computers and powerful and versatile instrumentation. Real-time facilities with hardware-in-the-loop extensions are a new tool that may be used very effectively to carry out event replication scenarios, and the testing of equipment and control boards in tightly controlled PQ environments.

1.5.1 Software tools

The software tools that power systems engineers have at their disposal to study PQ problems, and harmonic problems in particular, rely on complex mathematical models and powerful computer algorithms. An outline of the most successful methods used to calculate the steady-state response of non-linear electric circuits subjected to periodic excitations is presented below.

One of the earliest methods used to calculate the steady-state solution of non-linear circuits was to integrate the accompanying system of non-linear differential equations for as many cycles as required, until the transient response had disappeared from the solution, leaving only the periodic, steady-state response. In fact, any of the highly developed electromagnetic transient programs, such as EMTP [7] and PSCAD/EMTDC [8], may be used to such effect. This solution approach does not always yield satisfactory results because some power networks are lightly damped and because of difficulties in establishing suitable steady-state initial conditions.

Such difficulties provided the motivation for developing alternative, more reliable methods of solution. In power systems harmonics, methods were developed using harmonic balance techniques and shooting techniques.

The majority of power systems harmonics studies reported in the open literature have made use of harmonic balance techniques using Fourier series. Broadly speaking, two modelling approaches have emerged: one where the non-linear plant component is represented by a harmonic current source [9], and secondly where it is represented by either a harmonic Norton equivalent [10] or a harmonic admittance matrix. The former approach is amenable to Gauss-type iterative solutions. The latter approach is amenable to Newton-type iterative solutions, and takes place in the harmonic domain where the cross-couplings between frequencies is explicitly represented. This is a more elegant and comprehensive modelling approach, and it is the one adopted throughout this book.

1.5.2 Hardware tools

PQ covers a wide variety of transient disturbances, which range from transient over-voltages of extremely short duration, such as spikes, to long-term interruptions, which may last for several hours, with harmonics considered to be a steady-state phenomenon. Nowadays, advanced instruments exist that can monitor a number of PQ events with great fidelity, and in real time [11]. In a number of harmonic applications, a harmonic analyser with FFT calculation capabilities may be used to carry out the required measurements.

Off-the-shelf harmonic analysers may be utilised in power harmonic applications but care must be exercised because they are not designed specifically for power frequency sampling [12] and they also may not measure the three phases simultaneously. Alternatively, the development of data acquisition systems based on the virtual instrument concept, such as LabVIEW, yield quite satisfactory results [13][18].

1.5.3 Hardware-in-the-loop

The trend among manufacturers, large utilities and research organisations is to adopt real-time digital simulators to conduct complex power systems modelling and analysis. These simulators solve in real-world time the mathematical equations that govern the dynamic behaviour of the power system, providing continuous output conditions that accurately represent real network conditions. The real-time performance enables actual physical hardware to be mixed with computer models to replicate the total power network under investigation. The closed-loop operation of these simulators provides a very powerful tool for extensively evaluating and accurately testing new and existing equipment under normal and abnormal operating conditions to verify the equipment's performance. These simulators may also be used in an open-loop fashion for past event replication simulations.

PQ is an area where real-time simulators can provide a powerful solution. A case in point is the development of a flexible power waveform generator for investigating PQ-related issues, using an open-platform, multi-purpose real-time simulator [14]. Incorporated in the power waveform environment are voltage amplifiers, which provide in-service voltage levels to be amplified to the equipment under test. The real-time environment enables equipment to be tested in a large number of PQ environments, many of which cannot be performed by any other means, or would not be permitted in the real system.

1.6 Avenues of Research on Harmonic Analysis

Comprehensive harmonic models exist for most of the conventional power systems equipment. These include both linear and non-linear power plant components. The current research drive is in the development of advanced harmonic domain models of FACTS and custom power equipment – in particular, equipment that derives its working principle on the use of voltage source inverter technology.

Operational matrices [15] and orthogonal series offer a more general and systematic mathematical tool for the solution of linear and non-linear time-varying systems, compared with conventional harmonic domain and time domain methods. Operational matrix methods take the approach of transforming the differential equations that describe the time-varying system into algebraic equations. This approach is quite general in the sense that most orthogonal series possess the operational properties of integration, differentiation, product and coefficient matrices. To the best of our knowledge, the following orthogonal series have been used in the periodic, steady-state solution of electrical power circuits: complex and real Fourier, Hartley and Walsh [16]. It is likely that other orthogonal series with interesting properties, which may yield further insight into the periodic, steady-state behaviour of electrical power circuits, may be exploited. Notable among these is the family of wavelet transforms.

An issue of perhaps greater significance is the realisation that there is no need to constrain the use of the harmonic domain method to the solution of periodic, steady-state problems. Subtle but key extensions of the method allow for the direct computation of the transient response in addition to the periodic, steady-state response. The newly conceptualised method may be seen as an extension of the conventional harmonic domain. We would expect to see a marked increase of research activity on this method for it is a natural approach for conducting dynamic and steady-state studies of the evolution of harmonics in circuits containing non-linear components, including FACTS and custom power devices [17]. Preliminary investigations show that during the transient period, the extended harmonic domain yields cleaner harmonic information than solutions that rely on the combined use of conventional time domain and FFT solutions.

Nonetheless, an increase in research activity based on time domain's methods is expected, not only because they have a very traditional appeal among power systems engineers, but also because they lend themselves quite naturally to decomposed structures, which enable the use of parallel processing techniques for speedier computations. Current research in this area concentrates on limit cycle, real-time simulations and modelling of FACTS and custom power modelling.

1.7 Bibliography

1. E.L. Owen, "A History of Harmonics in Power Systems", *IEEE Industry Applications Magazine,* January/February 1998, pp. 6–12.
2. J.B.J. Fourier, *Théorie Analytique de la Chaluer*, Paris, 1822.
3. G. T. Heydt, *Electric Power Quality*, Stars in a Circle Publications, Scottsdale, AZ, 1991.
4. A. Domijan, G.T. Heydt, A.P.S. Meliopoulos, S.S. Venkata, S. West, "Directions of Research on Electric Power Quality", *IEEE Transactions on Power Delivery*, Vol. 8, No. 1, January 1993. pp. 429–436.
5. N.G. Hingorani, "High Power Electronics and Flexible AC Transmission Systems", *IEEE*

Power Engineering Review, July 1988, pp. 3–4.

6. N.G. Hingorani, "Introducing Custom Power", *IEEE Spectrum*, June 1995. pp. 41–48.
7. H.W. Dommel, Electromagnetic Transient Program (EMTP) Rule Book, EPRI, EL 642-1, Vol. 1,2, June 1989.
8. Manitoba HVDC Research Centre, PSCAD/EMTDC: Electromagnetic Transients Program Including DC Systems, 1994.
9. H.W. Dommel, A. Yan, S. Wei, "Harmonics from Transformer Saturation", *IEEE Transactions on Power Systems*, Vol. PWRD-1, No. 2, April 1986, pp. 209–215.
10. A. Semlyen, E. Acha, J. Arrillaga, "Newton-Type Algorithms for the Harmonic Phasor Analysis of Non-Linear Power Circuits in Periodical Steady-State with Special Reference to Magnetic Non-Linearities", *IEEE Transactions on Power Delivery*, Vol. 3, No. 3, July 1988, pp. 1090–1098.
11. J. Arrillaga, N.R. Watson, S. Chen, *Power System Quality Assessment*, John Wiley & Sons, Chichester, 2000.
12. R.C. Dugan, M.F. McGranaghan, H.W. Beaty, *Electrical Power System Quality*, McGraw Hill,New York, 1996.
13. E. Acha, N.P. Johnson, L.L. Loh, P. Miller, "The Impact of Energy Saving Apparatus on Power Quality", *Proceedings of ICHQP 1998*, 14–16 October, Athens, Greece, Vol. 1, pp. 36–41.
14. E. Acha, O. Anaya, J. Parle, M. Madrigal, "Real-Time Simulator for Power Quality Disturbance Applications", *Proceedings of ICHQP 2000*, Vol. III, Orlando, Florida, 1–4 October, pp. 763–768.
15. J.J. Rico, E. Acha, M. Madrigal, "The Study of Inrush Current Phenomenon using Operational Matrices", *IEEE Transactions on Power Delivery*, paper PE-004PRD(09-2000).
16. J.J. Rico, "Steady State Modelling of Non-Linear Power Plant Components", *PhD Thesis*, University of Glasgow, 1997.
17. J.J. Rico, M. Madrigal, E. Acha, "Dynamic Harmonic Evolution using the Extended Harmonic Domain", Submitted to *IEEE Transactions on Power Delivery*, June 2000.
18. N.P. Johnson, T.L. Tan, P. Miller, E. Acha, "A Laboratory Model for Harmonic Measurements in Windpower Generators", *IEEE Power Engineering Review*, Vol. 20, No. 3, March 2000, pp. 48–49.
19. IEEE Industry Applications Society/Power Engineering Society, *IEEE Recommended Practices and Requirements for Harmonic Control in Electrical Power Systems*, IEEE Std 519-1992, April 1993.
20. W. Shepherd, P. Zand, *Energy Flow and Power Factor in Nonsinusoidal Circuits*, Cambridge University Press, 1979.
21. J. Arrillaga, B.C. Smith, N.R. Watson, A.R. Wood, *Power System Harmonic Analysis*, John Wiley & Sons, Chichester, 1997.
22. Y.H. Song, A.T. Johns, *Flexible AC Transmission Systems (FACTS)*, The Institution of Electrical Engineers, London, 1999.
23. N.G. Hingorani, L. Gyugyi, *Understanding FACTS: Concepts and Technology of Flexible AC Transmission Systems*, IEEE Press, New York, 2000.

2

Orthogonal Series Expansions

2.1 Introduction

Orthogonal series expansions play a key role in solving a wide range of problems encountered in physics and engineering. Fourier, Hartley, Walsh, etc., have been used to approximate non-linear characteristics with good results. They can also be used to solve differential equations, resorting only to algebraic means. Because of their very interesting characteristics and high efficiency as integration solvers, they have been researched widely in the fields of system identification, model reduction and control.

In this book the use of orthogonal series expansions is twofold: (1) to extract the harmonic content of non-linear characteristics when subjected to periodic excitations and (2) to provide a generalised frame of reference where all the linear and non-linear elements of the electric circuit are represented together for unified, iterative, harmonic solutions. In principle at least, any orthogonal series expansion may be used to achieve these two aims but in this book only the Fourier and Hartley transforms will be used. Most of the attention will be centred on the use of the complex Fourier transform because its use is more widespread, but the real Fourier transform also receives attention.

A basic understanding of the theory behind orthogonal series expansions is essential for learning the concepts of harmonic domain modelling and understanding the applications presented in the first, second and third parts of the book. In this chapter, the most relevant aspects of Fourier and Hartley series are reviewed. Owing to the very practical nature of the second and third parts of the book, it is unavoidable that there is a bias towards solution by computer. The discrete versions of the Fourier and Hartley transforms are stepping stones towards the efficient algorithms associated with them, namely the fast Fourier transform (FFT) and the fast Hartley transform (FHT). In addition, the concept of convolution is also addressed because of its central role in harmonic domain solutions.

2.2 Orthogonal Functions

Definition of orthogonality

A set of functions $\phi_n(t)$ and $\phi_m(t)$ are said to be orthogonal over the interval $0 < t < T_0$, if the following basic property is observed:

$$\int_0^{T_0} \phi_n(t)\phi_m(t)\mathrm{d}t \begin{cases} = 0 & \forall n \neq m \\ \neq 0 & \forall n = m \end{cases} \tag{2.1}$$

For the case when the integral is equal to zero then $\phi_n(t)$ and $\phi_m(t)$ are orthogonal functions, i.e. $\phi_n(t) \perp \phi_m(t)$.

Moreover, the functions are also said to be orthonormal if the following more restrictive conditions are met:

$$\int_0^{T_0} \phi_n(t)\phi_m(t)\mathrm{d}t = \begin{cases} 0 & \forall n \neq m \\ 1 & \forall n = m \end{cases} \tag{2.2}$$

The Fourier and Hartley series are examples of orthonormal functions.

Example 2-1: Show that the functions $\cos \omega_0 t$ and $\sin \omega_0 t$ are orthogonal functions.

$$\begin{aligned} \int_0^{T_0} \cos \omega_0 t \sin \omega_0 t \mathrm{d}t &= \frac{1}{2}\int_0^{T_0} \sin 2\omega_0 t \mathrm{d}t = -\frac{1}{4\omega_0}\cos 2\omega_0 t\Big|_0^{T_0} \\ &= -\frac{1}{4\omega_0}\left(\cos 2\frac{2\pi}{T_0}T_0 - \cos(0)\right) = -\frac{1}{4\omega_0}(\cos 4\pi - 1) = 0 \end{aligned}$$

Example 2-2: Verify that the set of functions $\cos n\omega_0 t$ from $n = 1$ to ∞ form an orthogonal set.

The product $\cos n\omega_0 t \cos m\omega_0 t$ is given as

$$\cos n\omega_0 t \cos m\omega_0 t = \frac{1}{2}\cos(n+m)\omega_0 t + \frac{1}{2}\cos(n-m)\omega_0 t$$

Since n and m are positive integers then for $n = m$ the integral of $\cos(n+m)\omega_0 t$ over the interval $[0, T_0]$ is equal to zero, and due to the fact that $\cos(n-m)\omega_0 t$ is equal to one, then

$$\int_0^{T_0} \cos n\omega_0 t \cos m\omega_0 t \mathrm{d}t = \frac{T_0}{2} \quad \forall n = m \tag{2.3}$$

For $n \neq m$, the integrals of $\cos(n+m)\omega_0 t$ and $\cos(n-m)\omega_0 t$ are both equal to zero,

$$\int_0^{T_0} \cos n\omega_0 t \cos m\omega_0 t \mathrm{d}t = 0 \quad \forall n \neq m \tag{2.4}$$

The results (2.3) and (2.4) show that the set of functions $\{\cos n\omega_0 t\}_{n=1}^{\infty}$ form an orthogonal set. Following a similar procedure, it can be demonstrated that the same properties of orthogonality are preserved by the set of functions $\{\sin n\omega_0 t\}_{n=1}^{\infty}$ and $\{\mathrm{e}^{\mathrm{j}n\omega_0 t}\}_{n=1}^{\infty}$.

2.3 Periodic Functions

A periodic function $f(t)$ with frequency f_0 Hz and period T_0 seconds is defined as

$$f(t) = f(t + T_0) \tag{2.5}$$

where $T_0 = 1/f_0$. Figure 2.1 shows examples of periodic functions.

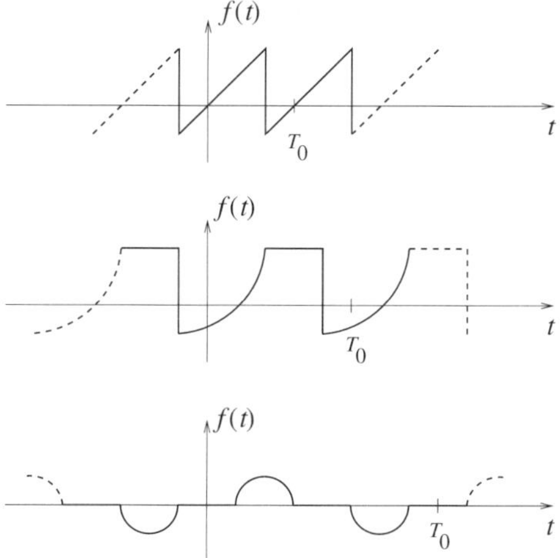

Figure 2.1 Periodic functions

Example 2-3: Find the period of the function $f(t) = \cos t/3 + \sin t/4$ using definition (2.5). Introducing the period T_0 into $f(t)$

$$\cos t/3 + \sin t/4 = \cos(t + T_0)/3 + \sin(t + T_0)/4$$

Since $\cos(\theta + 2\pi m) = \cos\theta$ for any integer m, writing $2\pi m$ for $T_0/3$ and $2\pi n$ for $T_0/4$ we have $T_0 = 6\pi m = 8\pi n$ where m and n are integers.

It follows that when $m = 4$ and $n = 3$, we obtain the smallest value of T_0, i.e. $T_0 = 24\pi$.

2.3.1 *Even and odd functions*

A periodic function $f(t)$ is said to be *even* if the following relation holds true:

$$f(t) = f(-t) \tag{2.6}$$

and it is said to be an *odd* function if

$$f(t) = -f(-t) \tag{2.7}$$

An arbitrary function $f(t)$ can be represented as

$$f(t) = f_{\text{even}}(t) + f_{\text{odd}}(t) \tag{2.8}$$

where

$$f_{\text{even}}(t) = \frac{1}{2}f(t) + \frac{1}{2}f(-t) \tag{2.9}$$

$$f_{\text{odd}}(t) = \frac{1}{2}f(t) - \frac{1}{2}f(-t) \tag{2.10}$$

It is not difficult to see that $f_{\text{even}}(t)$ and $f_{\text{odd}}(t)$ are, respectively, even and odd functions. This means that any arbitrary function $f(t)$ can be represented as the sum of even and odd components. It should be mentioned that the functions $\cos n\omega_0 t$ and $\sin n\omega_0 t$ are, respectively, even and odd functions of time.

Example 2-4: Find the integral of the product of two periodic functions, $f(t)$ and $g(t)$, over the period T_0.

$$\begin{aligned}
\frac{2}{T_0}\int_{-T_0/2}^{T_0/2} f(t)g(t)\mathrm{d}t &= \frac{2}{T_0}\int_{-T_0/2}^{0} f(t)g(t)\mathrm{d}t + \frac{2}{T_0}\int_{0}^{T_0/2} f(t)g(t)\mathrm{d}t \\
&= \frac{2}{T_0}\int_{0}^{T_0/2} f(-t)g(-t)\mathrm{d}t + \frac{2}{T_0}\int_{0}^{T_0/2} f(t)g(t)\mathrm{d}t \\
&= \frac{2}{T_0}\int_{0}^{T_0/2} \left[f(-t)g(-t) + f(t)g(t)\right]\mathrm{d}t \\
&= \begin{cases} \dfrac{4}{T_0}\displaystyle\int_{0}^{T_0/2} f(t)g(t)\mathrm{d}t & \text{if } f(t) \text{ and } g(t) \text{ are both even} \\ & \text{or both odd functions} \\ 0 & \text{otherwise} \end{cases}
\end{aligned}$$

When both periodic functions have even and odd parts,

$$\frac{2}{T_0}\int_{-T_0/2}^{T_0/2} f(t)g(t)\mathrm{d}t = \frac{4}{T_0}\int_{0}^{T_0/2} f_{\mathrm{even}}(t)g_{\mathrm{even}}(t)\mathrm{d}t + \frac{4}{T_0}\int_{0}^{T_0/2} f_{\mathrm{odd}}(t)g_{\mathrm{odd}}(t)\mathrm{d}t$$

These results show that even and odd functions are orthogonal functions. For example, $\cos\omega_0 t$ and $\sin\omega_0 t$ are, respectively, even and odd functions. Hence they are orthogonal functions.

2.3.2 *Half-wave symmetry*

A periodic function $f(t)$ has half-wave symmetry if it satisfies the following property:

$$f(t) = -f\left(t + \frac{T_0}{2}\right) \tag{2.11}$$

Example 2-5: Find the integral of the product of two functions, $f(t)$ and $g(t)$, where $f(t)$ has half-wave symmetry and:

(i) $g(t) = a$,
(ii) $g(t) = \cos n\omega_0 t$,
(iii) $g(t) = \sin n\omega_0 t$ and
(iv) $g(t) = \mathrm{e}^{-\mathrm{j}n\omega_0 t}$,
where $n = 1, 2, 3, \ldots$ and $\omega_0 = \frac{2\pi}{T_0}$.

In general form,

$$\begin{aligned}
\frac{2}{T_0}\int_{-T_0/2}^{T_0/2} f(t)g(t)\mathrm{d}t &= \frac{2}{T_0}\int_{-T_0/2}^{0} f(t)g(t)\mathrm{d}t + \frac{2}{T_0}\int_{0}^{T_0/2} f(t)g(t)\mathrm{d}t \\
&= \frac{2}{T_0}\int_{-T_0/2+T_0/2}^{0+T_0/2} f\left(t+\frac{T_0}{2}\right) g\left(t+\frac{T_0}{2}\right)\mathrm{d}t
\end{aligned}$$

$$+\frac{2}{T_0}\int_0^{T_0/2} f(t)g(t)\mathrm{d}t$$

If $f(t) = -f\left(t + \frac{T_0}{2}\right)$ then,

$$\frac{2}{T_0}\int_{-T_0/2}^{T_0/2} f(t)g(t)\mathrm{d}t = \frac{2}{T_0}\int_0^{T_0/2} f(t)\left[g(t) - g\left(t + \frac{T_0}{2}\right)\right]\mathrm{d}t$$

(i) $g(t) = a$:

$$\frac{2}{T_0}\int_0^{T_0/2} f(t)\left[g(t) - g\left(t + \frac{T_0}{2}\right)\right]\mathrm{d}t = \frac{2}{T_0}\int_0^{T_0/2} f(t)\left[a - a\right]\mathrm{d}t = 0$$

(ii) $g(t) = \cos n\omega_0 t$:

$$\begin{aligned}
\frac{2}{T_0}\int_0^{T_0/2} f(t)\left[g(t) - g\left(t + \frac{T_0}{2}\right)\right]\mathrm{d}t &= \frac{2}{T_0}\int_0^{T_0/2} f(t)\left[\cos n\omega_0 t - \cos(n\omega_0 t + n\pi)\right]\mathrm{d}t \\
&= \frac{2}{T_0}\int_0^{T_0/2} f(t)\left[\cos n\omega_0 t - (-1)^n \cos n\omega_0 t)\right]\mathrm{d}t \\
&= \frac{2}{T_0}\int_0^{T_0/2} f(t)\cos n\omega_0 t\left[1 - (-1)^n\right]\mathrm{d}t \\
&= \begin{cases} 0 & n = 2, 4, 6, \ldots \\ \dfrac{4}{T_0}\displaystyle\int_0^{T_0/2} f(t)\cos n\omega_0 t\mathrm{d}t & n = 1, 3, 5, \ldots \end{cases}
\end{aligned}$$

(iii) $g(t) = \sin n\omega_0 t$:

$$\begin{aligned}
\frac{2}{T_0}\int_0^{T_0/2} f(t)\left[g(t) - g\left(t + \frac{T_0}{2}\right)\right]\mathrm{d}t &= \frac{2}{T_0}\int_0^{T_0/2} f(t)\sin n\omega_0 t\left[1 - (-1)^n\right]\mathrm{d}t \\
&= \begin{cases} 0 & n = 2, 4, 6, \ldots \\ \dfrac{4}{T_0}\displaystyle\int_0^{T_0/2} f(t)\sin n\omega_0 t\mathrm{d}t & n = 1, 3, 5, \ldots \end{cases}
\end{aligned}$$

(iv) $g(t) = \mathrm{e}^{-\mathrm{j}n\omega_0 t}$:

$$\begin{aligned}
\frac{2}{T_0}\int_0^{T_0/2} f(t)\left[g(t) - g\left(t + \frac{T_0}{2}\right)\right]\mathrm{d}t &= \frac{2}{T_0}\int_0^{T_0/2} f(t)\mathrm{e}^{-\mathrm{j}n\omega_0 t}\left[1 - (-1)^n\right]\mathrm{d}t \\
&= \begin{cases} 0 & n = 2, 4, 6, \ldots \\ \dfrac{4}{T_0}\displaystyle\int_0^{T_0/2} f(t)\mathrm{e}^{-\mathrm{j}n\omega_0 t}\mathrm{d}t & n = 1, 3, 5, \ldots \end{cases}
\end{aligned}$$

2.3.3 Quarter-wave symmetry

If a periodic function $f(t)$ has half-wave symmetry and is an even function, then $f(t)$ is said to have even *quarter-wave* symmetry.

If a periodic function $f(t)$ has half-wave symmetry and is an odd function, then $f(t)$ is said to have odd *quarter-wave* symmetry.

2.3.4 Hidden symmetry

If a periodic function $f(t)$ is shifted with respect to the t axis by a constant, then it is said to have *hidden* symmetry.

2.4 The Fourier Series

2.4.1 Trigonometric form

A periodic signal $f(t)$ of period T_0 can be expanded into a trigonometric Fourier series of the form

$$f(t) = \frac{1}{2}a_0 + \sum_{n=1}^{\infty} (a_n \cos n\omega_0 t + b_n \sin n\omega_0 t) \tag{2.12}$$

where $\omega_0 = \frac{2\pi}{T_0}$ and the coefficients of the series are given as

$$a_0 = \frac{2}{T_0} \int_{-T_0/2}^{T_0/2} f(t) dt \tag{2.13}$$

$$a_n = \frac{2}{T_0} \int_{-T_0/2}^{T_0/2} f(t) \cos n\omega_0 t dt \tag{2.14}$$

$$b_n = \frac{2}{T_0} \int_{-T_0/2}^{T_0/2} f(t) \sin n\omega_0 t dt \tag{2.15}$$

Equation (2.12) can also be represented as

$$f(t) = A_0 + \sum_{n=1}^{\infty} A_n \cos (n\omega_0 t + \theta_n) \tag{2.16}$$

where

$$A_0 = \frac{1}{2}a_0 \tag{2.17}$$

$$A_n = \sqrt{a_n^2 + b_n^2} \tag{2.18}$$

$$\theta_n = \arctan\left(-\frac{b_n}{a_n}\right) \tag{2.19}$$

In electric circuit theory, the coefficient A_0 is referred to as the DC component, A_n and θ_n are the magnitude and angle of the nth harmonic component.

Figure 2.2 shows the graphical representation of the Fourier series. This figure shows how a rectangular waveform can be regarded as the summation of sinusoidal waves of different frequencies plus a constant term. It is customary to represent the set of sinusoidal waveforms as phasors of different harmonic frequencies, like those shown at the far end of the figure.

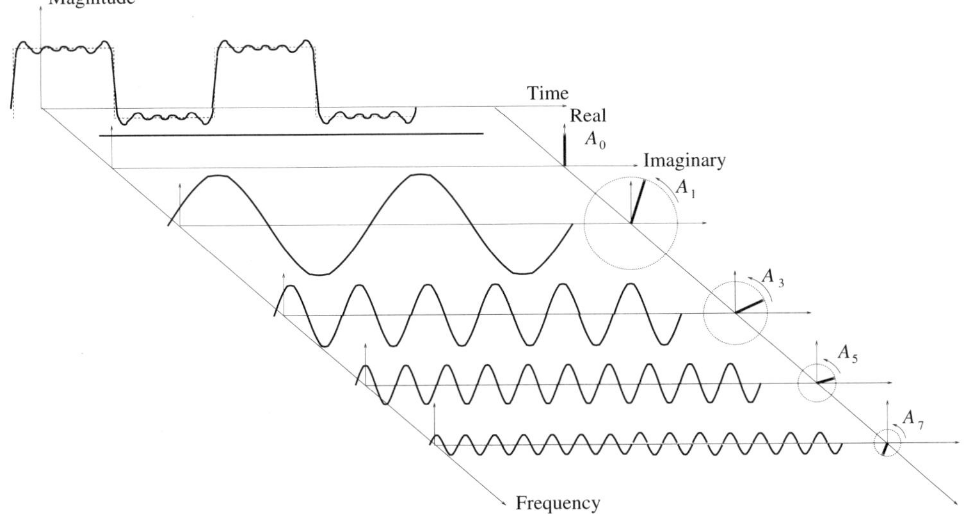

Figure 2.2 Fourier representation of a square waveform as the sum of a series of sinusoidal waves

Example 2-6: Find the Fourier series of the function

$$f(t) = \begin{cases} -1 & -\frac{T_0}{2} < t < 0 \\ 1 & 0 < t < \frac{T_0}{2} \end{cases}$$

This function corresponds to the periodic train of square waves shown in Figure 2.3.

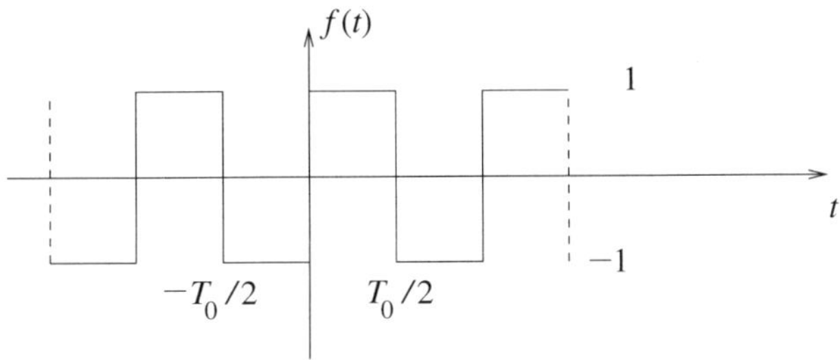

Figure 2.3 Periodic square waves

The series coefficients are given as

$$\begin{aligned} a_0 &= \frac{2}{T_0}\int_{-T_0/2}^{T_0/2} f(t)\mathrm{d}t = \frac{2}{T_0}\int_{-T_0/2}^{0}(-1)\mathrm{d}t + \frac{2}{T_0}\int_{0}^{T_0/2}(1)\mathrm{d}t = 0 \\ a_n &= \frac{2}{T_0}\int_{-T_0/2}^{T_0/2} f(t)\cos n\omega_0 t\mathrm{d}t \end{aligned}$$

$$
\begin{aligned}
&= \frac{2}{T_0}\int_{-T_0/2}^{0}(-1)\cos n\omega_0 t dt + \frac{2}{T_0}\int_{0}^{T_0/2}(1)\cos n\omega_0 t dt \\
&= -\frac{2}{n\omega_0 T_0}\cos n\pi + \frac{2}{n\omega_0 T_0}\cos n\pi = 0 \\
b_n &= \frac{4}{n\omega_0 T_0}(1-\cos n\pi) = \begin{cases} 0 & \text{for } n \text{ even} \\ \dfrac{4}{n\pi} & \text{for } n \text{ odd} \end{cases}
\end{aligned}
$$

If $A_n = b_n$ and $\theta_n = -\pi/2$ then the periodic function $f(t)$ can be represented by its trigonometric Fourier series,

$$
f(t) \approx \frac{4}{\pi}\cos(\omega_0 t - \pi/2) + \frac{4}{3\pi}\cos(3\omega_0 t - \pi/2) + \frac{4}{5\pi}\cos(5\omega_0 t - \pi/2) + \cdots
$$

For instance, it can be seen from Figure 2.3 that $f(t)$ is an even function with half-wave symmetry, so $a_0 = a_n = 0$. Also $b_n = 0$ for all n even. Examples 2-4 and 2-5 show these properties of the Fourier series.

2.4.2 *Exponential form*

The harmonic n is well represented by its coefficients a_n and b_n,

$$
f_n(t) = a_n \cos n\omega_0 t + b_n \sin n\omega_0 t \tag{2.20}
$$

In complex notation form, using Euler's formula,

$$
\begin{aligned}
e^{j\alpha} &= \cos\alpha + j\sin\alpha \\
\cos\alpha &= \frac{1}{2}e^{j\alpha} + \frac{1}{2}e^{-j\alpha} \\
\sin\alpha &= \frac{1}{2j}e^{j\alpha} - \frac{1}{2j}e^{-j\alpha}
\end{aligned}
$$

and substituting $\cos\alpha$ and $\sin\alpha$ in (2.20) it can be represented as

$$
\begin{aligned}
f_n(t) &= \frac{a_n}{2}\left(e^{jn\omega_0 t} + e^{-jn\omega_0 t}\right) - j\frac{b_n}{2}\left(e^{jn\omega_0 t} - e^{-jn\omega_0 t}\right) \\
&= \frac{1}{2}(a_n - jb_n)e^{jn\omega_0 t} + \frac{1}{2}(a_n + jb_n)e^{-jn\omega_0 t} \\
&= C_n e^{jn\omega_0 t} + C_{-n}e^{-jn\omega_0 t}
\end{aligned} \tag{2.21}
$$

where

$$
\begin{aligned}
C_n &= \frac{1}{2}(a_n - jb_n) \\
C_n &= C_{-n}^*
\end{aligned} \tag{2.22}
$$

where $*$ means complex conjugate. It should be noted that $n = 0$ is a special case,

$$C_0 = \frac{a_0}{2} \tag{2.23}$$

The nth harmonic index in (2.21) can also be expressed in polar form,

$$C_n = |C_n| \angle\theta_n = \frac{1}{2} A_n \angle\theta_n \tag{2.24}$$

Equation (2.12) can be rewritten as

$$f(t) = \sum_{n=-\infty}^{\infty} C_n \mathrm{e}^{\mathrm{j}n\omega_0 t} \tag{2.25}$$

which leads to the Fourier series in its complex form.

It can also be shown that the complex coefficients can be obtained from solving the following integral:

$$C_n = \frac{1}{T_0} \int_{-T_0/2}^{T_0/2} f(t)\mathrm{e}^{-\mathrm{j}n\omega_0 t}\mathrm{d}t \tag{2.26}$$

This expression gives a discrete-frequency representation of the continuous, periodic function $f(t)$, shown graphically in Figure 2.4. For this function, the coefficient C_n has only the real part and the waveform has hidden symmetry, i.e. $a_0 \neq 0$. It should be noted that the function is even with half-wave symmetry, i.e. $b_n = 0$. Also, $a_n = 0$ for $n = 2, 4, 6, \ldots$.

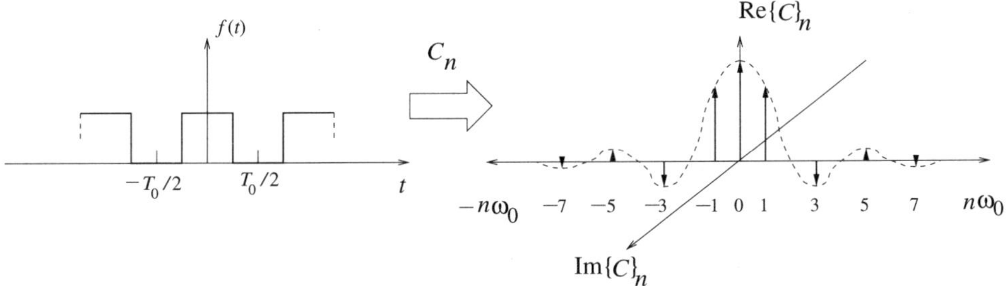

Figure 2.4 Graphical representation of the Fourier series coefficients

By inspecting (2.18) and (2.19) we can see that the magnitude and angle of the harmonic index n have the following properties:

$$\begin{aligned} A_n &= A_{-n} \\ \theta_n &= -\theta_{-n} \end{aligned} \tag{2.27}$$

These equations show that the magnitude A_n is the even part and the angle θ_n is the odd part of the harmonic n. Similarly, the real and imaginary parts of C_n are the even and odd parts, respectively.

2.4.3 *Summary of waveform symmetry in a Fourier series*

The property introduced by waveform symmetry in a Fourier series is summarised in the next table.

Waveform symmetry	Definition	Fourier coeff. diff. from zero for $n = 1, 2, 3, ...$	Fourier coeff. equal to zero for $n = 1, 2, 3, ...$
Even	$f(-t) = f(t)$	a_n	b_n
Odd	$f(-t) = -f(t)$	b_n	a_n
Half-wave	$f(t) = -f\left(t + \frac{T_0}{2}\right)$	a_{2n-1} b_{2n-1}	a_{2n} b_{2n}
Even Quarter $-$ wave	$f(-t) = f(t)$ and $f(t) = -f\left(t + \frac{T_0}{2}\right)$	a_{2n-1}	a_{2n} b_n
Odd Quarter $-$ wave	$f(-t) = -f(t)$ and $f(t) = -f\left(t + \frac{T_0}{2}\right)$	b_{2n-1}	b_{2n} a_n

2.5 The Fourier Transform

2.5.1 *Fundamental equation*

In the limit, when the period T_0 approaches infinity, equation (2.26) yields an infinite number of infinitesimally close frequency components. It is at this point that the discrete-frequency variable $n\omega_0$ merges into a continuum, becoming a function $F(\omega)$ of the continuous-frequency variable ω, and the Fourier transform is obtained:

$$F(\omega) = \int_{-\infty}^{\infty} f(t)\mathrm{e}^{-\mathrm{j}\omega t}\mathrm{d}t \tag{2.28}$$

This expression gives a continuous-frequency representation $F(\omega)$ of the continuous, non-periodic function $f(t)$ and the inverse Fourier transform of the $F(\omega)$ is given as

$$f(t) = \frac{1}{2\pi}\int_{-\infty}^{\infty} F(\omega)\mathrm{e}^{\mathrm{j}\omega t}\mathrm{d}\omega \tag{2.29}$$

Example 2-7: With reference to the function $f(t)$ in Figure 2.5(a), we have that the Fourier transform can be obtained as follows:

$$\begin{aligned}
F(\omega) &= \int_{-\infty}^{\infty} f(t)\mathrm{e}^{-\mathrm{j}\omega t}\mathrm{d}t \\
&= \int_{-\infty}^{-T_0/2} (0)\mathrm{e}^{-\mathrm{j}\omega t}\mathrm{d}t + \int_{-T_0/2}^{T_0/2} (1)\mathrm{e}^{-\mathrm{j}\omega t}\mathrm{d}t + \int_{T_0/2}^{\infty} (0)\mathrm{e}^{-\mathrm{j}\omega t}\mathrm{d}t \\
&= \int_{-T_0/2}^{T_0/2} \mathrm{e}^{-\mathrm{j}\omega t}\mathrm{d}t = -\frac{1}{\mathrm{j}\omega}\left(\mathrm{e}^{-\mathrm{j}\omega T_0/2} - \mathrm{e}^{\mathrm{j}\omega T_0/2}\right) = \frac{2\pi}{\omega_0}\frac{\sin\frac{\omega\pi}{\omega_0}}{\frac{\omega\pi}{\omega_0}}
\end{aligned}$$

This result is shown graphically in Figure 2.5(b).

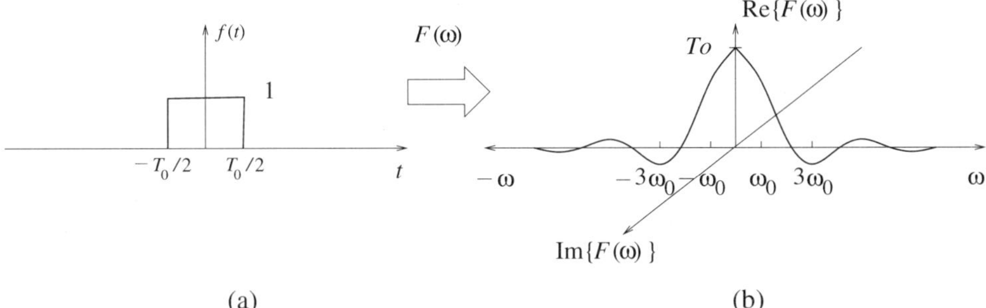

Figure 2.5 Graphical representation of the Fourier transform

2.5.2 *Discrete Fourier transform*

The discrete Fourier transform (DFT) is obtained from the Fourier transform of a sampled function $f_{\mathrm{d}}(t)$ in the interval $[0, T_0]$. To illustrate this point consider the function $f(t)$ shown in Figure 2.6, where the sampled function $f_{\mathrm{s}}(t)$ is given by the product of $f(t)$ and a sampled waveform $s_T(t)$,

$$f_s(t) = f(t)s_T(t) = \Delta T \sum_{n=-\infty}^{\infty} f(t)\delta(t - n\Delta T) \tag{2.30}$$

where $\delta(.)$ is the impulse function.

The Fourier transform of (2.30) will be

$$F[k\omega_0] = \frac{1}{T_0}\int_0^{T_0} f_s(t)\mathrm{e}^{-\mathrm{j}k\omega_0 t}\mathrm{d}t = \frac{\Delta T}{T_0}\int_0^{T_0}\sum_{n=-\infty}^{\infty} f(t)\delta(t - n\Delta T)\mathrm{e}^{-\mathrm{j}k\omega_0 t}\mathrm{d}t \tag{2.31}$$

If $g(t) = f(t)\mathrm{e}^{-\mathrm{j}k\omega_0 t}$ and the infinite summation is expressed as an integral,

$$F[k\omega_0] = \frac{\Delta T}{T_0}\int_0^{T_0}\int_{-\infty}^{\infty} g(t)\delta(t - \tau)\mathrm{d}t \tag{2.32}$$

where $\tau = n\Delta T$.

Since

$$\int_{-\infty}^{\infty} g(t)\delta(t - \tau)\mathrm{d}t = g(\tau)$$

equation (2.32) becomes

$$F[k\omega_0] = \frac{\Delta T}{T_0}\int_0^{T_0} g(\tau)\mathrm{d}t = \frac{\Delta T}{T_0}\int_0^{T_0} f[n\Delta T]\,\mathrm{e}^{-\mathrm{j}k\omega_0 n\Delta T}\mathrm{d}t \tag{2.33}$$

A window function $w(t)$ is used to select the length of the signal needed to fit the transform, where $f_{\mathrm{d}}(t) = f_{\mathrm{s}}(t)w(t)$, and the number of samples N is given by $N = \frac{T_0}{\Delta T}$. The integral of (2.33) becomes a summation,

$$F[k\omega_0] = \frac{1}{N}\sum_{n=0}^{N-1} f[n\Delta T]\,\mathrm{e}^{-\mathrm{j}k\omega_0 n\Delta T} \quad k = 0, 1, 2, \ldots, N-1 \tag{2.34}$$

Equation (2.34) is the Fourier transform of the sequence $f[n\Delta T]$, which is commonly known as the discrete Fourier transform (DFT). Using a similar procedure, the inverse discrete Fourier transform (IDFT) of the sequence $F[k\omega_0]$ will be

$$f[n\Delta T] = N\sum_{k=0}^{N-1} F[k\omega_0]\,\mathrm{e}^{-\mathrm{j}k\omega_0 n\Delta T} \quad n = 0, 1, 2, \ldots, N-1 \tag{2.35}$$

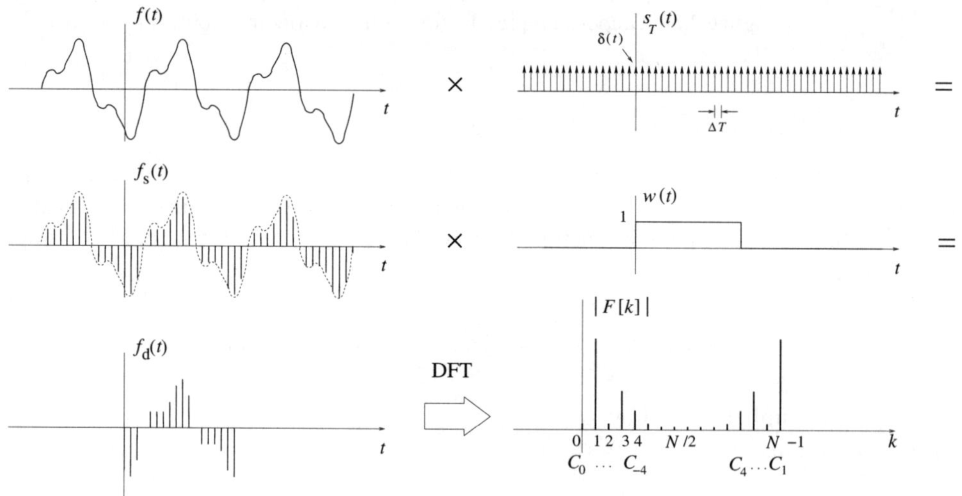

Figure 2.6 Sampled waveform and its discrete Fourier transform

Since $\omega_0 = 2\pi/T_0$ and $\Delta T = T_0/N$, equations (2.34) and (2.35) can be rewritten as

$$F[k] = \frac{1}{N}\sum_{n=0}^{N-1} f[n]\,\mathrm{e}^{-\mathrm{j}kn2\pi/N} \quad k = 0, 1, 2, \ldots, N-1 \tag{2.36}$$

$$f[n] = N\sum_{k=0}^{N-1} F[k]\,\mathrm{e}^{-\mathrm{j}kn2\pi/N} \quad n = 0, 1, 2, \ldots, N-1 \tag{2.37}$$

It should be noted that $F[k]$ is the counterpart of C_n calculated in equation (2.26) for periodic continuous functions. Also, $f[n]$ is the counterpart of $f(t)$ calculated in (2.25) for a periodic continuous function.

Example 2-8: The continuous function $f(t) = 3\cos(\omega_0 t + 120.32°) + \cos(3\omega_0 t - 85.94°)$ is sampled over one full period. The number of samples is 16 and a window with $T_0 = 16.7$ ms is selected. It should be noted that this is one full cycle of $f(t)$ at a base frequency $f_0 = 60$ Hz. The function is shown in Figure 2.7 where the sample values and the harmonic content, as given by a DFT using (2.36), are given in Table 2.1.

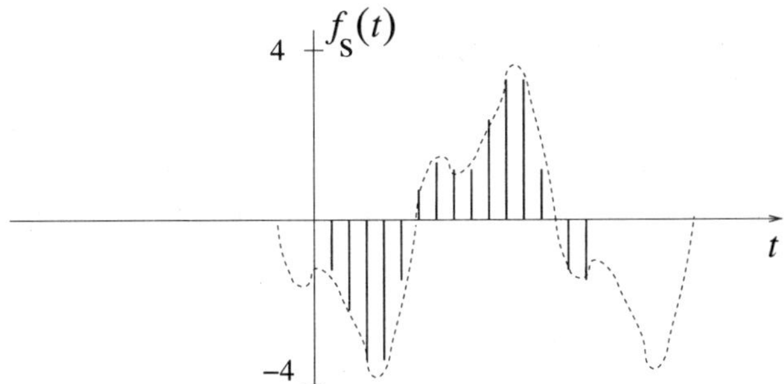

Figure 2.7 Sampled function

Table 2.1 Discrete Fourier transform and its relationship with the Fourier series

$f[n]$	k	$\|F[k]\|$	θ_k	A_n	θ_n	n
−1.4438	0	0.0928	180.0000	−0.0928	0.0000	0
−1.4672	1	1.4451	133.8798	2.8901	−133.8798	−1
−2.4142	2	0.1898	−52.9912	0.3796	52.9912	−2
−3.5786	3	0.5179	−49.9296	1.0358	49.9296	−3
−3.3340	4	0.0878	136.6834	0.1757	−136.6834	−4
−1.3909	5	0.0299	145.3872	0.0598	−145.3872	−5
0.6924	6	0.0168	155.7509	0.0337	−155.7509	−6
1.4738	7	0.0123	167.4691	0.0247	−167.4691	−7
1.3759	8	0.0112	−180.0000	0.0112	180.0000	
1.8386	9	0.0123	−167.4691	0.0247	167.4691	7
3.0958	10	0.0168	−155.7509	0.0337	155.7509	6
3.6990	11	0.0299	−145.3872	0.0598	145.3872	5
2.4749	12	0.0878	−136.6834	0.1757	136.6834	4
0.2110	13	0.5179	49.9296	1.0358	−49.9296	3
−1.2785	14	0.1898	52.9912	0.3796	−52.9912	2
−1.4386	15	1.4451	−133.8798	2.8901	133.8798	1

The comparison between the results given by the DFT and the Fourier harmonic coefficients $A_n\angle\theta_n$ yields useful insight. We have that $f(t)$ is a series of the form (2.16) and that, in this example, only the fundamental and the third harmonic exist. Their magnitudes and angles are $3\angle120.32°$ and $1\angle-85.94°$, respectively.

It is interesting to see that the DFT has given values of $2.8901\angle133.87°$ and $1.0358\angle-49.92°$ for the fundamental and third harmonic, respectively, as well as values different from zero for many other harmonics. Clearly, these values differ from the exact solution because of the small number of samples selected. Table 2.2 shows the result for the case when 256 samples are used, leading to a more accurate solution. Figure 2.8 shows the harmonic spectrum magnitude, a popular way of presenting the harmonic information produced by the DFT.

Table 2.2 Harmonic content with *N*=256

A_n	θ_n	n
0.00001	0.0000	DC
0.0018	90.3889	7
0.0027	90.6929	6
0.0048	91.6514	5
0.0124	93.6404	4
1.0083	−82.6403	3
0.0354	−75.4858	2
2.9912	121.5614	1

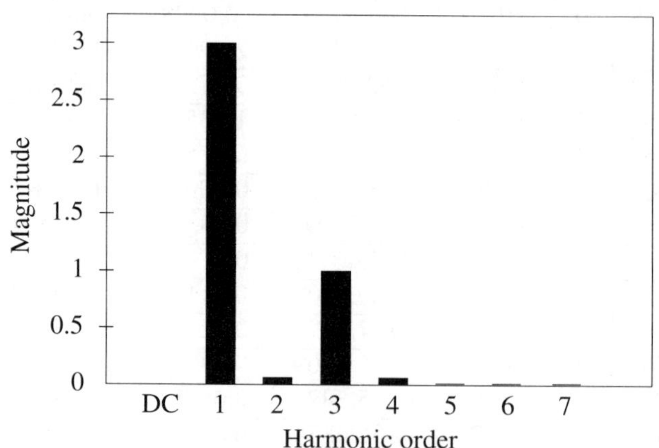

Figure 2.8 Harmonic content of $f[n]$ with 256 samples

A function in MATLAB™ to obtain the DFT is given below:

```
function F=dft(f,N)
% F : discrete Fourier transform
% f : discrete values in time domain
% N : number of points
```

```
for k=0:(N-1)
  F(k+1)=0;
  for n=0:(N-1)
    F(k+1)=f(n+1)*exp(-i*n*k*2*pi/N)+F(k+1);
  end
  F(k+1)=F(k+1)/N;
end
```

A function in MATLAB™ to obtain the IDFT is given below:

```
function f=idft(F,N)
% f : inverse discrete Fourier transform
% F : discrete values in the frequency domain
% N : number of points
for k=0:(N-1)
  f(k+1)=0;
  for n=0:(N-1)
    f(k+1)=F(n+1)*exp(i*n*k*2*pi/N)+f(k+1);
  end
  f(k+1)=N*f(k+1);
end
```

2.5.3 *Fast Fourier transform*

The fast Fourier transform (FFT) is an algorithm that solves the DFT in a highly efficient manner.

Equation (2.36) may be rewritten in a simpler form if the relation $W = \mathrm{e}^{-\mathrm{j}2\pi/N}$ is used:

$$F[k] = \frac{1}{N} \sum_{n=0}^{N-1} f[n] W^{kn} \quad k = 0, 1, 2, \ldots, N-1 \tag{2.38}$$

This leads to an expression in matrix form

$$\begin{bmatrix} F[0] \\ F[1] \\ F[2] \\ \vdots \\ F[N-1] \end{bmatrix} = \begin{bmatrix} W^0 & W^0 & W^0 & \cdots & W^0 \\ W^0 & W^1 & W^2 & \cdots & W^{(N-1)} \\ W^0 & W^2 & W^4 & \cdots & W^{2(N-1)} \\ \vdots & \vdots & \vdots & \ddots & \vdots \\ W^0 & W^{(N-1)} & W^{2(N-1)} & \cdots & W^{(N-1)(N-1)} \end{bmatrix} \times \begin{bmatrix} f[0] \\ f[1] \\ f[2] \\ \vdots \\ f[N-1] \end{bmatrix} \tag{2.39}$$

or, in compact form,

$$\mathbf{F} = \mathbf{W}\mathbf{f} \tag{2.40}$$

By way of example, we have the values of W^{kn} shown in Figure 2.9 for the case when $N = 8$. From this figure it is easy to see that all the $kn = 7 \times 7 = 49$ operations required to obtain matrix **W** can in fact be achieved with only three values: $1, \frac{1}{2} + \mathrm{j}\frac{1}{2}, \mathrm{j}1$. The remaining 46 values are the conjugate or negative values of those three values. This simplifying philosophy is at the heart of the FFT. The only restriction in achieving this degree of simplification is that the algorithm is limited to using samples of power of two, i.e. $N = 2^m$.

It should be noted that to obtain the Fourier transform of a discrete function, using (2.36), we would require N^2 multiplications to produce the result whereas using the FFT we would require only $\frac{1}{2}N \log_2 N$ multiplications to achieve a similar result. Figure 2.10 shows a comparison between the number of multiplications required by these two methods as a function of the number of samples.

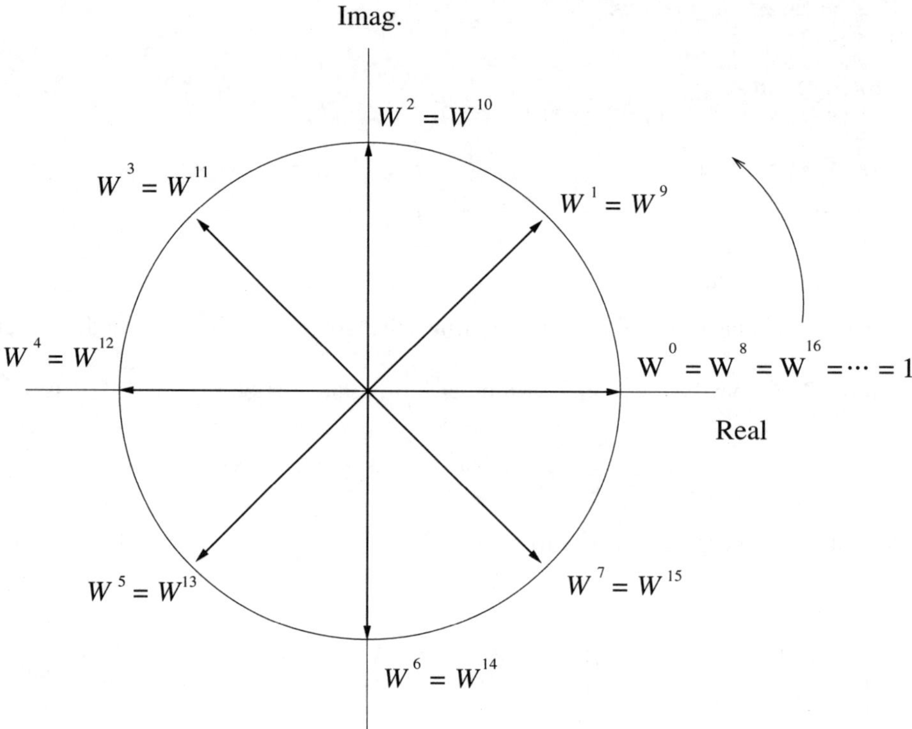

Figure 2.9 Values of W for N=8

2.6 Convolutions

The convolution of two functions is defined by an integral operand for continuous functions or by a sum for the case of discrete functions.

2.6.1 *Time convolutions*

The convolution of two continuous functions $x(t)$ and $h(t)$ is expressed by means of the following integration:

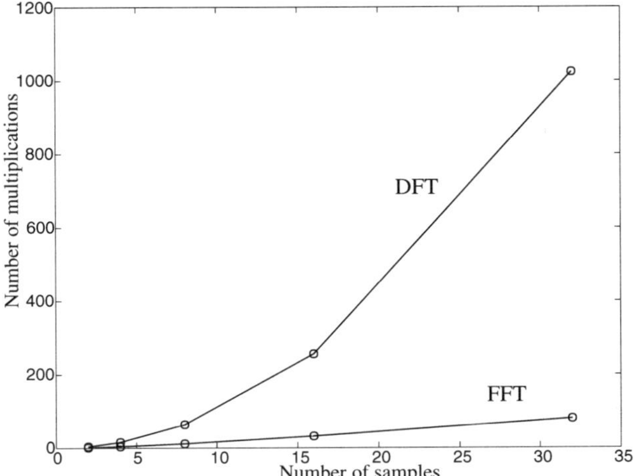

Figure 2.10 Comparison of multiplications required by DFT and FFT

$$y(t) = \int_{-\infty}^{\infty} x(t-\tau)h(\tau)\mathrm{d}\tau \tag{2.41}$$

or, in symbolic notation,

$$y(t) = x(t) * h(t) \tag{2.42}$$

The discrete-time convolution of $x[n\Delta T]$ and $h[n\Delta T]$ is expressed by the sum

$$y[n\Delta T] = \sum_{m=-\infty}^{\infty} x[n\Delta T - m\Delta T]h[m\Delta T] \tag{2.43}$$

or

$$y[n\Delta T] = x[n\Delta T] * h[n\Delta T] \tag{2.44}$$

2.6.2 *Frequency convolutions*

The convolution of continuous functions $F(\omega)$ and $H(\omega)$ is expressed by the integral

$$Y(\omega) = \int_{-\infty}^{\infty} F(\omega - \nu)H(\omega)\mathrm{d}\nu \tag{2.45}$$

or

$$Y(\omega) = F(\omega) \otimes H(\omega) \tag{2.46}$$

The discrete-frequency convolution of $F[n\omega_0]$ and $H[n\omega_0]$ is calculated by the summation

$$Y[n\omega_0] = \sum_{m=-\infty}^{\infty} F[n\omega_0 - m\omega_0]H[m\omega_0] \tag{2.47}$$

or

$$Y[n\omega_0] = F[n\omega_0] \otimes H[n\omega_0] \tag{2.48}$$

2.7 The Hartley Transform

2.7.1 From Fourier to Hartley series

Starting from the trigonometric Fourier series, with the angular frequency defined as ν_0 instead of ω_0,

$$f(t) = \frac{1}{2}a_0 + \sum_{n=1}^{\infty}(a_n \cos n\nu_0 t + b_n \sin n\nu_0 t) \tag{2.49}$$

this series can be rewritten as

$$f(t) = \frac{1}{2}a_0 + \sum_{n=1}^{\infty}\left\{\frac{1}{2}(a_n + b_n)\,\mathrm{cas}\, n\nu_0 t + \frac{1}{2}(a_n - b_n)\,\mathrm{cas}(-n\nu_0 t)\right\} \tag{2.50}$$

where the function $\mathrm{cas}\,\theta$ is defined as

$$\mathrm{cas}\,\theta = \cos\theta + \sin\theta \tag{2.51}$$

Since $a_{-n} = a_n$ and $b_{-n} = -b_n$ then (2.50) becomes

$$f(t) = \frac{1}{2}\sum_{n=1}^{\infty}(a_n + b_n)\,\mathrm{cas}\, n\nu_0 t \tag{2.52}$$

or

$$f(t) = \sum_{n=1}^{\infty} S_n\,\mathrm{cas}\, n\nu_0 t \tag{2.53}$$

where

$$S_n = \frac{1}{2}(a_n + b_n) \tag{2.54}$$

Substituting (2.14) and (2.15) into (2.54) gives

$$S_n = \frac{1}{2}\left\{\frac{2}{T_0}\int_{-T_0/2}^{T_0/2} f(t)\cos\nu_0 t\mathrm{d}t + \frac{2}{T_0}\int_{-T_0/2}^{T_0/2} f(t)\sin\nu_0 t\mathrm{d}t\right\} \tag{2.55}$$

$$S_n = \frac{1}{T_0}\int_{-T_0/2}^{T_0/2} f(t)\mathrm{cas}\,\nu_0 t\mathrm{d}t \tag{2.56}$$

We have now obtained the Hartley series defined by equations (2.53) and (2.56). The coefficients a_n and b_n can be obtained in trigonometric form,

$$\begin{aligned} a_0 &= 2S_0 \\ a_n &= S_{-n} + S_n \\ b_n &= -S_{-n} + S_n \end{aligned} \tag{2.57}$$

and, in complex form,

$$C_n = \frac{1}{2}(S_{-n} + S_n) - \mathrm{j}\frac{1}{2}(-S_{-n} + S_n) \tag{2.58}$$

2.7.2 *Properties of the* cas $(\cdot)$ *function*

Arguably the two most important properties of the cas $(\cdot)$ function are its mathematical elegance and the fact that all its operations take place in the real domain rather than the complex domain. The properties in Tables 2.3 and 2.4 are very useful in understanding and using the Hartley series efficiently. Table 2.3 summarises the simplifying properties of the Hartley coefficients as a function of symmetry. Table 2.4 gives the most relevant trigonometric properties of the cas $(\cdot)$ function.

Table 2.3 Simplification of the Hartley and Fourier series coefficients simplification

Symmetry	Hartley coefficients	Fourier coefficients
$f(t) = f(-t)$	$S_n = S_{-n}$	$C_n = C_{-n},\ \Im\{C_n\} = 0$
$f(t) = -f(-t)$	$S_n = -S_{-n}$	$C_n = -C_{-n},\ \Re\{C_n\} = 0$
$f(t) = -f\left(t + \frac{T_0}{2}\right)$	$S_n = S_{-n} = 0$ for n even	$C_n = C_{-n} = 0$ for n even
none	no simplification	$C_n = C^*_{-n}$

Table 2.4 Trigonometric properties of the cas $(\cdot)$ function

$\operatorname{cas}\alpha = \cos\alpha + \sin\alpha$	$\operatorname{cas}(-\alpha) = \cos\alpha - \sin\alpha$
$\sin\alpha = \frac{1}{2}\operatorname{cas}\alpha - \frac{1}{2}\operatorname{cas}(-\alpha)$	$\operatorname{cas}(\pm\alpha) = \sqrt{2}\cos(\alpha \mp \pi/4)$
$\operatorname{cas}\alpha\operatorname{cas}(-\alpha) = \cos 2\alpha$	$\operatorname{cas}^2\alpha - \operatorname{cas}^2(-\alpha) = 2\sin 2\alpha$
$\operatorname{cas}^2\alpha + \operatorname{cas}^2(-\alpha) = 2$	$\operatorname{cas}^2\alpha\operatorname{cas}^2\alpha = 2 + 2\sin 2\alpha$
$\operatorname{cas} 2\alpha = \operatorname{cas}^2\alpha - 2\sin^2\alpha$	$\operatorname{cas}(\alpha + \pi/2) = \operatorname{cas}(-\alpha)$
$\sin\alpha\operatorname{cas}\beta = \operatorname{cas}(\alpha - \beta) - \operatorname{cas}(-\alpha - \beta)$	$\cos\alpha = \frac{1}{2}\operatorname{cas}\alpha + \frac{1}{2}\operatorname{cas}(-\alpha)$
$\cos\alpha\operatorname{cas}\beta = \operatorname{cas}(-\alpha + \beta) + \operatorname{cas}(\alpha + \beta)$	$\operatorname{cas}(\alpha + \beta) = \cos\beta\operatorname{cas}\alpha + \sin\beta\operatorname{cas}(-\alpha)$
$\operatorname{cas}\alpha\operatorname{cas}\beta = \frac{1}{2}[\operatorname{cas}(\alpha + \beta) + \operatorname{cas}(\alpha - \beta) + \operatorname{cas}(-\alpha + \beta) - \operatorname{cas}(-\alpha - \beta)]$	$\mathrm{e}^{\mathrm{j}\alpha} = \frac{1}{2}[\operatorname{cas}\alpha + \operatorname{cas}(-\alpha)] + \frac{1}{2}\mathrm{j}[\operatorname{cas}\alpha - \operatorname{cas}(-\alpha)]$
$\dfrac{\mathrm{d}\operatorname{cas}\alpha t}{\mathrm{d}t} = \alpha\operatorname{cas}(-\alpha t)$	$\int \operatorname{cas}\alpha t\,\mathrm{d}t = -\frac{1}{\alpha}\operatorname{cas}(-\alpha t)$
$\dfrac{\mathrm{d}\operatorname{cas}(-\alpha t)}{\mathrm{d}t} = -\alpha\operatorname{cas}\alpha t$	$\int \operatorname{cas}(-\alpha t)\mathrm{d}t = \frac{1}{\alpha}\operatorname{cas}\alpha t$

2.7.3 *The Hartley transform*

In the limit, when the period T_0 approaches infinity, equation (2.56) leads to an infinite number of infinitesimally close frequency components. It is at this point that the discrete-frequency variable $n\nu_0$ merges into a continuum becoming a function $H(\nu)$ of the continuous-frequency variable ν and the Hartley transform is obtained:

$$H(\nu) = \int_{-\infty}^{\infty} f(t)\text{cas}\,\nu t\text{d}t \tag{2.59}$$

This expression gives a continuous-frequency representation $H(\nu)$ of the continuous, non-periodic function $f(t)$. The inverse Hartley transform is given by

$$f(t) = \frac{1}{2\pi}\int_{-\infty}^{\infty} H(\nu)\text{cas}\,\nu t\text{d}\nu \tag{2.60}$$

2.7.4 *Properties of the Hartley transform*

In order to show some very useful properties of the Hartley and Fourier transforms, $f(t)$ is separated into even and odd parts:

$$\begin{aligned}
H(\nu) &= \int_{-\infty}^{\infty} f(t)\cos\nu t\text{d}t + \int_{-\infty}^{\infty} f(t)\sin\nu t\text{d}t \\
&= \int_{-\infty}^{\infty} f_{\text{even}}(t)\cos\nu t\text{d}t + \int_{-\infty}^{\infty} f_{\text{odd}}(t)\cos\nu t\text{d}t \\
&\quad + \int_{-\infty}^{\infty} f_{\text{even}}(t)\sin\nu t\text{d}t + \int_{-\infty}^{\infty} f_{\text{odd}}(t)\sin\nu t\text{d}t \\
&= \int_{-\infty}^{\infty} f_{\text{even}}(t)\text{cas}\,\nu t\text{d}t + \int_{-\infty}^{\infty} f_{\text{odd}}(t)\text{cas}\,\nu t\text{d}t \\
&= H_{\text{even}}(\nu) + H_{\text{odd}}(\nu)
\end{aligned} \tag{2.61}$$

From the Fourier transform (2.28) we have

$$\begin{aligned}
F(\omega) &= \int_{-\infty}^{\infty} f(t)\cos\omega t\text{d}t - \text{j}\int_{-\infty}^{\infty} f(t)\sin\omega t\text{d}t \\
&= \Re\{F(\omega)\} + \text{j}\Im\{F(\omega)\}
\end{aligned} \tag{2.62}$$

Comparing (2.61) and (2.62) gives the following relations:

$$H(\nu) = (\Re\{F(\omega)\} - \Im\{F(\omega)\})|_{\omega=\nu} \tag{2.63}$$

$$F(\omega) = (H_{\text{even}}(\nu) - \text{j}H_{\text{odd}}(\nu))|_{\nu=\omega} \tag{2.64}$$

A summary of the main theorems for the Fourier and Hartley transforms are given in Table 2.5.

It should be noted that convolutions in the Hartley domain are greatly simplified if the even and odd nature of Hartley functions is observed, as shown in Table 2.6.

2.7.5 *Discrete Hartley transform*

Similarly to the DFT, the discrete Hartley transform (DHT) and its inverse IDHT also exist and are given by

Table 2.5 Properties of the Hartley and Fourier transforms

Theorem	$f(t)$	$F(\omega)$	$H(\nu)$
Similarity	$f(t/T_0)$	$\|T_0\|\,F(T_0\omega)$	$\|T_0\|\,H(T_0\nu)$
Addition	$f_1(t)+f_2(t)$	$F_1(\omega)+F_2(\omega)$	$H_1(\nu)+H_2(\nu)$
Negative	$f(-t)$	$F(-\omega)$	$H(-\nu)$
Shift	$f(t-T_0)$	$\mathrm{e}^{-\mathrm{j}\omega T_0}F(\omega)$	$\sin\nu T_0 H(-\nu)+\cos\nu T_0 H(\nu)$
Modulation	$f(t)\cos\omega_0 t$	$\frac{1}{2}F(\omega-\omega_0)$ $+\frac{1}{2}F(\omega+\omega_0)$	$\frac{1}{2}H(\nu-\nu_0)+\frac{1}{2}H(\nu+\nu_0)$
Convolution	$f_1(t)*f_2(t)$	$F_1(\omega)F_2(\omega)$	$\frac{1}{2}[H_1(\nu)H_2(\nu)+H_1(\nu)H_2(-\nu)$ $+H_1(-\nu)H_2(\nu)-H_1(-\nu)H_2(-\nu)]$
Autocorrelation	$f_1(t)\star f_2(t)$	$\|F(\omega)\|^2$	$\frac{1}{2}[H^2(\nu)+H^2(-\nu)]$
Product	$f_1(t)f_2(t)$	$F_1(\omega)\otimes F_2(\omega)$	$\frac{1}{2}[H_1(\nu)\otimes H_2(\nu)+H_1(\nu)\otimes H_2(-\nu)$ $+H_1(-\nu)\otimes H_2(\nu)-H_1(-\nu)\otimes H_2(-\nu)]$
Derivative	$f'(t)$	$\mathrm{j}\omega F(\omega)$	$-\nu H(-\nu)$
Second derivative	$f''(t)$	$-\omega^2F(\omega)$	$-\nu^2H(\nu)$

Table 2.6 Convolutions

$H_1(\nu)$	$H\{f_1(t)*f_2(t)\}$
Even	$H_1(\nu)H_2(\nu)$
Odd	$H_1(\nu)H_2(-\nu)$

$$H[k]=\frac{1}{N}\sum_{n=0}^{N-1}f[n]\,\mathrm{cas}(nk2\pi/N)\quad k=0,1,2,\ldots,N-1 \tag{2.65}$$

$$f[n]=\sum_{k=0}^{N-1}H[k]\,\mathrm{cas}(nk2\pi/N)\quad n=0,1,2,\ldots,N-1 \tag{2.66}$$

It should be noted that $H[k]$ is the counterpart of S_n in (2.56). Also, $f[n]$ is the counterpart of $f(t)$ in (2.53).

Example 2-9: Find the DHT for the case presented in Example 2-8. Upon conversion, using equation (2.57), these results should be the same as those given by the DFT, but in this case the number of multiplications and storage requirements are only 25% of those required by the DFT. The DFT requires N^2 complex operations, whereas the DHT only requires N^2 real operations. The results are shown in Table 2.7.

A function in MATLAB™ to obtain the DHT is given below:

```
function H=dht(f,N)
% H : discrete Hartley transform
% f : discrete values in time domain
% N : number of points
for k=0:(N-1)
  H(k+1)=0;
  for n=0:(N-1)
    H(k+1)=f(n+1)*(cos(n*k*2*pi/N)+sin(n*k*2*pi/N))+H(k+1);
```

Table 2.7 Discrete Hartley transform and Fourier series results

k	$H[k]$	a_n	b_n	A_n	θ_n	n
0	−0.0928	−0.0928	0.0000	−0.0928	0.0000	0
1	−2.0432	−2.0033	−2.0832	2.8901	−133.8798	−1
2	0.2658	0.2285	0.3031	0.3796	52.9912	−2
3	0.7297	0.6668	0.7926	1.0358	49.9296	−3
4	−0.1242	−0.1278	−0.1205	0.1757	−136.6834	−4
5	−0.0416	−0.0492	−0.0340	0.0598	−145.3872	−5
6	−0.0223	−0.0307	−0.0138	0.0337	−155.7509	−6
7	−0.0147	−0.0241	−0.0053	0.0247	−167.4691	−7
8	−0.0112	−0.0112	−0.0112	0.0112	180.0000	
9	−0.0094	−0.0241	0.0053	0.0247	167.4691	7
10	−0.0084	−0.0307	0.0138	0.0337	155.7509	6
11	−0.0076	−0.0492	0.0340	0.0598	145.3872	5
12	−0.0036	−0.1278	0.1205	0.1757	136.6834	4
13	−0.0629	0.6668	−0.7926	1.0358	−49.9296	3
14	−0.0373	0.2285	−0.3031	0.3796	−52.9912	2
15	0.0399	−2.0033	2.0832	2.8901	133.8798	1

```
  end
  H(k+1)=H(k+1)/N;
end
```

A function in MATLAB™ to obtain the IDHT is given below:

```
function f=idht(H,N)
% f : inverse discrete Hartley transform
% H : discrete values in the Hartley domain
% N : number of points
for k=0:(N-1)
  f(k+1)=0;
  for n=0:(N-1)
    f(k+1)=H(n+1)*(cos(n*k*2*pi/N)+sin(n*k*2*pi/N))+f(k+1);
  end
end
```

2.7.6 *Fast Hartley transform*

Similarly to the FFT algorithm found in Fourier theory, the fast transform also exists in Hartley theory, the only major difference being that the FHT requires $\frac{1}{2}N\log_2 N$ real operations as opposed to the $\frac{1}{2}N\log_2 N$ complex operations. Another advantage of the FHT over the FFT is that it only requires half the memory storage for the same number of samples.

A function in MATLAB™ to obtain the FHT is given below [4]:

```
function H=FHT(f,N);
P=log2(N);
if (round(P)-P)~0
 fprintf('The number of samples needs to be power of 2');
 break;
```

```
end
for j=1:N,
 H(j)=0;
end
% permutation
for j=1:(P-1),
 X=N/(2^j);
 for k=0:2^(j-1)-1,
  m=1+2*k*X;
  r=m;
  for n=0:X-1,
   H(m)=f(2*m-1-(r-1));
   H(X+m)=f(2*m-(r-1));
   m=m+1;
  end
  f=H;
 end
end
% step 1
j=1;
for k=1:N/2,
 H(j)=f(j)+f(j+1);
 H(j+1)=f(j)-f(j+1);
 j=2*k+1;
end
f=H;
% step 2
j=1;
for k=1:N/4,
 H(j)=f(j)+f(j+2);
 H(j+1)=f(j+1)+f(j+3);
 H(j+2)=f(j)-f(j+2);
 H(j+3)=f(j+1)-f(j+3);
 j=4*k+1;
end
f=H;
% step 3,4,...,P
for s=3:P,
 m=2^s;
 for j=1:m,
  C(j)=cos(2*pi*(j-1)/m);
  S(j)=sin(2*pi*(j-1)/m);
 end
 for k=1:N/m,
  j=1;
  q=j+(k-1)*m;
  H(q)=f(q)+C(j)*f(m/2+q);
  H(m/2+q)=f(q)+C(m/2+j)*f(m/2+q);
  for j=2:m/2,
   q=j+(k-1)*m;
   H(q)=f(q)+C(j)*f(m/2+q)+S(j)*f((2*k-1)*m+2-q);
   H(m/2+q)=f(q)+C(m/2+j)*f(m/2+q)+S(m/2+j)*f((2*k-1)*m+2-q);
  end
 end
 f=H;
end
```

```
H=H/N;
end
```

A function in MATLAB™ to obtain the IFHT is given below:

```
function f=IFHT(H,N)
P=log2(N);
if (round(P)-P)~0
 fprintf('The number of samples needs to be power of 2');
 break;
end
f=fht(H);
f=N*f;
end
```

2.8 Summary

This chapter has presented an overview of orthogonal series expansions with emphasis on Fourier and Hartley analysis. The most important properties of periodic functions have received attention, e.g. even and odd periodicity and half-wave symmetry. Because of their popularity in waveform analysis of electric circuits, the trigonometric and exponential Fourier series have been covered thoroughly. A relatively large portion of the chapter has been dedicated to the Fourier transform where the discrete version of the transform has led to a treatment of the FFT, a very ingenious computer algorithm that has made Fourier analysis applicable to a wide range of engineering problems. A similar treatment has been given to the lesser known Hartley transform, that makes use only of the real plane as oppossed to the complex plane. The all-important concept of convolution has been introduced in both the time and the frequency domains. This is a very useful operation that has opened up the possibility for conducting direct harmonic domain computations in the frequency domain, instead of alternating between time and frequency domain representations. The concept of convolution in both the Fourier and the Hartley frequency domains is used extensively throughout the book. It is important to remark that for the remainder of the book, the periodic, continuous function $f(t)$ is analysed by means of the discrete, non-periodic function $\mathbf{F}(\omega)$, where the latter is a vector formed with the harmonic content of $f(t)$. This action is achieved with the use of both the Fourier and the Hartley series. Equations (2.26), (2.36) and (2.47) are used to carry out complex Fourier operations. Equivalent equations are used to real Fourier and Hartley operations.

2.9 Bibliography

1. P. Kraniauskas, *Transforms in Signals and Systems*, Addison-Wesley, Wokingham, England, 1993.
2. R.N. Bracewell, *The Fourier Transform and its Application*, McGraw-Hill, New York, 1965.
3. O. Brigham, *The Fast Fourier Transform and its Applications*, Prentice Hall, 1988.
4. R.N. Bracewell, *The Hartley Transform*, Oxford University Press, 1986.

3

Electric Circuit Analysis Under Non-sinusoidal Conditions

3.1 Introduction

The apparent power, measured in kVA, delivered by an electric utility consists of active power, measured in kW, and reactive power, measured in kVAR. Active power is the power that performs actual work and reactive power has traditionally been associated with energy storage in a magnetic field. The reactive power may be defined as the orthogonal component of the apparent power,

$$\text{kVAR} = \sqrt{\text{kVA}^2 - \text{kW}^2}$$

Reactive power plays a key role in the operation of electrical power networks, where the inductive nature of most electric equipment leads to a very significant consumption of reactive power. This is mainly in industrial installations where the use of motors and transformers is extensive.

The flow of reactive power through a watt-hour meter will not affect the rate of the meter, but the flow in the power system will result in increased energy losses on both the utility and the industrial site. Accordingly, some utilities charge for the use of kVAR by using a penalty factor, or a kVA demand charge. Others utilities charge a direct fee to customers who operate at a power factor less than a specified target. It is argued that this charge is to recover the cost for lost energy and the additional conductor and transformer capacity required to carry the required kVAR. In addition to being responsible for energy losses, reactive powers flow can also cause excessive voltage drops. Consumption of kVAR from the utility and associated cost may be avoided by using local reactive compensation using bank of capacitors installed at the customer's premises, reducing the kVA delivered by the electric utility, and improving the power factor.

However, non-linear loads and electronic equipment such as electric and induction furnaces and motor drives, generate harmonics which increase the root mean square (RMS) current, and therefore increase kVA consumption and decrease power factor. It should be remarked that harmonics do not necessarily increase kVAR consumption. Under these conditions the use of a bank of capacitors is not enough to restore power factor close to unity. The two orthogonal components of the apparent power do not provide sufficient information to improve power factor compensation when harmonics are present in the electrical system.

Since non-linear loads are an essential component of today's electrical systems, reactive power compensation methods require more complete studies based on harmonic modelling and analysis and metering practices in order to obtain better solutions.

A review of the fundamental concepts of electric circuits operating under non-sinusoidal conditions is presented in this chapter. These concepts will be presented using harmonic domain techniques with emphasis on complex Fourier series. A review of phasor analysis and symmetrical components for steady-state circuit analysis are also presented. The chapter closes by addressing harmonic propagation in linear circuits.

3.2 Key Periodic Electrical Relations

Traditionally, the following active and reactive power expressions

$$P = V_{\text{rms}} I_{\text{rms}} \cos\theta$$

$$Q = V_{\text{rms}} I_{\text{rms}} \sin\theta$$

have been used in electrical engineering applications. In these expressions, $V_{\text{rms}} = V_{\text{peak}}/\sqrt{2}$, $I_{\text{rms}} = V_{\text{peak}}/\sqrt{2}$ and θ is the phase angle between the voltage and the current.

It is important to understand that these expressions are well suited to the study of electric circuits operating with sinusoidal voltage and current, and that they will not apply in cases when the voltage and current waveforms are non-sinusoidal. This is an issue of a great importance since, as time goes on, with the composition of the load shifting from quasi-linear to highly non-linear, the use of such simplified expressions becomes more difficult to justify.

Electric power in non-sinusoidal circuits has been a subject of much study since 1927 when Budeanu separated the apparent power into three orthogonal components [12],

$$S^2 = P^2 + Q_B^2 + D^2$$

where reactive power was defined as $Q_B = \sum_{n=1}^{\infty} V_n I_n \sin\theta_n$, where V_n and I_n are the voltage and current at the harmonic n and θ_n the phase angle between them.

Since the time and over the years, different power terms and definitions have appeared in the literature. Some of these terms yield insight but not others. Tariffs and reactive power compensation applications have provided the motivation for continuous work in this contentious are of electrical power systems. To date, there seems to be no unified view on what set of power terms, definitions or standards are to be used in cases of non-sinusoidal operation. This issue has a significant bearing on how the utility bills the customer, and each utility seems to have found a working solution to the problem. Also, these power terms have played an important role in reactive power compensation applications.

A more academic angle on the subject of power flows in non-sinusoidal circuits is pursued in this chapter. We start by examining the fundamental principle of instantaneous power, which clearly has a physical meaning for it satisfies the principle of conservation of energy in an electric circuit operating under either sinusoidal and non-sinusoidal operation.

Instantaneous power

From the physical and AC electric circuit theory viewpoints, the product of the voltage and the current represents the instantaneous power delivered by the source,

$$p(t) = v(t)i(t) \tag{3.1}$$

where the voltage $v(t)$ and current $i(t)$ are periodic functions, with period T_0,

$$v(t) = v(t + T_0) \tag{3.2}$$

$$i(t) = i(t + T_0) \tag{3.3}$$

The integral of the instantaneous power gives the average power, which is directly related to the energy delivered over the life span of the source. The instantaneous and average power both have physical meaning and satisfy the principle of conservation of energy. Other power concepts, such as apparent power, and reactive power do not have physical meaning and, in general, do not conform to the principle of conservation of energy.

Average or active power

The active power P is defined as the instantaneous power consumed in a period T_0,

$$P = \frac{1}{T_0} \int_{-T_0/2}^{T_0/2} p(t) \mathrm{d}t \tag{3.4}$$

It is this power that will be read by a wattmeter, and its units are watts (W).

RMS values

The RMS voltage and the current are defined as

$$V_{\mathrm{rms}} = \sqrt{\frac{1}{T_0} \int_{-T_0/2}^{T_0/2} v^2(t) \mathrm{d}t} \tag{3.5}$$

$$I_{\mathrm{rms}} = \sqrt{\frac{1}{T_0} \int_{-T_0/2}^{T_0/2} i^2(t) \mathrm{d}t} \tag{3.6}$$

which have units of volts and amperes, respectively. It should be noted that the active power given by the RMS voltage and current values in an AC circuit should equal those of a DC circuit of comparable rating.

From a mathematical viewpoint, the RMS is the norm 2 value. It has the geometric interpretation of a distance in the real space.

Apparent power

The apparent power S is defined as

$$S = V_{\mathrm{rms}} I_{\mathrm{rms}} \tag{3.7}$$

which has units of volt amperes (VA).

Power factor

The power factor (PF) is the ratio by which the apparent power must be multiplied to obtain the active power,

$$\mathrm{PF} = \frac{P}{S} \tag{3.8}$$

The power factor gives the energy efficiency of an electric circuit.

3.2.1 *Non-sinusoidal functions*

The above definitions of instantaneous power, active power, RMS values, apparent power and power factor are valid for any shape of the voltage and current waveforms as long as they are periodic. They provide a good starting point for deriving useful expressions for the study of electric circuits operating with periodic, non-sinusoidal voltages and currents.

The periodic voltage and current expressed by means of their complex Fourier series are

$$v(t) = \sum_{m=-\infty}^{\infty} V_m \mathrm{e}^{\mathrm{j}m\omega_0 t} \tag{3.9}$$

$$i(t) = \sum_{n=-\infty}^{\infty} I_n \mathrm{e}^{\mathrm{j}n\omega_0 t} \tag{3.10}$$

These relations will be substituted in the various functions introduced in the previous section.

Active power

Substitution of the periodic relations (3.9) and (3.10) in (3.1) leads to an instantaneous power relation which is also periodic:

$$p(t) = \sum_{m=-\infty}^{\infty} \sum_{n=-\infty}^{\infty} V_m I_n \mathrm{e}^{\mathrm{j}(m+n)\omega_0 t} \tag{3.11}$$

Applying the definition of active power given by (3.4) to (3.11) leads to the following expression:[1]

$$P = \sum_{m=-\infty}^{\infty} V_m I_{-m} \tag{3.12}$$

where the active power has units of watts. It should be noted that the integral operation yields a value different from zero only when $m = -n$.

RMS values

Using (3.9) we have that $v^2(t)$ is given by

---

[1] The active power is given by the *inner product* of the complex vectors, $P = \sum V_n^* I_n = V^* I = |V^* I|$, where * means transposed complex conjugate.

$$v^2(t) = \sum_{m=-\infty}^{\infty} \sum_{m'=-\infty}^{\infty} V_m V_{m'} \mathrm{e}^{\mathrm{j}(m+m')\omega_0 t}$$

where the integral of $v^2(t)$ is different from zero only when $m = -m'$. Then, the RMS value is given by[2]

$$V_{\text{rms}} = \sqrt{\sum_{m=-\infty}^{\infty} V_m V_{-m}} \tag{3.13}$$

or

$$V_{\text{rms}} = \sqrt{\sum_{m=-\infty}^{\infty} |V_m|^2} \tag{3.14}$$

By the same token,

$$I_{\text{rms}} = \sqrt{\sum_{n=-\infty}^{\infty} |I_n|^2} \tag{3.15}$$

When the harmonics are included in the RMS value, the RMS is known as "true" RMS.

Apparent power

Using the results in (3.14) and (3.15), and the apparent power definition in (3.7), we have a very useful expression

$$S = \sqrt{\sum_{m=-\infty}^{\infty} \sum_{n=-\infty}^{\infty} |V_m|^2 |I_n|^2} \tag{3.16}$$

that is developed further below.

Orthogonal components of the apparent power

The apparent power, given by (3.16), is decomposed into three orthogonal components in order to generate additional information for power factor compensation studies.

Using expression (3.12) in (3.16), and squaring, gives

$$S^2 = S^2 - P^2 + P^2 = \sum_{m=-\infty}^{\infty} \sum_{n=-\infty}^{\infty} \left(|V_m|^2 |I_n|^2 - V_m I_{-m} V_n I_{-n}\right) + P^2 \tag{3.17}$$

---

[2] The RMS values correspond to the norm 2 for complex vectors:

$$||V||_2 = \left(|V_{-n}|^2 + \cdots + |V_0|^2 + \cdots + |V_n|^2\right)^{\frac{1}{2}} = (V^* V)^{\frac{1}{2}}$$

where

$$P^2 = \sum_{m=-\infty}^{\infty} \sum_{n=-\infty}^{\infty} V_m I_{-m} V_n I_{-n} \tag{3.18}$$

The first term on the rhs of (3.17) may be written as follows:

$$\sum_{m=-\infty}^{\infty} \sum_{n=-\infty}^{\infty} \left(|V_m|^2 |I_n|^2 - V_m I_{-m} V_n I_{-n} \right) = \sum_{m=-\infty}^{\infty} \left(|V_m|^2 |I_m|^2 - V_m I_{-m} V_m I_{-m} \right) + \sum_{m=-\infty}^{\infty} \sum_{n=-\infty,\ |n| \neq |m|}^{\infty} \left(|V_m|^2 |I_n|^2 - V_m I_{-m} V_n I_{-n} \right) \tag{3.19}$$

Substituting this result in (3.17) gives

$$\begin{aligned} S^2 &= \sum_{m=-\infty}^{\infty} \left(|V_m|^2 |I_m|^2 - V_m I_{-m} V_m I_{-m} \right) \\ &\quad + \sum_{m=-\infty}^{\infty} \sum_{n=-\infty,\ |n| \neq |m|}^{\infty} \left(|V_m|^2 |I_n|^2 - V_m I_{-m} V_n I_{-n} \right) + P^2 \end{aligned} \tag{3.20}$$

This is a key expression which shows that in addition to the active power term, P, the apparent power can be divided into two other kinds of orthogonal components:

1. The first term on the rhs of equation (3.20) is termed reactive power, Q_{H},

$$Q_{\mathrm{H}}^2 = \sum_{m=-\infty}^{\infty} \left(|V_m|^2 |I_m|^2 - V_m I_{-m} V_m I_{-m} \right) \tag{3.21}$$

This term is given by the multiplication of harmonic voltages and currents of the same frequency.

2. The second term on the rhs of (3.20) is termed distortion power, D_{H},

$$D_{\mathrm{H}}^2 = \sum_{m=-\infty}^{\infty} \sum_{n=-\infty,\ |n| \neq |m|}^{\infty} \left(|V_m|^2 |I_n|^2 - V_m I_{-m} V_n I_{-n} \right) \tag{3.22}$$

This distortion power is given by the multiplication of harmonic voltages and currents of different frequencies.

It should be noted that the reactive power and the distortion power are very similar in form. Also, the following power equality should be satisfied at any given moment:

$$S^2 = P^2 + Q_{\mathrm{H}}^2 + D_{\mathrm{H}}^2 \tag{3.23}$$

where S, P, Q_{H}, and D_{H} have units of VA, W, VAR and volt amperes distortion (VAD), respectively. Figure 3.1 shows the geometric representation of these powers.

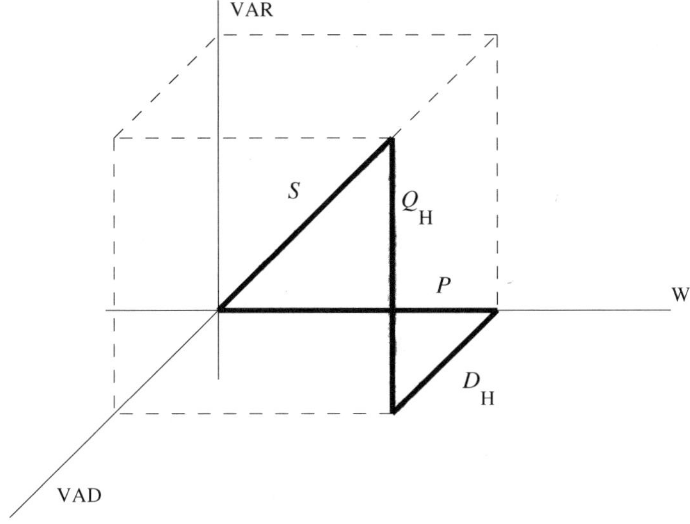

Figure 3.1 Power cube

Power factor

Substituting equations (3.12) and (3.16) into equation (3.8) gives a general expression for the power factor:[3]

$$\mathrm{PF} = \frac{\sum\limits_{n=-\infty}^{\infty} V_m I_{-m}}{\sqrt{\sum\limits_{m=-\infty}^{\infty} \sum\limits_{n=-\infty}^{\infty} |V_m|^2 \, |I_n|^2}} \tag{3.24}$$

Total harmonic distortion

The total harmonic distortion (THD) for the voltage and the current is given as

$$\mathrm{THD}_V = \sqrt{\frac{\sum\limits_{m=2}^{\infty} |V_m|^2}{|V_1|^2}} \times 100\% \tag{3.25}$$

$$\mathrm{THD}_I = \sqrt{\frac{\sum\limits_{n=2}^{\infty} |I_n|^2}{|I_1|^2}} \times 100\% \tag{3.26}$$

---

[3] The power factor represents the *Cauchy–Schwartz inequality* for complex vectors $|V^* I| \leq ||V||_2 \, ||I||_2$ or $\frac{|V^* I|}{||V||_2 ||I||_2} \leq 1$, where $|V^* I|$ is the active power, and $||V||_2 \, ||I||_2$ is the apparent power.

Example 3-1: Use the instantaneous voltage and current relations given in Section 3.2.1 to find the relevant power quality information for the following voltage and current set:

$$\mathbf{V} = \begin{bmatrix} V_{-4} \\ V_{-3} \\ V_{-2} \\ V_{-1} \\ V_0 \\ V_1 \\ V_2 \\ V_3 \\ V_4 \end{bmatrix} = \begin{bmatrix} 0 \\ 0 \\ 0 \\ 1 \\ 0 \\ 1 \\ 0 \\ 0 \\ 0 \end{bmatrix} ; \quad \mathbf{I} = \begin{bmatrix} I_{-4} \\ I_{-3} \\ I_{-2} \\ I_{-1} \\ I_0 \\ I_1 \\ I_2 \\ I_3 \\ I_4 \end{bmatrix} = \begin{bmatrix} 0 \\ -0.15 \\ 0 \\ 0.5 \\ 0 \\ 0.5 \\ 0 \\ -0.15 \\ 0 \end{bmatrix}$$

From inspection of these vectors we can see that the voltage is a sinusoidal, even function and the current is a non-sinusoidal, even function.

The power quality parameters are

$$\begin{aligned}
V_{\text{rms}} &= \sqrt{|V_1|^2 + |V_{-1}|^2} = \sqrt{|1|^2 + |1|^2} = 1.4141 \\
I_{\text{rms}} &= \sqrt{|I_{-3}|^2 + |I_{-1}|^2 + |I_1|^2 + |I_3|^2} \\
&= \sqrt{|-0.15|^2 + |0.5|^2 + |0.5|^2 + |-0.15|^2} = 0.7382 \\
S &= V_{\text{rms}} I_{\text{rms}} = (1.4141)(0.7382) = 1.0440 \\
P &= V_{-1} I_1 + V_1 I_{-1} = (1)(0.5) + (1)(0.5) = 1 \\
Q_{\text{H}} &= \sqrt{|V_{-1}|^2 |I_1|^2 - V_{-1} I_1 V_{-1} I_1 + |V_1|^2 |I_{-1}|^2 - V_1 I_{-1} V_1 I_{-1}} \\
&= \sqrt{|1|^2 |0.5|^2 - (1)(0.5)(1)(0.5) + |1|^2 |0.5|^2 - (1)(0.5)(1)(0.5)} = 0 \\
D_{\text{H}} &= \sqrt{\begin{aligned}&|V_{-1}|^2 |I_{-3}|^2 - V_{-1} I_1 V_{-3} I_3 + |V_{-1}|^2 |I_3|^2 - V_{-1} I_1 V_3 I_{-3} \\ &+ |V_1|^2 |I_{-3}|^2 - V_1 I_{-1} V_{-3} I_3 + |V_1|^2 |I_3|^2 - V_1 I_{-1} V_3 I_{-3}\end{aligned}} \\
&= \sqrt{\begin{aligned}&|1|^2 |-0.15|^2 - (1)(0.5)(0)(-0.15) + |1|^2 |-0.15|^2 - (1)(0.5)(0)(-0.15) \\ &+ |1|^2 |-0.15|^2 - (1)(0.5)(0)(-0.15) + |1|^2 |-0.15|^2 - (1)(0.5)(0)(-0.15)\end{aligned}} \\
&= 3 \\
\text{PF} &= \frac{1}{1.0440} = 0.9578 \\
\text{THD}_V &= \sqrt{\frac{0}{(1)^2}} \times 100\% = 0 \\
\text{THD}_I &= \sqrt{\frac{(0.15)^2}{(0.5)^2}} \times 100\% = 30\%
\end{aligned}$$

This example bears some resemblance to the situation present in energy saving lamps, TVs, and computers. where the voltage and current have equal symmetry, i.e., they are in phase, but the current is non-sinusoidal, resulting in $Q_{\mathrm{H}} = 0$ and $D_{\mathrm{H}} \neq 0$, leading to a low power factor. In practice, the power factor will be much lower than 0.9578 because the distortion power will be much higher.

Example 3-2: A more involved case is presented below, where the harmonic currents and voltages both contain harmonic distortion:

$$\mathbf{V} = \begin{bmatrix} 0 \\ 0.5 \\ 0 \\ -1+\mathrm{j}0.5 \\ 0 \\ -1-\mathrm{j}0.5 \\ 0 \\ 0.5 \\ 0 \end{bmatrix} ; \mathbf{I} = \begin{bmatrix} 0 \\ 0.5+\mathrm{j}0.2 \\ 0 \\ -0.5+\mathrm{j}0.5 \\ 0 \\ -0.5-\mathrm{j}0.5 \\ 0 \\ 0.5-\mathrm{j}0.2 \\ 0 \end{bmatrix}$$

The following power quality parameters are obtained:

$$V_{\mathrm{rms}} = 1.7321 \qquad P = 2 \qquad \mathrm{PF} = 0.9186$$

$$I_{\mathrm{rms}} = 1.2570 \qquad Q_{\mathrm{H}} = 0.5385 \qquad \mathrm{THD}_V = 44.7214\%$$

$$S = 2.1772 \qquad D_{\mathrm{H}} = 0.6708 \qquad \mathrm{THD}_I = 76.1577\%$$

It should be noted that the values of P, Q_{H} and D_{H} satisfy the relation $S^2 = P^2 + Q_{\mathrm{H}}^2 + D_{\mathrm{H}}^2$.

The function in MATLAB™ used to calculate the power quality parameters is given below:

```
function [S,P,Q,D,Vrms,Irms,PF,VTHD,ITHD]=spqd(V,I,h)
% V      : Vector V=[-h ... -1 0 1 ... h]
% I      : Vector I=[-h ... -1 0 1 ... h]
% h      : number of harmonics
Vrms  = sqrt(sum(abs(V).^2));
Irms  = sqrt(sum(abs(I).^2));
S     = Vrms*Irms;
P     = sum(V.*conj(I));
Q     = sqrt(sum(abs(V).^2.*abs(I).^2-V.*conj(I).*V.*conj(I)));
D     = sqrt(S^2-P^2-Q^2);
PF    = P/S;
VTHD  = sqrt(sum(abs(V(h+3:2*h+1)).^2)/abs(V(h+2).^2))*100;
ITHD  = sqrt(sum(abs(I(h+3:2*h+1)).^2)/abs(I(h+2).^2))*100;
```

3.2.2 Sinusoidal functions

Particular cases of the general expressions derived in Section 3.2 become readily available when the voltage and current waveforms are purely sinusoidal, i.e.

$$v(t) = V_{\max} \cos \omega_0 t \tag{3.27}$$

$$i(t) = I_{\max} \cos(\omega_0 t + \theta) \tag{3.28}$$

Using the complex exponential form of $v(t)$ and $i(t)$,

$$v(t) = \frac{V_{\max}}{2}\mathrm{e}^{-\mathrm{j}\omega_0 t} + \frac{V_{\max}}{2}\mathrm{e}^{\mathrm{j}\omega_0 t}$$

and

$$i(t) = \frac{I_{\max}}{2}(\cos\theta - \mathrm{j}\sin\theta)\mathrm{e}^{-\mathrm{j}\omega_0 t} + \frac{I_{\max}}{2}(\cos\theta + \mathrm{j}\sin\theta)\mathrm{e}^{\mathrm{j}\omega_0 t}$$

leads to the following vectors of harmonic coefficients:

$$\begin{aligned} \begin{bmatrix} V_{-2} \\ V_{-1} \\ V_0 \\ V_1 \\ V_2 \end{bmatrix} &= \frac{V_{\max}}{2}\begin{bmatrix} 0 \\ 1 \\ 0 \\ 1 \\ 0 \end{bmatrix} \\ \begin{bmatrix} I_{-2} \\ I_{-1} \\ I_0 \\ I_1 \\ I_2 \end{bmatrix} &= \frac{I_{\max}}{2}\begin{bmatrix} 0 \\ \cos\theta - \mathrm{j}\sin\theta \\ 0 \\ \cos\theta + \mathrm{j}\sin\theta \\ 0 \end{bmatrix} \end{aligned} \tag{3.29}$$

These harmonic coefficients, in turn, are used in the basic definitions given in Section 3.2.1 to derive fundamental frequency expressions for the following electrical parameters.

Active power

Using equation (3.12), the active power P is given by

$$P = V_{-1}I_1 + V_1 I_{-1} = \frac{V_{\max}I_{\max}}{2}\cos\theta \tag{3.30}$$

RMS values

Substituting the harmonics content of voltage and current into (3.14) and (3.15) gives the V_{rms} and I_{rms} expressions, respectively:

$$\begin{aligned} V_{\text{rms}} &= \sqrt{|V_{-1}|^2 + |V_1|^2} = \frac{V_{\max}}{\sqrt{2}} \\ I_{\text{rms}} &= \sqrt{|I_{-1}|^2 + |I_1|^2} = \frac{I_{\max}}{\sqrt{2}} \end{aligned} \tag{3.31}$$

Apparent power

The apparent power becomes readily available from the above result:

$$S = V_{\text{rms}}I_{\text{rms}} = \frac{V_{\max}I_{\max}}{2} \tag{3.32}$$

Reactive power

From equation (3.21) we have that the reactive power is

$$\begin{aligned} Q_{\mathrm{H}}^2 &= |V_{-1}|^2 |I_1|^2 - V_{-1} I_1 V_{-1} I_1 + |V_1|^2 |I_{-1}|^2 - V_1 I_{-1} V_1 I_{-1} \\ &= \left(\frac{V_{\max} I_{\max}}{2} \sin\theta \right)^2 \\ Q_{\mathrm{H}} &= Q = \frac{V_{\max} I_{\max}}{2} \sin\theta \end{aligned} \tag{3.33}$$

where Q_{H} for sinusoidal conditions equals the reactive power term Q, an expression well known in classic electric circuit theory.

Distortion power

In this case, only the fundamental voltage and current values are different from zero. Accordingly, we have that the inner summation of the distortion power given by equation (3.22) yields a value of zero, i.e. $D_{\mathrm{H}} = 0$. As expected, the following relation is satisfied:

$$S^2 = P^2 + Q^2 \tag{3.34}$$

Figure 3.2 shows the geometric representation of the three powers in (3.34), where the inductive and capacitive nature of the reactive power is emphasised.

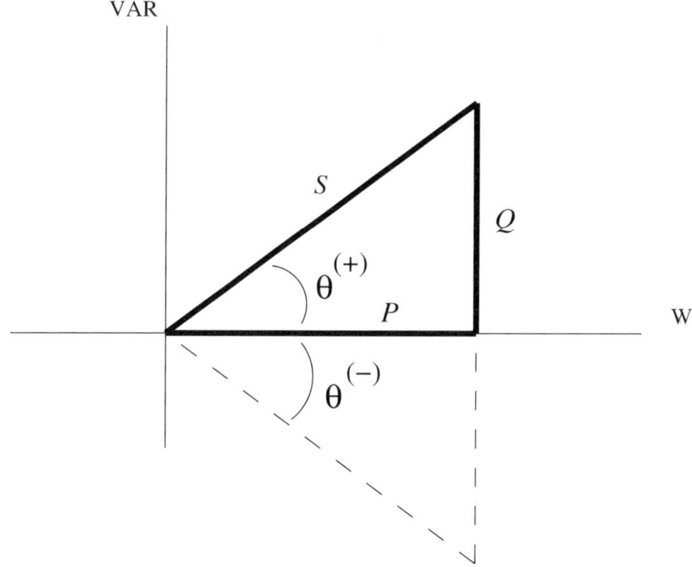

Figure 3.2 Power triangle

Power factor

It is not difficult to see that for cases of sinusoidal voltage and current waveforms, a straightforward substitution of (3.30) and (3.32) into (3.8) leads to the well-known power factor expression

$$\mathrm{PF} = \cos\theta \tag{3.35}$$

When harmonics are involved in the analysis, the shift angle between voltage and current at the fundamental frequency, i.e. θ_1, is used to define the displacement power factor (DPF) given by

$$\mathrm{DPF} = \cos\theta_1 \tag{3.36}$$

In general DPF > PF, and being equal for sinusoidal conditions.

The following useful expression can be derived from inspection of the power triangle in Figure 3.2:

$$\mathrm{PF} = \cos\left(\arctan\left(\frac{Q}{P}\right)\right) \tag{3.37}$$

The power factor will vary in the interval

$$0 \leq \mathrm{PF} \leq 1 \tag{3.38}$$

PF will take lagging (−) and leading (+) values to reflect the sign of the angle θ. Lagging power factors will exist in inductive circuits and leading power factors in capacitive circuits.

3.2.3 Symmetry and power definitions

Key properties of the various electrical parameters introduced above are emphasised in this section by refering to four cases of the symmetry of voltage and current functions.

1. Sinusoidal voltage and current waveforms.

Figure 3.3 shows the behaviour of the various power terms as a function of the phase angle difference between the voltage and the current. At zero phase shift both voltage and current have odd symmetry, i.e. $V_n = -V_{-n}$ and $I_n = -I_{-n}$, giving $P = S$, $Q_{\mathrm{H}} = 0$ and $D_{\mathrm{H}} = 0$. This case corresponds to a resistive load.

At 90° phase shift the current has even symmetry, i.e. $I_n = -I_{-n}$, and the voltage keeps its odd symmetry giving $P = 0$, $Q_{\mathrm{H}} = S$ and $D_{\mathrm{H}} = 0$. This case corresponds to an inductive load.

This analysis corresponds to a linear load operating under sinusoidal conditions. As expected, the power factor decreases from one to zero as the phase shift difference increases.

2. The voltage is sinusoidal and the current is non-sinusoidal.

Figure 3.4 shows the behaviour of the various power terms as a function of the phase angle between the voltage and the current. It can be seen that at zero phase shift both voltage and current have odd symmetry but the current is non-sinusoidal, giving $P < S$, $Q_{\mathrm{H}} = 0$ and $D_{\mathrm{H}} \neq 0$, resulting in PF < 1. This operating condition is typical of current waveforms found in TVs, computers, energy saving lamps and printers.

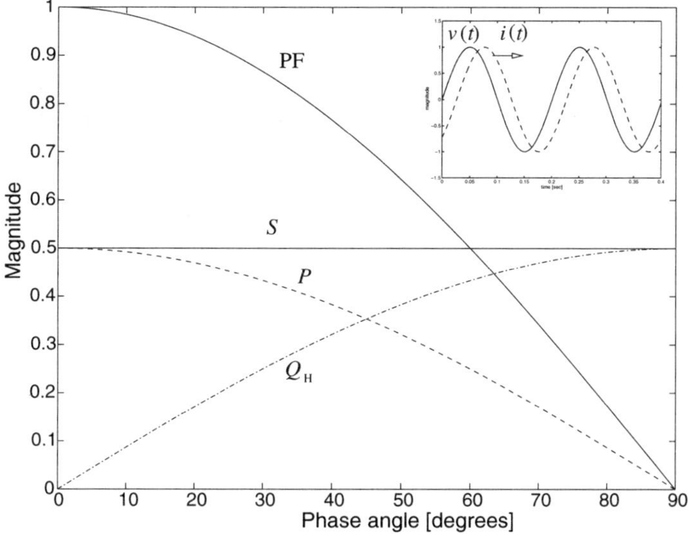

Figure 3.3 Sinusoidal voltage, sinusoidal current

At 90° the current and voltage have even and odd symmetry, respectively, giving $P = 0$, $Q_H < S$ and $D_H \neq 0$, resulting in PF = 0. This operating condition is typical of non-linear reactors, induction furnaces and motor drives. In general, this condition corresponds to a non-linear load operating under sinusoidal voltage excitation. It should be noted that the distortion power remains constant throughout the phase shift sweep.

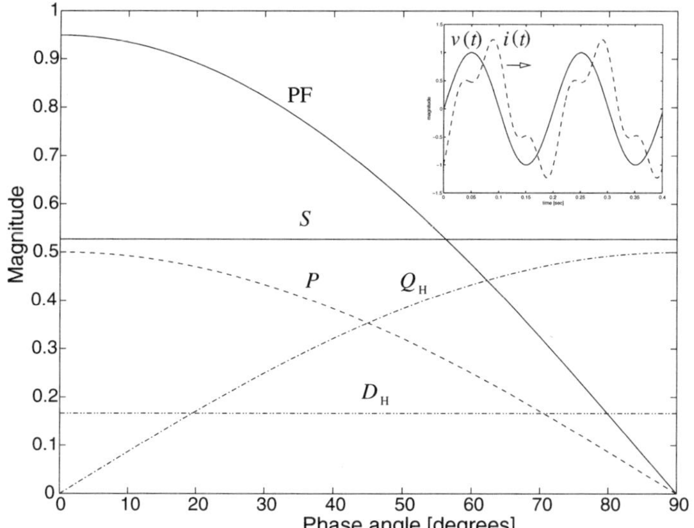

Figure 3.4 Sinusoidal voltage, non-sinusoidal current

3. The voltage and current are both non-sinusoidal, with identical harmonic content.

Figure 3.5 shows the behaviour of the various power terms as a function of the phase angle difference between the voltage and the current. It can be seen that at zero phase shift both voltage and current have odd symmetry with identical harmonic content, giving $P = S$, $Q_{\mathrm{H}} = 0$ and $D_{\mathrm{H}} = 0$, resulting in $\mathrm{PF} = 1$.

At 90° the current has even symmetry and the voltage odd symmetry, giving $P = 0$, $Q_{\mathrm{H}} < S$ and $D_{\mathrm{H}} \neq 0$, resulting in $\mathrm{PF} = 0$. The behaviour of D_{H} is not constant in this case. The former condition involves a resistive circuit whereas the latter would involve an inductive circuit.

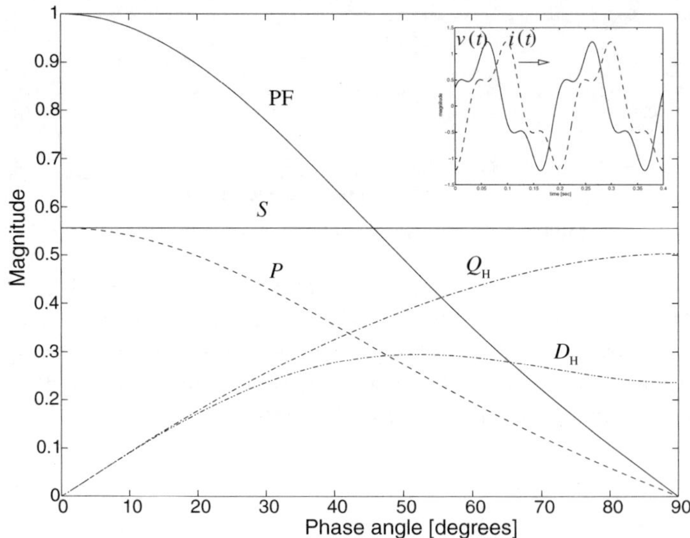

Figure 3.5 Non-sinusoidal voltage, non-sinusoidal current

4. The voltage and current are both non-sinusoidal and have different harmonic contents.

Figure 3.6 shows the behaviour of the various power terms as a function of the phase angle difference between the voltage and the current. It can be seen that at zero phase shift both voltage and current have odd symmetry but different harmonic content, giving $P < S, Q \neq 0$ and $D \neq 0$, resulting in $\mathrm{PF} < 1$.

At 90° the current and the voltage have, respectively, even and odd symmetry, giving $P \neq 0, Q_{\mathrm{H}} < S$ and $D_{\mathrm{H}} \neq 0$, resulting in $\mathrm{PF} > 0$ where D_{H} shows a non-constant behaviour. This is the more general case corresponding to a circuit with non-linear load operating under non-sinusoidal conditions.

3.3 Phasor Analysis

The use of complex numbers is a popular resort in the solution of electric circuits operating under steady-state, sinusoidal conditions. A systematic treatment of this approach is called phasor analysis. It is based on Euler's identity,

$$|A|\,\mathrm{e}^{\mathrm{j}\alpha} = |A|\cos\alpha + \mathrm{j}\,|A|\sin\alpha$$

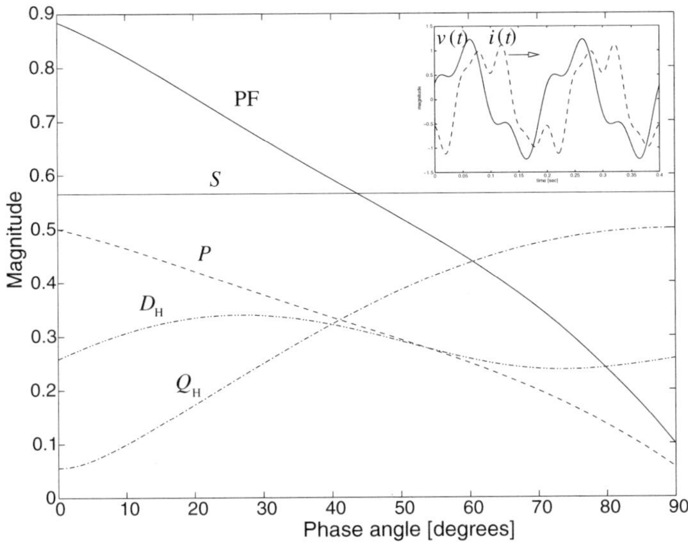

Figure 3.6 Non-sinusoidal voltage, non-sinusoidal current

where the sinusoidal voltage and current can be represented by

$$v(t) = |V| \cos(\omega_0 t + \theta_v) = |V| \Re \left\{ \mathrm{e}^{\mathrm{j}\theta_v} \mathrm{e}^{\mathrm{j}\omega_0 t} \right\} \tag{3.39}$$

$$i(t) = |I| \cos(\omega_0 t + \theta_i) = |I| \Re \left\{ \mathrm{e}^{\mathrm{j}\theta_i} \mathrm{e}^{\mathrm{j}\omega_0 t} \right\} \tag{3.40}$$

The steady-state response of passive elements is given by

$$v(t) = Ri(t)$$

$$v(t) = L \frac{\mathrm{d}i(t)}{\mathrm{d}t}$$

$$v(t) = \frac{1}{C} \int i(t) \mathrm{d}t$$

can be expressed as

$$|V| \Re \left\{ \mathrm{e}^{\mathrm{j}\theta_v} \mathrm{e}^{\mathrm{j}\omega_0 t} \right\} = R \, |I| \Re \left\{ \mathrm{e}^{\mathrm{j}\theta_i} \mathrm{e}^{\mathrm{j}\omega_0 t} \right\} \tag{3.41}$$

$$|V| \Re \left\{ \mathrm{e}^{\mathrm{j}\theta_v} \mathrm{e}^{\mathrm{j}\omega_0 t} \right\} = \mathrm{j}\omega_0 L \, |I| \Re \left\{ \mathrm{e}^{\mathrm{j}\theta_i} \mathrm{e}^{\mathrm{j}\omega_0 t} \right\} \tag{3.42}$$

$$|V| \Re \left\{ \mathrm{e}^{\mathrm{j}\theta_v} \mathrm{e}^{\mathrm{j}\omega_0 t} \right\} = \frac{1}{\mathrm{j}\omega_0 C} \, |I| \Re \left\{ \mathrm{e}^{\mathrm{j}\theta_i} \mathrm{e}^{\mathrm{j}\omega_0 t} \right\} \tag{3.43}$$

By removing the terms $\mathrm{e}^{\mathrm{j}\omega_0 t}$ and $\Re\{\bullet\}$ from the lhs and rhs of these equations, the phasor quantities are obtained:

$$|V| \, \mathrm{e}^{\mathrm{j}\theta_v} = R \, |I| \, \mathrm{e}^{\mathrm{j}\theta_i} \tag{3.44}$$

$$|V| \, \mathrm{e}^{\mathrm{j}\theta_v} = \mathrm{j}X_L \, |I| \, \mathrm{e}^{\mathrm{j}\theta_i} \tag{3.45}$$

$$|V|\,\mathrm{e}^{\mathrm{j}\theta_v} = -\mathrm{j}X_C\,|I|\,\mathrm{e}^{\mathrm{j}\theta_i} \tag{3.46}$$

where $X_L = \omega_0 L$ and $X_C = \frac{1}{\omega_0 C}$.

From these expressions, the *impedance* term is defined,

$$Z = \frac{|V|\,\mathrm{e}^{\mathrm{j}\theta_v}}{|I|\,\mathrm{e}^{\mathrm{j}\theta_i}} = \frac{|V|}{|I|}\mathrm{e}^{\mathrm{j}(\theta_v - \theta_i)} = |Z|\,\mathrm{e}^{\mathrm{j}\theta_z} = R + \mathrm{j}X \tag{3.47}$$

and the *admittance* is the inverse relationship of the impedance, i.e. $Y = \frac{1}{|Z|}\mathrm{e}^{-\mathrm{j}\theta_z} = G + \mathrm{j}B$, where R, X, G and B are termed *resistance*, *reactance*, *conductance* and *susceptance*, respectively.

The use of phasors $|V|\,\mathrm{e}^{\mathrm{j}\theta_v}$, $|I|\,\mathrm{e}^{\mathrm{j}\theta_i}$ and $|Z|\,\mathrm{e}^{\mathrm{j}\theta_z}$ lends to simple calculations in nodal and mesh circuit analysis.

Using phasors, the complex power is defined by the product of the voltage and the complex conjugate current,

$$U = VI^* = |V|\,|I|\,\mathrm{e}^{\mathrm{j}(\theta_v - \theta_i)} = |U|\,\mathrm{e}^{\mathrm{j}\theta_u} \tag{3.48}$$

where the following relations hold for cases of sinusoidal voltage and current: $|U| = 2S$, $\Re\{U\} = 2P$, $\Im\{U\} = -2Q$ and $\cos\theta_u = \mathrm{PF}$. Moreover, if $|V|$ and $|I|$ are the RMS values of $v(t)$ and $i(t)$, respectively, then $|U| = S$, $\Re\{U\} = P$ and $\Im\{U\} = -Q$.

It should be remarked that only the voltage and current have time domain representations, with the impedance and complex power having been defined in the complex Fourier domain at a specific angular frequency ω_0. When different frequencies are involved in the circuit, then phasor analysis is carried out at each frequency of interest, and by using superposition the final time domain representation of voltages and currents can be obtained.

3.4 Symmetrical Components

Charles L. Fortescue [15] formulated the theory of symmetrical components, a frame of reference where an unbalanced three-phase phasor system (*abc*) can be resolved into three balanced phasor systems (012). The three sets are:

1. Zero sequence components “0”, consisting of three phasors equal in magnitude and with zero phase displacement from each other.
2. Positive sequence components “1”, consisting of three phasors of equal magnitude, shifted from each other by 120° in phase, and having the same phase sequence as the original set (*abc*).
3. Negative sequence components “2”, consisting of three phasors of equal magnitude, shifted from each other by 120° in phase, and having the opposite phase sequence as the original set (*acb*).

The matrix relation that transforms phase quantities into symmetrical components quantities is

$$\begin{bmatrix} I_0 \\ I_1 \\ I_2 \end{bmatrix} = \frac{1}{3}\begin{bmatrix} 1 & 1 & 1 \\ 1 & a & a^2 \\ 1 & a^2 & a \end{bmatrix}\begin{bmatrix} I_a \\ I_b \\ I_c \end{bmatrix}$$

and the inverse relation is

$$\begin{bmatrix} I_a \\ I_b \\ I_c \end{bmatrix} = \begin{bmatrix} 1 & 1 & 1 \\ 1 & a^2 & a \\ 1 & a & a^2 \end{bmatrix} \begin{bmatrix} I_0 \\ I_1 \\ I_2 \end{bmatrix}$$

where $a = 1\angle 120^\circ$ and $a^2 = 1\angle -120^\circ$.

3.4.1 *Three-phase sinusoidal systems*

A balanced three-phase sinusoidal system is given by

$$i_a(t) = I \cos \omega_0 t$$

$$i_b(t) = I \cos(\omega_0 t - 120^\circ)$$

$$i_c(t) = I \cos(\omega_0 t + 120^\circ)$$

or, in phasor quantities,

$$I_a = I\angle 0^\circ, I_b = I\angle -120^\circ \text{ and } I_c = I\angle 120^\circ$$

The phasor representation is expressed with ease in the symmetrical components frame of reference,

$$\begin{bmatrix} I_0 \\ I_1 \\ I_2 \end{bmatrix} = \frac{1}{3} \begin{bmatrix} 1 & 1 & 1 \\ 1 & a & a^2 \\ 1 & a^2 & a \end{bmatrix} \begin{bmatrix} I_a \\ I_b \\ I_c \end{bmatrix} = \begin{bmatrix} 0 \\ I \\ 0 \end{bmatrix}$$

For unbalanced three-phase sinusoidal quantities the zero and negative sequences may differ from zero, depending on the nature of the imbalance and the circuit connection.

In delta and floating star connections the following basic principle stands: $I_a + I_b + I_c = 0$, which means that no zero sequence can exist in these three-phase connections. Conversely, in a grounded star circuit the basic principle is $I_a + I_b + I_c = I_n$, where I_n is different from zero if unbalanced conditions are present. The I_n current returns to supply via the neutral wire, and it is related to the zero sequence current.

3.4.2 *Three-phase non-sinusoidal systems*

The current in a balanced[4], three-phase non-sinusoidal system is given by

$$\begin{aligned} i_a(t) &= I \cos \omega_0 t + I_h \cos(h\omega_0 t + \phi_h) \\ i_b(t) &= I \cos(\omega_0 t - 120^\circ) + I_h \cos(h(\omega_0 t - 120^\circ) + \phi_h) \\ i_c(t) &= I \cos(\omega_0 t + 120^\circ) + I_h \cos(h(\omega_0 t + 120^\circ) + \phi_h) \end{aligned}$$

and in symmetrical components form:

---

[4] Understanding a three-phase balanced system as $i_a(t) = i_a(t + T_0)$, $i_b(t) = i_a(t + T_0 - 120^\circ)$ and $i_c(t) = i_a(t + T_0 + 120^\circ)$.

1. At fundamental frequency,

$$\begin{bmatrix} I_0 \\ I_1 \\ I_2 \end{bmatrix} = \frac{1}{3}\begin{bmatrix} 1 & 1 & 1 \\ 1 & a & a^2 \\ 1 & a^2 & a \end{bmatrix}\begin{bmatrix} I\angle 0^\circ \\ I\angle -120^\circ \\ I\angle 120^\circ \end{bmatrix} = \begin{bmatrix} 0 \\ I \\ 0 \end{bmatrix}$$

2. At the general harmonic h,

$$\begin{bmatrix} I_0 \\ I_1 \\ I_2 \end{bmatrix} = \frac{1}{3}\begin{bmatrix} 1 & 1 & 1 \\ 1 & a & a^2 \\ 1 & a^2 & a \end{bmatrix}\begin{bmatrix} I_h\angle\phi_h \\ I_h\angle(-h120^\circ + \phi_h) \\ I_h\angle(h120^\circ + \phi_h) \end{bmatrix}$$

Three possible cases emerge from this general condition, the cases of positive, negative and zero sequence harmonics.

The harmonics $h = 3n + 1$, for n integer, e.g. $h = 4,\ 7,\ 10,\ \ldots$, have the same behaviour as the positive sequence fundamental quantity.

$$\begin{bmatrix} I_0 \\ I_1 \\ I_2 \end{bmatrix} = \begin{bmatrix} 0 \\ I_h\angle\phi_h \\ 0 \end{bmatrix}$$

The triplen harmonics $h = 3n$, for n integer, e.g. $h = 3,\ 6,\ 9,\ \ldots$, have the same behaviour as the zero sequence fundamental quantity.

$$\begin{bmatrix} I_0 \\ I_1 \\ I_2 \end{bmatrix} = \begin{bmatrix} I_h\angle\phi_h \\ 0 \\ 0 \end{bmatrix}$$

The harmonics $h = 3n - 1$, for n integer, e.g. $h = 2,\ 5,\ 8,\ \ldots$, have the same behaviour as the negative sequence fundamental quantity.

$$\begin{bmatrix} I_0 \\ I_1 \\ I_2 \end{bmatrix} = \begin{bmatrix} 0 \\ 0 \\ I_h\angle\phi_h \end{bmatrix}$$

In a delta-connected circuit, the zero sequence component is forced to zero since the summation of the three line currents should be zero, i.e. $I_{a_h} + I_{b_h} + I_{c_h} = 0$. Under balanced conditions, in a delta-connected circuit, the harmonics $h = 3,\ 6,\ 9$ do not exist.

For example, in a delta-connected circuit, if the phase c is open, then the harmonic phasors are

$$\begin{aligned} I_{a_h} &= I_h\angle\phi_h \\ I_{b_h} &= I_h\angle(180^\circ + \phi_h) \\ I_{c_h} &= 0 \end{aligned}$$

and in symmetrical components form

$$
\begin{aligned}
I_0 &= 0 \\
I_1 &= \frac{I_h}{\sqrt{3}}\angle(-30^\circ + \phi_h) \\
I_2 &= \frac{I_h}{\sqrt{3}}\angle(30^\circ + \phi_h)
\end{aligned}
$$

As expected, the result indicates that the zero sequence component is zero, but the harmonic h flows in the line, where h may be any harmonic, including the triplens. The same situation occurs in a floating star connection.

In a grounded star circuit, I_n is zero only if the system is balanced and with no triplen harmonics since $I_a + I_b + I_c = I_n$. Under balanced non-sinusoidal conditions the triplen harmonics ($h = 3, 6, 9, \ldots$) flow through the neutral wire. Under unbalanced conditions zero sequence current at the fundamental and harmonic frequencies will circulate through the neutral wire.

3.5 Linear Circuit Analysis

The steady-state analysis of linear circuits operating under sinusoidal and non-sinusoidal conditions may be carried out using phasor analysis, but in the latter case the circuit is solved at each frequency of interest as opposed to only the fundamental frequency. The inductors and capacitors have linear frequency dependence, i.e. $X_L = \mathrm{j}h\omega_0 L$ and $X_C = -\mathrm{j}\frac{1}{h\omega_0 C}$, whereas the resistor may be assumed to remain constant. With these three passive elements and the sources represented by their harmonic content, a linear circuit analysis can be carried out.

In general, a linear circuit operating under non-sinusoidal conditions is well represented by the following linear system of equations:

$$
\begin{bmatrix} I_h^1 \\ I_h^2 \\ \vdots \\ I_h^j \\ \vdots \\ I_h^N \end{bmatrix} = \begin{bmatrix} Y_h^{1,1} & Y_h^{1,2} & \cdots & Y_h^{1,j} & \cdots & Y_h^{1,N} \\ Y_h^{2,1} & Y_h^{2,2} & \cdots & Y_h^{2,j} & \cdots & Y_h^{2,N} \\ \vdots & \vdots & \ddots & \vdots & \ddots & \vdots \\ Y_h^{j,1} & Y_h^{j,2} & \cdots & Y_h^{j,j} & \cdots & Y_h^{j,N} \\ \vdots & \vdots & \ddots & \vdots & \ddots & \vdots \\ Y_h^{N,1} & Y_h^{N,2} & \cdots & Y_h^{N,j} & \cdots & Y_h^{N,N} \end{bmatrix} \begin{bmatrix} V_h^1 \\ V_h^2 \\ \vdots \\ V_h^j \\ \vdots \\ V_h^N \end{bmatrix} \tag{3.49}
$$

where the current I_h^j is the phasor current at frequency h injected at node j, i.e. $I_h^j = |I_h|\angle\theta_h$, $Y_h^{i,j}$ is the equivalent admittance at frequency h between nodes i and j, V_h^j is the phasor voltage at frequency h at node j, and N is the number of nodes of the electric network. Equation (3.49) in compact form is given by

$$
\mathbf{I}_h = \mathbf{Y}_h \mathbf{V}_h \tag{3.50}
$$

The linear equation (3.50) is solved at each frequency h and the final result is obtained by superposition of effects.

The inverse of the admittance matrix $\mathbf{Y}_h$ gives the impedance matrix $\mathbf{Z}_h$,

$$\mathbf{Z}_h = \begin{bmatrix} Z_h^{1,1} & Z_h^{1,2} & \cdots & Z_h^{1,j} & \cdots & Z_h^{1,N} \\ Z_h^{2,1} & Z_h^{2,2} & \cdots & Z_h^{2,j} & \cdots & Z_h^{2,N} \\ \vdots & \vdots & \ddots & \vdots & \ddots & \vdots \\ Z_h^{j,1} & Z_h^{j,2} & \cdots & Z_h^{j,j} & \cdots & Z_h^{j,N} \\ \vdots & \vdots & \ddots & \vdots & \ddots & \vdots \\ Z_h^{N,1} & Z_h^{N,2} & \cdots & Z_h^{N,j} & \cdots & Z_h^{N,N} \end{bmatrix} \tag{3.51}$$

where the impedance $Z_h^{j,j}$ is known as the driving point impedance of node j at different frequencies.

Example 3-3: Find the driving point impedance of the three-node circuit shown in Figure 3.7.

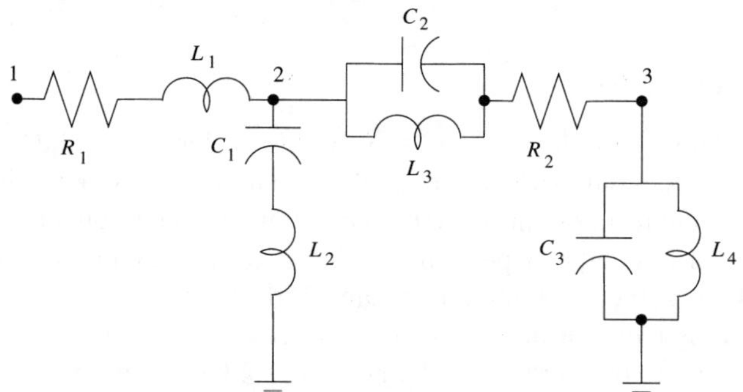

Figure 3.7 Passive circuit

The first step is to determine the equivalent admittance at each frequency of interest for the three-node circuit:

$$\mathbf{Y}(\mathrm{j}h\omega_0) = \begin{bmatrix} Y^{1,1} & Y^{1,2} & Y^{1,3} \\ Y^{2,1} & Y^{2,2} & Y^{2,3} \\ Y^{3,1} & Y^{3,2} & Y^{3,3} \end{bmatrix} = \begin{bmatrix} y_{12} & -y_{12} & 0 \\ -y_{12} & y_{12}+y_{23}+y_{20} & -y_{23} \\ 0 & -y_{23} & y_{23}+y_{30} \end{bmatrix}$$

where

$$\begin{aligned} y_{12} &= \frac{1}{R_1 + \mathrm{j}h\omega_0 L_1} \\ y_{20} &= \frac{1}{\mathrm{j}h\omega_0 L_2 - \mathrm{j}\dfrac{1}{h\omega_0 C_1}} \\ y_{23} &= \frac{1}{R_2 + \dfrac{L_3/C_2}{\mathrm{j}h\omega_0 L_3 - \mathrm{j}\dfrac{1}{h\omega_0 C_2}}} \end{aligned}$$

$$y_{30} = \frac{\mathrm{j}h\omega_0 L_4 - \mathrm{j}\dfrac{1}{h\omega_0 C_3}}{L_4/C_3}$$

and then the equivalent impedance matrices are obtained by inverting the admittance matrices, i.e. $\mathbf{Z}(\mathrm{j}h\omega_0) = \mathbf{Y}^{-1}(\mathrm{j}h\omega_0)$.

The data used in the circuit are

$$R_1 = 0.1\,\Omega \qquad R_2 = 0.2\,\Omega \qquad L_1 = 20\,\mathrm{mH}$$

$$L_2 = 11.3\,\mathrm{mH} \quad L_3 = 2\,\mathrm{mH} \qquad L_4 = 0.69\,\mathrm{mH}$$

$$C_1 = 100\,\mu\mathrm{H} \quad C_2 = 200\,\mu\mathrm{H} \quad C_3 = 300\,\mu\mathrm{H}$$

and $f_0 = 50$ Hz.

Figure 3.8 shows the driving point impedance as seen from node 1, i.e. element $Z^{1,1}(\mathrm{j}h\omega_0)$, where resonant points appear close to the third, fifth and seventh harmonic frequencies.

To achieve a good resolution for the driving point impedance, it is recommended that the impedances are obtained at every 5 Hz.

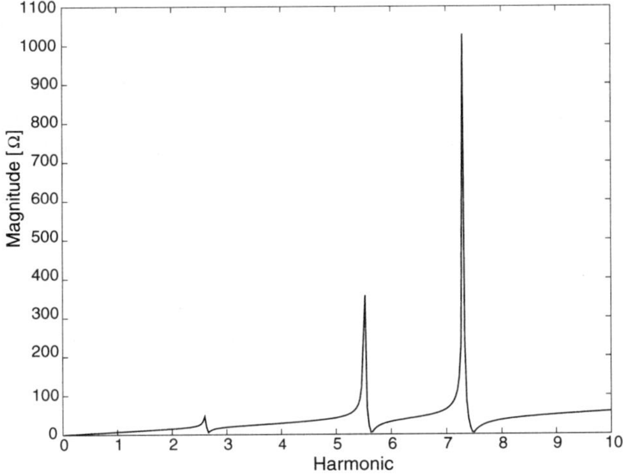

Figure 3.8 Driving point impedance as seen from node 1

Using the following harmonic injection currents in node 1: $I^1_{h=1} = 1\angle 0°$ A, $I^1_{h=3} = 0.3\angle 0°$ A, $I^1_{h=5} = 0.2\angle 0°$ A and $I^1_{h=7} = 0.14\angle 0°$ A, the nodal voltages given in Table 3.1 are obtained by solving equation (3.50). It is observed from Figure 3.8 and the results in Table 3.1 that resonant points can lead to high harmonic voltages if the injected currents excite the circuit at that particular frequency.

3.5.1 Harmonic propagation studies in large networks

The method of harmonic current injection is widely used to carry out harmonic propagation studies in industrial, distribution and power systems. The salient features of the method are outlined below:

Table 3.1 Nodal voltages in volts

Node \ h	1	3	5	7
1	$7.1918\angle 87.50°$	$5.6667\angle 89.69°$	$8.4581\angle 89.86°$	$9.0412\angle 89.91°$
2	$0.9266\angle 76.71°$	$0.0118\angle 89.96°$	$2.1749\angle 89.99°$	$2.8837\angle 89.99°$
3	$0.2284\angle 89.58°$	$0.0025\angle 93.04°$	$0.0198\angle 90.04°$	$2.8741\angle 89.99°$

- Main operating point: A conventional power flow solution is carried out to derive fundamental frequency information for the voltage magnitudes and angles at all nodes of the network.
- Passive elements: All the passive elements are considered to behave linearly with frequency, e.g. resistors, inductors and capacitors. The following characteristics are observed by these elements:

$$\begin{aligned} R &= \text{constant} \\ X_L(h) &= \mathrm{j}hX_L \\ X_C(h) &= -\mathrm{j}\frac{X_C}{h} \end{aligned}$$

  where X_L and X_C are the rated inductive and capacitive reactances, respectively.
- Overhead lines and cables: These are represented by the exact equivalent π-circuit at different frequencies where long-line effects, line imbalances and line transpositions should be taken into account. A nominal π-circuit, scaled up by the harmonic order should only be used in cases of very short transmission lines and low harmonic order.
- Generators: These are considered to be linear elements whose harmonic impedance is derived similarly to passive elements, using the following impedance:

$$Z_{\mathrm{g}} = R\sqrt{h} + \mathrm{j}X_{\mathrm{d}}''h$$

  where R is derived from the generator power losses and X_{d}'' is the generator subtransient reactance. If required, the generator's saturation and frequency conversion effects could be represented by means of harmonic current and voltage injection sources, respectively.
- Transformers: These are considered to be linear elements whose harmonic impedance is derived similarly to passive elements, using the following impedance:

$$Z_{\mathrm{t}} = R\sqrt{h} + \mathrm{j}X_{\mathrm{t}}h$$

  where R is derived from the transformer power losses and X_{t} is the transformer's short-circuit reactance. The connection of the transformer should be taken into account when zero sequence harmonics are present in the network. The transformer's saturation effect could be represented by means of a harmonic current injection source.
- Capacitor banks: These are considered to be passive elements, where:

$$X_C(h) = -\mathrm{j}\frac{V_{\mathrm{LL}}^2}{hQ_{3\phi}}$$

  V_{LL} is the RMS line voltage in kV and $Q_{3\phi}$ is the three-phase reactive power in MVAR.
- Linear loads: These may be represented by three different models given by the CIGRE Working Group 36-05 [16]:

(i) Parallel $R - X_L$ equivalent, where $R = \frac{V_{\mathrm{LL}}^2}{P_{3\phi}}$ and $X_L(h) = \mathrm{j}\frac{hV_{\mathrm{LL}}^2}{Q_{3\phi}}$.

(ii) Parallel $R - X_L$ with $R(h) = \frac{V_{\mathrm{LL}}^2}{kP_{3\phi}}$ and $X_L(h) = \mathrm{j}\frac{V_{\mathrm{LL}}^2}{kQ_{3\phi}}$ where $k = 0.1h + 0.9$.

(iii) Parallel $R - X_L$ in series with X_S, where $R = \frac{V_{\mathrm{LL}}^2}{P_{3\phi}}$, $X_L(h) = \mathrm{j}\frac{hR}{6.7\left(\frac{Q_{3\phi}}{P_{3\phi}} - 0.74\right)}$ and $X_S(h) = \mathrm{j}0.073hR$.

- Non-linear loads: These are represented by either a harmonic current injection source or by a harmonic voltage source. Harmonic current injection sources are used to represent the harmonic contributions of static VAR compensators (SVCs), induction arc furnaces, rectifiers and electronic appliances. For example, an SVC is represented by the harmonic current injection given by $I_h = (\%_h)I_{1\phi}$, where $(\%_h)$ is a percentage of the current at fundamental frequency given by $I_{1\phi} = \frac{Q_{3\phi}}{\sqrt{3}V_{\mathrm{LL}}}\mathrm{e}^{\mathrm{j}(\theta \pm \pi/2)}$, where θ is the voltage angle obtained from a conventional power flow, and $\pi/2$ is required phase shift since the current leads or lags the voltage by $90°$. Harmonic voltage sources are arc furnaces, and pulse-width modulation (PWM) converters. For example, a PWM converter can be represented by a Thévenin equivalent given by the leakage reactance of the transformer and a harmonic voltage source.
- Transmission elements and linear loads are represented by impedances, at each harmonic h, with which the admittance matrix $\mathbf{Y}_h$ of the system is built. The relevant harmonic current and voltage sources $\mathbf{I}_h$ and $\mathbf{V}_h$ are injected into the partially inverted nodal admittance matrix and the remaining nodal harmonic voltages and currents are obtained.

A program written in MATLAB™ to calculate the driving point impedance (DPI) is given below:

```
% dpi.m
% Obtain the DPI in node (N1h,N2h) for the system in datadpi.m
clear all
datadpi;
z=(h1:dh:h2)*0;
k=0;
for h=h1:dh:h2
  Y=zeros(nN,nN);
  for n=1:nL
    node1=n1(n);
    node2=n2(n);
    y12=1/(rl(n)+j*h*xl(n));
    y00=j*h*b2(n);
    Y(node1,node1)=Y(node1,node1)+y12+y00;
    Y(node2,node2)=Y(node2,node2)+y12+y00;
    Y(node1,node2)=Y(node1,node2)-y12;
    Y(node2,node1)=Y(node2,node1)-y12;
  end
  for n=1:nG
    node1=ng(n);
    yg=1/(j*h*xg(n));
    Y(node1,node1)=Y(node1,node1)+yg;
  end
  for n=1:nN
    node1=nn(n);
```

```
      if PL(n)~=0
        R =V(n)^2/PL(n);
        Xl=j*h*R/(6.7*(QL(n)/PL(n)-0.74));
        Xs=j*h*0.073*R;
        yl=1/(R*Xl/(R+Xl)+Xs);
        Y(node1,node1)=Y(node1,node1)+yl;
      end
      yc=j*h*Qc(n)/(V(n)^2);
      Y(node1,node1)=Y(node1,node1)+yc;
    end
    k=k+1;
    Z=inv(Y);
    z(k)=Z(N1h,N2h);
  end
```

Example 3-4: The classic test system of Figure 3.9 [6] has been modified to include one static VAR compensator (SVC) connected at node 5 to provide voltage support at that node. The base values are 100 kV and 100 MVA. The line parameters are given in Table 3.2 where a nominal π-circuit is used to simplify the example as much as possible. The maximum amplitudes of harmonic current injected by the SVC are given in Table 3.3. The fundamental power flow solution is given in Tables 3.4 and 3.5. Table 3.4 shows the nodal voltage magnitudes and angles, the nodal powers injected by the generators and that consumed by the loads. Table 3.5 gives the power flows.

The SVC is considered to be delta connected, hence no zero sequence harmonic current is injected into the system.

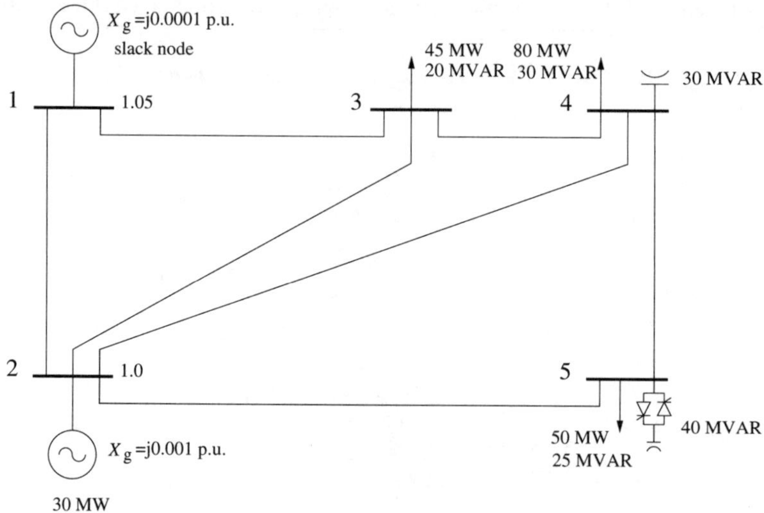

Figure 3.9 Power system diagram with SVC

Figure 3.10 shows the passive equivalent circuit used to obtain the driving point impedance and the harmonic voltages throughout the network. The loads are represented as passive loads, as given by CIGRE's model (iii). The SVC is assumed to be a constant harmonic current

Table 3.2 Transmission line parameters [p.u.]

Line, p–q	Impedance, z_{pq}	Line charging, $y_{pq}/2$
1–2	0.02+j0.06	0.0+j0.030
1–3	0.08+j0.24	0.0+j0.025
2–3	0.06+j0.18	0.0+j0.020
2–4	0.06+j0.18	0.0+j0.020
2–5	0.04+j0.12	0.0+j0.015
3–4	0.01+j0.03	0.0+j0.010
4–5	0.08+j0.24	0.0+j0.025

Table 3.3 Maximum amplitudes of harmonic current in the SVC [18]

Harmonic	% of fundamental	Harmonic	% of fundamental
5	5.05	17	0.44
7	2.59	19	0.35
11	1.05	23	0.24
13	0.75	25	0.20

Table 3.4 Fundamental frequency power flow [p.u.]

Node	$\|V\|$	θ	P_g	Q_g	P_d	Q_d
1	1.0500	0.00000	1.5260	0.65840	0.00	0.00
2	1.0000	−2.6944	0.3000	−0.6512	0	0
3	0.9796	−6.2114	0	0	0.45	0.20
4	0.9776	−6.9232	0	0	0.80	0.30
5	0.9922	−6.6963	0	0	0.50	0.25

Table 3.5 Power flows [p.u.]

Node 1	Node 2	S_{12}	S_{21}
1	2	1.0087+j0.5250	−0.9846−j0.5158
1	3	0.5173+j0.1333	−0.4960−j0.1210
2	3	0.3376−j0.0087	−0.3307−j0.0099
2	4	0.4022−j0.0149	−0.3925+j0.0049
2	5	0.5448−j0.1118	−0.5326+j0.1188
3	4	0.3768−j0.0691	−0.3752+j0.0545
4	5	−0.0323−j0.0727	0.0326+j0.0250

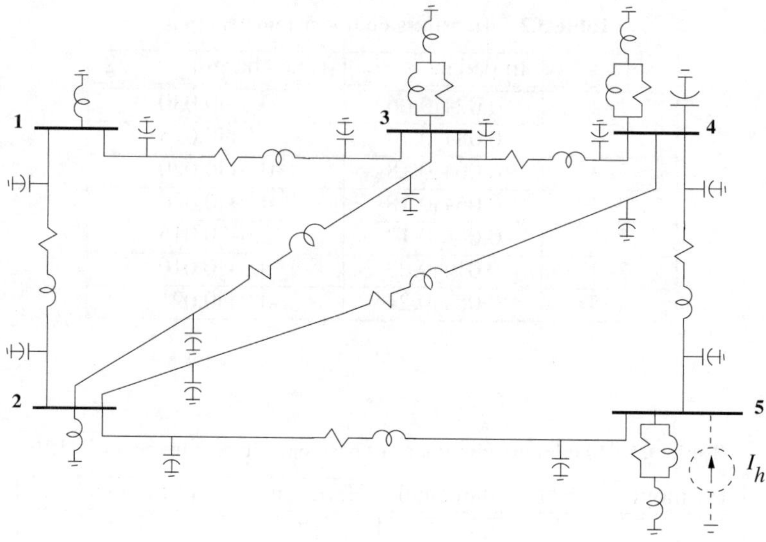

Figure 3.10 Passive circuit used to obtain the driving point impedance and harmonic propagation source.

The impedance of the capacitor bank at node 4 is given by

$$X_C = -\mathrm{j}\frac{(0.9776)^2}{h0.30} = -\mathrm{j}\frac{3.1857}{h}$$

and the load in node 4 is represented by the equivalent impedance

$$Z_L = \frac{RX_L}{R + X_L} + X_S$$

where

$$R = \frac{(0.9776)^2}{0.80} = 1.1946$$

$$X_L = \mathrm{j}h1.1946 \frac{1}{6.7\left(\dfrac{0.30}{0.80} - 0.74\right)} = -\mathrm{j}h0.4885$$

$$X_S = \mathrm{j}0.073h(1.1946) = \mathrm{j}h0.0872$$

Similarly, the load in node 3 has the following parameters, $R = 2.1325$, $X_L = -\mathrm{j}h1.0769$ p.u. and $X_S = \mathrm{j}h0.1557$, and the load in node 5, $R = 1.9689$, $X_L = -\mathrm{j}h1.2245$ and $X_S = \mathrm{j}h0.1437$. The generators reactance are $X_{\mathrm{g}1} = \mathrm{j}h0.0001$ and $X_{\mathrm{g}2} = \mathrm{j}h0.001$. The transmission lines parameters are given in Table 3.6.

Figures 3.11(a) and (b) show the driving point impedance for the five-node system. Figure 3.11(c) shows the equivalent impedance between node 5 and the rest of the system. It should

Table 3.6 Transmission parameters at harmonic frequencies [p.u.]

Line, p–q	Series impedance	Shunt admittance
1–2	0.02+jh0.06	jh0.030
1–3	0.08+jh0.24	jh0.025
2–3	0.06+jh0.18	jh0.020
2–4	0.06+jh0.18	jh0.020
2–5	0.04+jh0.12	jh0.015
3–4	0.01+jh0.03	jh0.010
4–5	0.08+jh0.24	jh0.025

be noted that in this example the harmonic voltage h in node i, i.e. V_h^i for $i = 1, 2, \ldots, 5$, will due to the injection of harmonic current in node 5, i.e. I_h^5, since $V_h^i = Z_h^{i,5} I_h^5$. From these figures it can be seen that if the 21st harmonic current is injected at node 5; hence, we would expect the highest harmonic voltage at that frequency to appear at node 5, followed by nodes 3, 4, 2 and 1, in decreasing order of magnitude.

For the SVC, the fundamental frequency current is given by

$$I_{\text{svc1}} = \frac{0.40}{\sqrt{3} \times 0.9922} \mathrm{e}^{\mathrm{j}(-0.1169+\pi/2)} \Rightarrow 0.2328\angle 83.3037^\circ \text{ p.u.}$$

and the harmonic currents given as a percentage of this value according to Table 3.3.

The system is solved at each frequency of interest using $\mathbf{V}_h = \mathbf{Y}_h^{-1}\mathbf{I}_h$, where $\mathbf{I}_h$ has values different from zero only in entry 5 of the harmonic current vector. Table 3.7 shows the harmonics voltages as percentages of the fundamental.

Table 3.7 Harmonic voltage in the network

$V \setminus h$	5%	7%	11%	13%	17%	19%	23%	25%	THD%
$\lvert V\rvert^1$	0.00007	0.00005	0.00001	0.0000	0.0000	0.0000	0.0000	0.0000	0.00009
θ^1	108.12	58.543	−4.728	−16.52	−40.39	−66.91	162.95	142.81	
$\lvert V\rvert^2$	0.0057	0.0030	0.0015	0.0016	0.0020	0.0026	0.0022	0.0012	0.0080
θ^2	137.75	118.91	147.36	148.11	137.44	120.68	39.513	19.029	
$\lvert V\rvert^3$	0.1675	0.1235	0.0304	0.0188	0.0126	0.0134	0.0098	0.0056	0.2122
θ^3	99.919	50.809	−11.68	−22.73	−42.19	−61.16	−146.1	−169.1	
$\lvert V\rvert^4$	0.2107	0.1540	0.0360	0.0213	0.0123	0.0116	0.0060	0.0026	0.2650
θ^4	103.58	55.324	−6.193	−16.89	−35.55	−53.89	−136.2	−156.2	
$\lvert V\rvert^5$	0.5213	0.3302	0.2307	0.2217	0.2621	0.3293	0.2795	0.1512	0.8725
θ^5	148.76	145.76	150.37	148.30	136.65	119.83	38.673	18.173	

To show the impact that the load representation has on these kinds of studies, Figure 3.12 shows the driving point impedance when load model (i) is used. It is observed by comparing the results in Figures 3.11 and 3.12 that the load model affects very significantly the high frequencies of the system response.

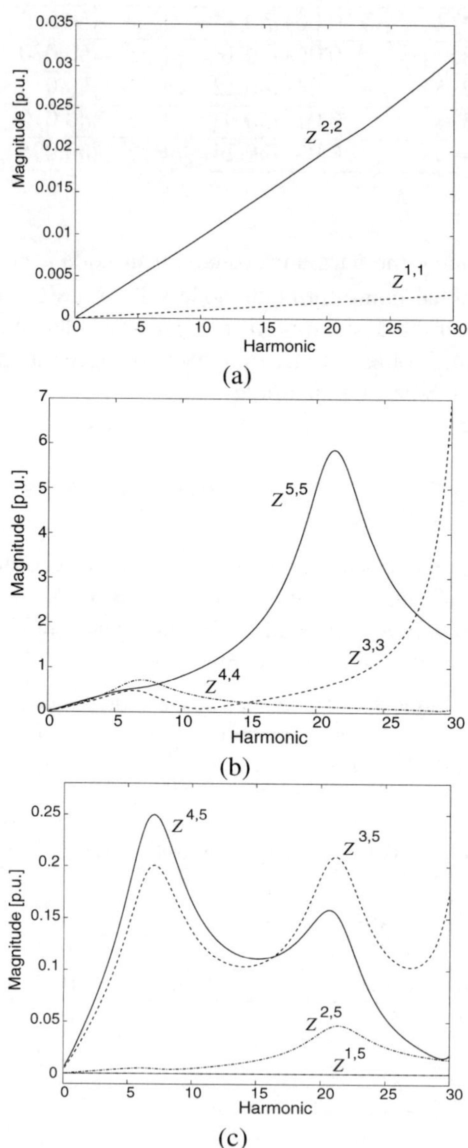

Figure 3.11 Driving point impedance of the power system using load model (iii)

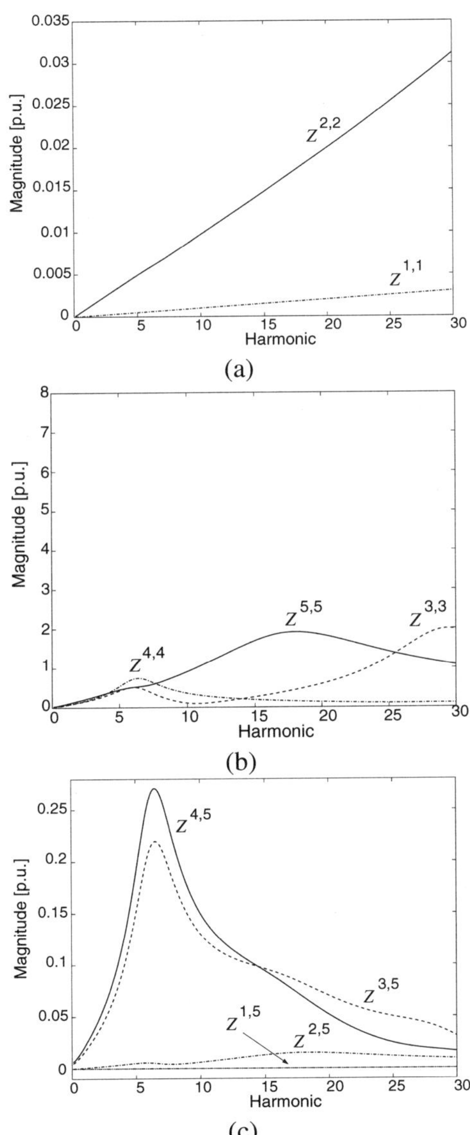

Figure 3.12 Driving point impedance of the power system using load model (i)

The data in MATLAB™ for Example 3-4 are given in the program below:

```
% datadpi.m
% data for Example 3-4
nN=5; % number of nodes
nL=7; % number of transmission elements
nG=2; % number of generators
h1=0.1; % first harmonic
dh=0.1; % harmonic step
h2=30; % final harmonic
N1h=5; % node 1 of Zeqh
N2h=5; % node 2 of Zeqh
nn=[      1;       2;       3;       4;       5]; % nodes
V =[1.0500; 1.0000; 0.9796; 0.9776; 0.9922]; % from power flow
A =[   0.0;-2.6944;-6.2114;-6.9232;-6.6963]; % voltage angles
PL=[      0;       0;    0.45;    0.80;    0.50]; % active power load
QL=[      0;       0;    0.20;    0.30;    0.25]; % reactive power load
Qc=[      0;       0;       0;    0.30;       0]; % Qc by shunt capacit
%  node 1 and 2 of the transmission elements
n1=[     1;      1;      2;      2;      2;      3;      4];
n2=[     2;      3;      3;      4;      5;      4;      5];
rl=[0.020; 0.080; 0.060; 0.060; 0.040; 0.010; 0.080]; % resistance
xl=[0.060; 0.240; 0.180; 0.180; 0.120; 0.030; 0.240]; % reactance
b2=[0.030; 0.025; 0.020; 0.020; 0.015; 0.010; 0.025]; % susceptance
ng=[      1;       2]; % node of connection of the generators
xg=[0.0001; 0.001]; % reactance of the generators
```

3.6 Passive Filters

Passive shunt filters are a very popular resort used to control the propagation of harmonic currents, and are normally designed as a series combination of reactors and capacitors. Passive filters are also referred to as sinks because they absorb the harmonic currents. They present a low-impedance path to the harmonic to which it is tuned.

Power harmonic filters are installed at the AC terminals of rectifiers, motor drives, uninterruptible power supplies, and other non-linear loads, to reduce voltage and current distortion to acceptable limits at the point of connection. Filter design is normally carried out assuming static operating conditions, although in practice filters may be operated dynamically, switching them on and off by sections, according to system requirements. Examples of applications where switched filters are used are elevator drives, adjustable speed pump drives, reactive power compensators and HVDC converter stations.

In systems with a number of filters, it is necessary to develop a switching strategy. A good practice is to switch off the higher order harmonic filters before the lower order ones. For switching-on operations the opposite would apply, but due care should be exercised because their response is strongly dependent on electric plant conditions. If banks of capacitors are provided with the filters, it makes sense to leave the filters permanently switched on, and to use the banks to provide voltage and power factor regulation. A suitable switching strategy may be based on a comprehensive transient analysis of the filtering system and the plant, taking a wide range of operating conditions into consideration.

3.6.1 Single-tuned first-order filters

These filters are mostly used at low harmonic frequencies. They are probably the most common shunt filters in use today and comprise a series *RLC* circuit, as shown in Figure 3.13(a), and are tuned to a particular harmonic frequency.

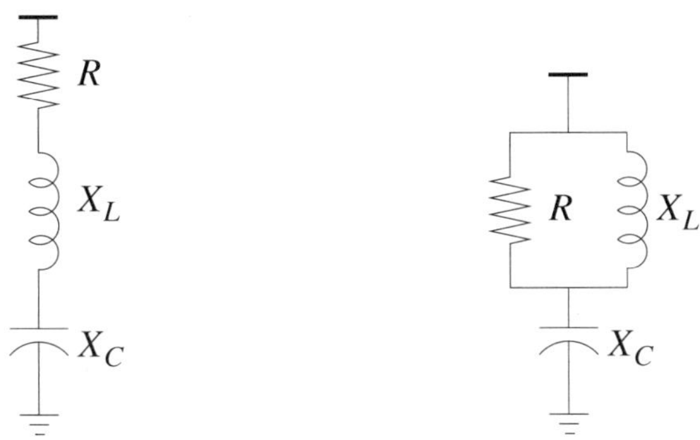

Figure 3.13 Common passive filter configuration

Some characteristics of first-order filters are as follows: (1) They act as a very low-impedance path at the frequency at which they are tuned. (2) When the source impedance is inductive, there is a resonant peak which always occurs at a frequency lower than the frequency at which the filter is tuned. (3) There is a sharp increase in the impedance below the tuned frequency due to the proximity of the resonant frequency. (iv) The impedance rises with frequency for frequencies above the tuning frequency.

The tuning characteristic of the filter is described by its quality factor Q given by

$$Q = \frac{hX_L}{R}$$

where R is the resistance of the filter, h is the tuned harmonic, and X_L is the rated reactance of the inductor. Typically, the value of R consists of only the resistance of the inductor, resulting in a large value of Q and a very sharp filtering characteristic. The value of the resistance may be obtained by selecting an appropriate value of quality factor, $20 < Q < 30$. Figure 3.14 shows a typical response for this kind of filter for two quality factors and the same tuned frequency f_t.

3.6.2 High-pass second-order filters

In cases with high-order harmonic currents, e.g. 11th, 13th and higher, high-pass filters are normally used. High-pass filters are named after the low-impedance characteristic that they observe above a corner frequency. The filter will shunt a large percentage of all harmonics at or above the corner frequency. Frequently, only one high-pass filter is used to eliminate a range of harmonics, whose corner frequency is located at the lowest harmonic that it is meant to eliminate. Two factors may discourage such an application. (1) The minimum impedance of the high-pass filter never achieves a value comparable to that of a single-tuned filter at tuned frequency. (2) The shunting of a percentage of all the system harmonics through a single filter may require the filter to be vastly over-rated from the fundamental frequency point of view. In contrast to the single-tuned filter, the quality factor of a high-pass filter is given by its inverse relation

$$Q = \frac{R}{hX_L}$$

Figure 3.14 shows the response of these kinds of filters for two different quality factors, where typical values of Q are $0.5 < Q < 2$.

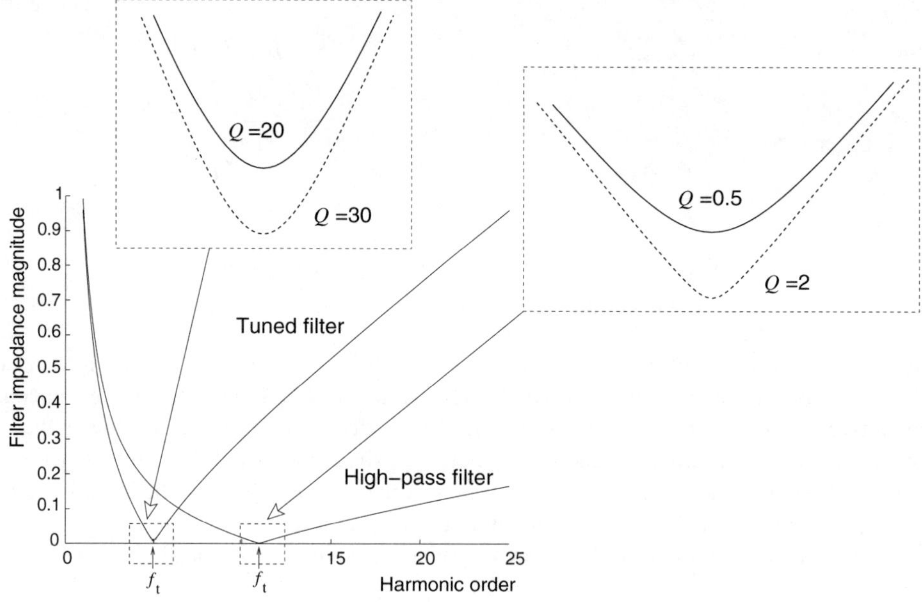

Figure 3.14 Typical frequency response of filters

3.6.3 Single-tuned first-order harmonic filter design

The design of the single-tuned harmonic filter shown in Figure 3.15 will be illustrated through a simple but commonly used calculation procedure [4,17]:

- Capacitor reactance X_C: The capacitor reactance may be obtained from the following relation:

$$X_C = \frac{V^2_{C_{\text{rated}}}}{Q_{C_{\text{rated}}}}$$

  where $V_{C_{\text{rated}}}$ is the line-to-line rated voltage of the capacitor. $Q_{C_{\text{rated}}}$ is the three-phase rated reactive power of the capacitor. The reactive power is based on a reactive power compensation study at fundamental frequency, with due regard to power factor improvement of the electric plant.
- The capacitor rated line current $I_{C_{\text{rated}}}$. This current is calculated using

$$I_{C_{\text{rated}}} = \frac{Q_{C_{\text{rated}}}}{\sqrt{3}V_{C_{\text{rated}}}}$$

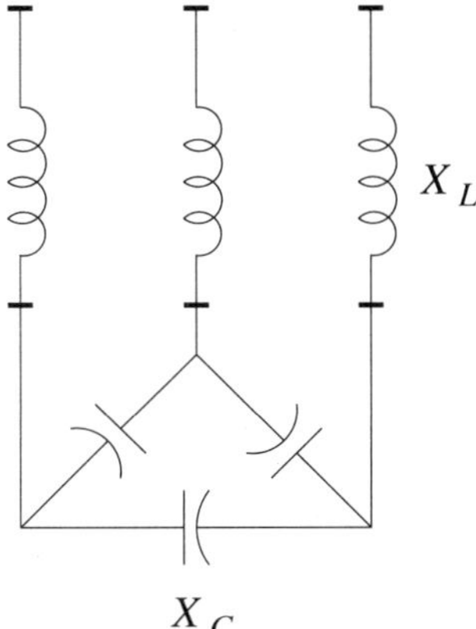

Figure 3.15 Three-phase single-tuned filter configuration at low voltage

- Tuned reactor reactance X_L. The reactance of the filter is obtained by

$$X_L = \frac{X_C}{(rh)^2}$$

where h is the harmonic to which the filter is tuned and r is an empirical factor smaller than one, giving rh a value slightly below the harmonic frequency of concern. This factor minimises the possibility of unwanted harmonic resonances, which might take place as system parameters change. A typical value of r, for the fifth harmonic, is 0.94.
- Reactive power delivered by the tuned filter:

$$Q_{\mathrm{F}} = \frac{V_{\mathrm{LL}}^2}{X_C - X_L}$$

where V_{LL} is the line-to-line voltage at the point of connection. Q_{F} is the reactive power delivered by the capacitor–inductor arrangement. Here, Q_{F} is greater than the reactive power delivered by the capacitor alone, giving a power factor compensation slightly higher than the value previously calculated.
- The line current in the filter at fundamental frequency is given by

$$I_{\mathrm{F}} = \frac{Q_{\mathrm{F}}}{\sqrt{3}V_{\mathrm{LL}}}$$

- The true RMS line current in the filter is

$$I_{\mathrm{F_{rms}}} = \sqrt{I_{\mathrm{F}}^2 + I_{\mathrm{h}}^2}$$

where I_{h} is the RMS harmonic line current to be filtered out.
- The fundamental frequency RMS voltage across the capacitor is given by

$$V_C = \sqrt{3}I_{\mathrm{F}}X_C$$

- The voltage drop across the capacitor due to the harmonic current I_{h} is given by

$$V_{C_\mathrm{h}} = \sqrt{3} I_\mathrm{h} \frac{X_C}{h}$$

- The true RMS voltage across the capacitor is

$$V_{C_\mathrm{rms}} = \sqrt{V_C^2 + V_{C_\mathrm{h}}^2}$$

- The maximum peak voltage across the capacitor is

$$V_{C_\mathrm{peak}} = \sqrt{2} V_C + \sqrt{2} V_{C_\mathrm{h}}$$

- Assuming that the totality of the apparent power delivered by the capacitor is reactive power, then the power delivered by the capacitor as part of the filter is

$$Q_{C_\mathrm{F}} = \sqrt{3} V_{C_\mathrm{rms}} I_{\mathrm{F}_\mathrm{rms}}$$

Finally is important to check that the capacitor is within the limits, prescribed by Table 3.8.

Table 3.8 ANSI/IEEE 18-1980 standard for shunt power capacitors

	Maximum limits
True RMS current	180%
True RMS voltage	110%
Maximum peak voltage	120%
Reactive power	135%

The main advantages of passive filters may be summarised as follows:

- Easy to design.
- Reliable operation.
- May be used as part of filter banks.
- Relatively inexpensive configuration.
- Act as a power factor improvement device.

Some of the disadvantages of passive filters are:

- Resonance peaks depend on system impedance.
- Tuned to only one harmonic frequency.
- Resonance problems may occur at different harmonic frequencies.
- Affected by capacitor ageing.

In spite of such drawbacks, passive filters are widely used since they provide an effective means of delivering harmonic current compensation and power factor compensation at the same time.

Example 3-5: A 1 MVA, 13.2/480 V power transformer with a short-circuit reactance of 10% supplies a 933 kW, 1405 A load at 480 V.

The load consists mainly of six pulse rectifiers which inject 30% of fifth harmonic current with respect to the fundamental frequency. The parameters of a fifth harmonic tuned filter are calculated to compensate the power factor and to absorb the fifth harmonic current.

Four basic steps will be followed in this example: (1) reactive power requirements, (2) resonance problem prediction, (3) calculation of the filter and (4) checking capacitor limits.

1. Reactive power requirements:

The kVA and kVAR supplied by the transformer are

$$S_{\mathrm{T}} = \sqrt{3} \times 0.480 \times 1405 = 1168 \text{ kVA and } Q_{\mathrm{T}} = \sqrt{1168^2 - 933^2} = 702.82 \text{ kVAR}$$

and the power factor is

$$\mathrm{PF} = \frac{933}{1168} = 0.798\ (-)$$

In order to compensate the power factor at a value of 0.95, a capacitor bank with a value of

$$Q_{\mathrm{C}} = P_L\,(\tan\theta_1 - \tan\theta_2) = 933(0.7552 - 0.3286) = 398.01 \text{ kVAR}$$

is required. Using a commercially available capacitor bank, we select a 350 kVAR at 480 V. The new kVAR and kVA supplied by the transformer are

$$Q_{\mathrm{T}} = 702.82 - 350 = 352.82 \text{ kVAR and } S_{\mathrm{T}} = \sqrt{933^2 + 352.82^2} = 997.82 \text{ kVA}$$

and the new power factor is

$$\mathrm{PF} = \frac{933}{997.5} = 0.935\ (-)$$

with a transformer line current of

$$I_{\mathrm{T}} = \frac{997.82}{\sqrt{3} \times 0.480} = 1199.8 \text{ A}$$

2. Resonance problem prediction:

In order to predict potential parallel resonance problems between the power transformer and the capacitor bank, the resonant frequency, f_{r}, and the short-circuit ratio (SCR) are calculated from the following relations:

$$f_{\mathrm{r}} = 2\pi f_0 \sqrt{\frac{\mathrm{MVA_{SC}}}{\mathrm{MVAR_C}}}$$

$$\mathrm{SCR} = \frac{\mathrm{MVA_{SC}}}{\mathrm{MW_{rect}}}$$

where $\mathrm{MVA_{SC}}$ is the short-circuit MVA at the point of connection of the capacitor and $\mathrm{MVAR_C}$ is the MVAR rating of the capacitor bank. $\mathrm{MW_{rect}}$ is the MW rating of the rectifier.

These values give a good indication of potential resonance problems. From a practical viewpoint, if the harmonic current generated by the non-linear load is close to f_{r} and the SCR is below 20, the resonant condition will drastically affect the operation of the capacitor bank.

For this example, we have that the MVA short-circuit level is

$$\mathrm{MVA_{SC}} = \frac{\mathrm{kV}^2}{X_\mathrm{T}} = \frac{100\,\mathrm{MVA_T}}{Z\%} = 10\ \mathrm{MVA} \quad \text{where} \quad X_\mathrm{T} = \frac{Z\%}{100}\frac{\mathrm{kV}^2}{\mathrm{MVA_T}}$$

The resonant harmonic frequency and SCR are given by

$$f_\mathrm{r} = 2\pi f_0\sqrt{\frac{10}{0.35}} = 2\pi f_0 \times 5.35 \quad \text{and} \quad \mathrm{SCR} = \frac{10}{0.993} = 10.72$$

The resonant harmonic frequency is very close to the fifth harmonic generated by the load, and the SCR is very low. These results suggest the use of a fifth harmonic tuned filter.

3. Calculation of the filter:

Proceeding with the calculation of the fifth harmonic tuned filter, the reactance of the capacitor bank and the reactance of the filter, using $r = 0.94$, are given by

$$X_C = \frac{0.480^2}{0.35} = 0.6582\,\Omega \quad \text{and} \quad X_L = \frac{0.6582}{4.7^2} = 0.0298\,\Omega$$

The reactive power delivered by the filter is

$$Q_\mathrm{F} = \frac{0.480^2}{0.6582 - 0.0298} = 366.6\ \mathrm{kVAR}$$

The rated current of the capacitor bank and the current in the filter are

$$I_{C_\mathrm{rated}} = \frac{350}{\sqrt{3} \times 0.480} = 420.98\,\mathrm{A} \quad \text{and} \quad I_\mathrm{F} = \frac{366.6}{\sqrt{3} \times 0.480} = 440.95\,\mathrm{A}$$

The fifth harmonic current generated by the load is 30% of the fundamental frequency,

$$I_1 = \frac{933}{\sqrt{3} \times 0.480} = 1122.22\,\mathrm{A} \quad \text{and} \quad I_5 = 0.3 \times 1122.22 = 336.67\,\mathrm{A}$$

The true RMS current in the filter is

$$I_{\mathrm{F_{rms}}} = \sqrt{440.95^2 + 336.67^2} = 554.74\,\mathrm{A}$$

The fundamental frequency and fifth harmonic voltage across the capacitor bank are

$$V_{C_1} = \sqrt{3} \times 440.95 \times 0.6582 = 502.69\,\mathrm{V} \text{ and } V_{C_5} = \sqrt{3} \times 336.67 \times \frac{0.6582}{5} = 76.76\,\mathrm{V}$$

The true RMS voltage across the capacitor is

$$V_{C_\mathrm{rms}} = \sqrt{502.69^2 + 76.76^2} = 508.52\,\mathrm{V}$$

The maximum peak voltage in the capacitor is given by

$$V_{C_\mathrm{peak}} = \sqrt{2} \times 502.69 + \sqrt{2} \times 76.76 = \sqrt{2} \times 579.45\,\mathrm{V}$$

The reactive power delivered by the capacitor as part of the filter is

$$Q_{C_\mathrm{F}} = \sqrt{3} \times 508.52 \times 554.74 = 488.6\,\mathrm{kVAR}$$

4. Checking capacitor limits:

The results show that the true RMS current in the capacitor is $554.74/420.98 = 131.77\%$, the true RMS voltage is $508.52/480 = 105.94\%$, the maximum peak voltage is $579.45/480 = 120.71\%$ and the reactive power is $488.6/350 = 139.6\%$. These parameters are given as a percentage of their respective rated values.

As indicated by these results, two capacitor limits have been exceeded, the maximum peak voltage and the reactive power limit. Accordingly, a new set of calculations is required, using a larger value of capacitor.

If a capacitor bank of 400 kVAR is selected as opposed to a 350 kVAR, the true RMS current is 125.96%, the true RMS voltage is 105.66%, the maximum peak voltage is 118.73% and the reactive power is 133.10%. These values comply with the limits given in Table 3.8.

The results give a filter reactor with a value of 0.0261 Ω, a fundamental frequency and fifth harmonic current in the filter of 606.04 A and 336.67 A, respectively, and a power factor of 0.956.

The reactor may be selected to be either an air-cored inductor or an iron-cored inductor but iron-cored inductors are preferred because of their smaller size. In such cases the knee point of the magnetising characteristic must be sufficiently high to prevent magnetic saturation at the maximum expected value of current. The overall filter design must be based on standards.

In this example the effect of the equivalent impedance of the system was not taken into account since the equivalent reactance of the system-transformer was comparable in magnitude to that of the transformer. Nevertheless, in practical applications, the effect of the system must be taken into account since the resonant peak changes with the system equivalent impedance. Figure 3.16 shows the effect of the capacitor on the driving point impedance at the point of common coupling (PCC), both when it is used on its own and when it is used as a part of a passive filter. This figure also shows how the equivalent impedance of the system affects the resonant peak, increasing the risk of excitation by a low-order harmonic frequency, such as the third harmonic, in the case of a weak system condition, i.e. small SCR.

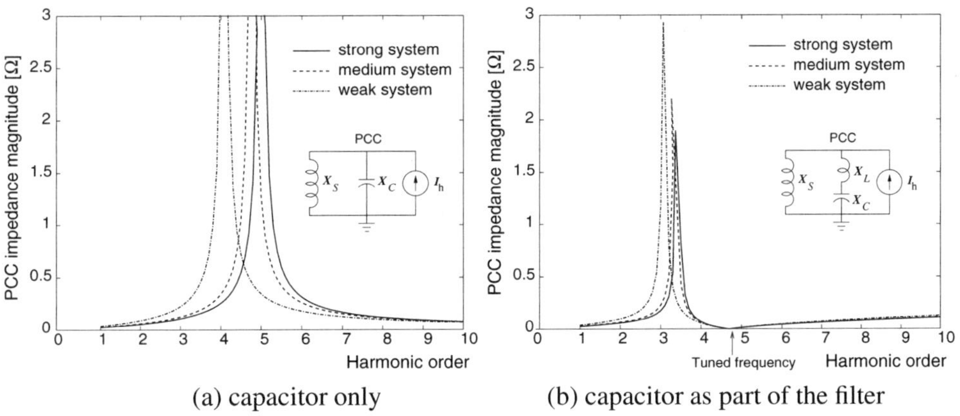

(a) capacitor only (b) capacitor as part of the filter

Figure 3.16 Driving point impedance at the point of common coupling

Passive filters are not too difficult to design but careful and extensive analysis is required because they may induce low-order resonant peaks in the power system. In actual applications a proper harmonic study is required, based on computer simulations and measurements.

3.7 Summary

This chapter has introduced a systematic treatment of all the relevant terms associated with electric power in circuits operating under non-sinusoidal and sinusoidal conditions. The treatment is based on the theory of Fourier series, as presented in the previous chapter, together with fundamental definitions found in electric circuit theory books. To conform with current thinking, the generic apparent power has been assumed to comprise active power, reactive power and distortion power. Using the generic expression for the apparent power and the active power, an expression for the "true" power factor is readily available.

It has been shown that the generic expressions for the various quantities simplify to well-known power expressions if only fundamental frequency voltages and currents are assumed to exist. Numeric examples have been used to illustrate the simplicity of the concepts involved.

A review of harmonic propagation in linear circuits, driving point impedances and passive filters has been presented, and the use of a harmonic propagation algorithm illustrated with the aid of a numeric example. The MATLAB™ code used to solve the example was given.

3.8 Bibliography

1. W. Shepherd, P. Zand, *Energy Flow and Power Factor in Nonsinusoidal Circuits*, Cambridge University Press, 1979.
2. G.T. Heydt, *Electric Power Quality*, Stars in a Circle Publications, Scottsdale, AZ, 1991.
3. J. Arrillaga, B.C. Smith, N.R. Watson, A.R. Wood, *Power System Harmonic Analysis*, John Wiley & Sons, Chichester, 1997.
4. R.C. Dugan, M.F. McGranaghan, H.W. Beaty, *Electrical Power System Quality*, McGraw Hill,New York, 1996.
5. J. Arrillaga, N.R. Watson, S. Chen, *Power System Quality Assessment*, John Wiley & Sons, Chichester, 2000.
6. G.W. Stagg, A.H. El-Abiad, *Computer Methods in Power System Analysis*, McGraw-Hill, New York, 1968.
7. C.A. Desoer, E.S. Kuh, *Basic Circuit Theory*, McGraw-Hill, New York, 1969.
8. L.S. Czarnecki, "Misinterpretations of Some Power Properties of Electric Circuits", *IEEE Transactions on Power Delivery*, Vol. 9, No. 4, October 1994, pp. 1760–1769.
9. P.S. Filipski, Y. Baghzouz, M.D. Cox, "Discussion of Power Definitions Contained in the IEEE Dictionary", *IEEE Transactions on Power Delivery*, Vol. 9, No. 3, July 1994, pp. 1237–1244.
10. A.E. Emanuel, "On the Definitions of Power Factor and Apparent Power in Unbalanced Polyphase Circuits with Sinusoidal Voltage and Currents", *IEEE Transactions on Power Delivery*, Vol. 8, No. 3, July 1993, pp. 841–852.
11. L. Czarnecki, "Physical Reasons of Currents RMS Value Increase in Power Systems with Nonsinusoidal Voltage", *IEEE Transactions on Power Delivery*, Vol. 8, No. 1, January 1993, pp. 437–447.
12. C. Budeanu, "Reactive and Fictitious Power", Publication No. 2 *of the Rumanian National Institute*, Bucharest, 1927.
13. L. Czarnecki, "What is Wrong with the Budeanu Concept of Reactive and Distortion Power and why it Should be Abandoned", *IEEE Transactions on Instrumentation and Measurement*, Vol. 36, No. 36, September 1987, pp. 834–837.
14. A.E. Emanuel, "Power in Nonsinusoidal Situations: A Review of Definitions and Physical

Meaning", *IEEE Transactions on Power Delivery*, Vol. 5, No. 3, July 1990, pp. 1377–1383.

15. C.L. Fortescue, "Method of Symmetrical Coordinates Applied to the Solution of Polyphase Networks", *Transactions of the AIEE*, Vol. 37, 1918, pp. 1027–1140.
16. CIGRE Working Group, 36-05, "Harmonics, Characteristic Parameters, Methods of Study, Estimates of Existing Values in the Networks", *Electra*, No. 77, July 1981, pp. 35–54.
17. IEEE Industry Applications Society/Power Engineering Society, *IEEE Recommended Practices and Requirements for Harmonic Control in Electrical Power Systems*, IEEE Std 519-1992, April 1993.
18. T.J.E. Miller, *Reactive Power Control in Electric Systems*, Wiley-Interscience, New York, 1982.
19. A.A. Mahmound (Editor), W.M. Grady, M.F. McGranaghan (Co-Editors), "Power Systems Harmonics", *IEEE Tutorial Course*, 1984.
20. IEEE Working Group on Nonsinusoidal Situations, "Effects on Meter Performance and Definitions of Power, Practical Definitions for Powers in Systems with Nonsinusoidal Waveforms and Unbalanced Loads: A discussion", *IEEE Transactions on Power Delivery*, Vol. 11, No. 1, January 1996, pp. 79–101.
21. Task Force on Harmonics Modeling and Simulation, "Modeling and Simulation of Propagation of Harmonics in Electric Power Networks, Part I: Concepts, Models, and Simulation Techniques", *IEEE Transactions on Power Delivery*, Vol. 11, No. 1, January 1996, pp. 452–465.
22. Task Force on Harmonics Modeling and Simulation, "Modeling and Simulation of Propagation of Harmonics in Electric Power Networks, Part II: Sample Systems and Examples", *IEEE Transactions on Power Delivery*, Vol. 11, No. 1, January 1996, pp. 466–474.
23. Task Force on Harmonics Modeling and Simulation, "Test Systems for Harmonics Modeling and Simulation", *IEEE Transactions on Power Delivery*, Vol. 14, No. 2, April 1999, pp. 579–587.
24. T.H. Ortmeyer, T. Hiyama, "Distribution System Harmonic Filter Planning", *IEEE Transactions on Power Delivery*, Vol. 11, No. 4, October 1996, pp. 2005–2012.
25. E.B. Makram, E.V. Subramaniam, A.A. Girgis, R. Catoe, "Harmonic Filter Design Using Actual Recorded Data", *IEEE Transactions on Industry Applications*, Vol. 29, No. 6, November/December 1993, pp. 1176–1183.
26. K.P. Lin, M.H. Lin, T.P. Lin, "An Advanced Computer Code for Single-Tuned Harmonic Filter Design", *IEEE Transactions on Industry Applications*, Vol. 34, No. 4, July/August 1998, pp. 640–648.
27. J.A. Bonner, W.M. Hurst, R.G. Rocamora, R.F. Dudley, M.R. Sharp, J.A. Twiss, "Selecting Rating for Capacitors and Reactors in Applications Involving Multiple Single-Tuned Filters", *IEEE Transactions on Power Delivery*, Vol. 10, No. 1, January 1995, pp. 547–555.

4

Harmonic Evaluation of Non-linear Functions

4.1 Introduction

A wide range of non-linear characteristics relating to physical, electrical equipment can be accurately represented by means of standard polynomial equations,

$$u = a_0 + a_1 x + a_2 x^2 + a_3 x^3 + \cdots + a_n x^n \tag{4.1}$$

rational polynomials,

$$w = \frac{b_0 + b_1 y + b_2 y^2 + b_3 y^3 + \cdots + b_p y^p}{c_0 + c_1 z + c_2 z^2 + c_3 z^3 + \cdots + c_q z^q} \tag{4.2}$$

and switching functions,

$$v = s \cdot y \tag{4.3}$$

where s is the switching function.

These representations are amenable to direct frequency domain evaluations via discrete convolutions, a more efficient alternative for performing harmonic domain calculations than techniques that resort to combined time and frequency domain representations. Most of the chapter is dedicated to direct frequency domain evaluations but time–frequency domain techniques are very important for some kinds of harmonic domain applications and they receive attention towards the end of the chapter.

Saturation and hysteresis phenomena are important classes of non-linearities exhibited by most electromagnetic components operating above their rated design values. As shown towards the end of the chapter, they are well represented by polynomial equations of various kinds. This is significant for circuit analysis problems because repeated convolutions provide an efficient way of performing polynomial evaluations in harmonic domain applications. On the other hand, switching functions are commonly used to describe the operation of power electronic devices, and, when evaluated in the harmonic domain via repeated convolutions, provide a very powerful tool for assessing the harmonic behaviour of these commutated devices. The value of switching functions lies not just in their simplicity but also in the fact that the polynomial representation of commutated device characteristics is difficult to achieve to satisfactory standards.

Several hand-worked examples of harmonic evaluations are presented in this chapter. They correspond to simple algebraic manipulations of complex exponential terms and also to full convolution-based methods.

4.2 Polynomial Evaluation by Hand Calculations

Harmonic domain evaluations by hand are only possible for very simple cases. Nevertheless, they provide insight into the mechanism of harmonic domain evaluation and some examples are presented below.

Example 4-1: Linearise the relation

$$u = y + y^3$$

around the base point

$$y_0 = \sin \omega_0 t$$

(a) Substituting the base operating point into the non-linear function gives

$$u_0 = \sin \omega_0 t + \sin^3 \omega_0 t \tag{4.4}$$

The powers of $\sin x$ are given as

$$\begin{aligned}
\sin^0 x &= 1 \\
\sin^1 x &= \sin x \\
\sin^2 x &= \frac{1}{2}(1 - \cos 2x) \\
\sin^3 x &= \frac{1}{4}(3 \sin x - \sin 3x) \\
\sin^4 x &= \frac{1}{8}(3 - 4 \cos 2x + \cos 4x) \\
\sin^5 x &= \frac{1}{16}(10 \sin x - 5 \sin 3x + \sin 5x) \\
\sin^6 x &= \frac{1}{32}(10 - 15 \cos 2x + 6 \cos 4x - \cos 6x) \\
\sin^7 x &= \frac{1}{64}(35 \sin x - 21 \sin 3x + 7 \sin 5x - \sin 7x)
\end{aligned}$$

Hence, the harmonic content of equation (4.4) is

$$u_0 = \sin \omega_0 t + \frac{1}{4}(3 \sin \omega_0 t - \sin 3\omega_0 t) = \frac{7}{4} \sin \omega_0 t - \frac{1}{4} \sin 3\omega_0 t \tag{4.5}$$

With reference to equation (2.12), the coefficients of the trigonometric series (4.5), in vector form, are

$$\mathbf{U}_0 = \left[\begin{array}{c} \mathbf{U}_0^s \\ \hline \mathbf{U}_0^c \end{array} \right] = \left[\begin{array}{c} b_1 \\ b_2 \\ b_3 \\ \hline a_0 \\ a_1 \\ a_2 \\ a_3 \end{array} \right] = \left[\begin{array}{c} 1.75 \\ 0 \\ -0.25 \\ \hline 0 \\ 0 \\ 0 \\ 0 \end{array} \right] \tag{4.6}$$

(b) Alternatively, equation (4.4) can be expressed in terms of complex harmonics,

$$u_0 = \left(\frac{e^{j\omega_0 t} - e^{-j\omega_0 t}}{2j} \right) + \left(\frac{e^{j\omega_0 t} - e^{-j\omega_0 t}}{2j} \right)^3 \tag{4.7}$$

since

$$\sin \omega_0 t = \left(\frac{e^{j\omega_0 t} - e^{-j\omega_0 t}}{2j} \right)$$

Equation (4.7) simplifies to the following:

$$u_0 = j\frac{1}{8}e^{j3\omega_0 t} - j\frac{7}{8}e^{j\omega_0 t} + j\frac{7}{8}e^{-j\omega_0 t} - j\frac{1}{8}e^{-j3\omega_0 t}$$

The coefficients of the complex exponential series, in vector form, are

$$\mathbf{U}_0 = \left[\begin{array}{c} C_{-3} \\ C_{-2} \\ C_{-1} \\ C_0 \\ C_1 \\ C_2 \\ C_3 \end{array} \right] = \left[\begin{array}{c} -j0.125 \\ 0 \\ j0.875 \\ 0 \\ -j0.875 \\ 0 \\ j0.125 \end{array} \right] \tag{4.8}$$

It should be noticed that solutions (4.6) and (4.8) are fully equivalent.

Example 4-2: A more elaborate example involving two inputs and one output is presented next. The non-linear relation

$$u = f(x, y) = xy^2(1 + y)$$

is linearised around the base point

$$\begin{aligned} x_0 &= \sin \omega_0 t + \sin 3\omega_0 t \\ y_0 &= \cos \omega_0 t \end{aligned}$$

or, in terms of complex harmonics,

$$x_0 = \frac{e^{j\omega_0 t} - e^{-j\omega_0 t}}{2j} + \frac{e^{j3\omega_0 t} - e^{-j3\omega_0 t}}{2j} \tag{4.9}$$

$$y_0 = \frac{e^{j\omega_0 t} + e^{-j\omega_0 t}}{2} \tag{4.10}$$

Substituting the base operating point, (4.9) and (4.10), into the non-linear relation gives

$$\begin{aligned} u_0 &= \left\{\left(\frac{e^{j\omega_0 t} - e^{-j\omega_0 t}}{2j}\right) + \left(\frac{e^{j3\omega_0 t} - e^{-j3\omega_0 t}}{2j}\right)\right\}\left(\frac{e^{j\omega_0 t} + e^{-j\omega_0 t}}{2}\right)^2 \\ &\quad \times \left\{1 + \left(\frac{e^{j\omega_0 t} + e^{-j\omega_0 t}}{2}\right)\right\} \\ &= j\frac{1}{16}e^{-j6\omega_0 t} + j\frac{1}{8}e^{-j5\omega_0 t} + j\frac{3}{16}e^{-j4\omega_0 t} + j\frac{3}{8}e^{-j3\omega_0 t} + j\frac{1}{8}e^{-j2\omega_0 t} + j\frac{1}{4}e^{-j\omega_0 t} \\ &\quad - j\frac{1}{4}e^{j\omega_0 t} - j\frac{1}{8}e^{j2\omega_0 t} - j\frac{3}{8}e^{j3\omega_0 t} - j\frac{3}{16}e^{j4\omega_0 t} - j\frac{1}{8}e^{j5\omega_0 t} - j\frac{1}{16}e^{j6\omega_0 t} \end{aligned}$$

or, in harmonic coefficients vector form,

$$\mathbf{F}_0 = j\frac{1}{16}\left[\begin{array}{ccccccccccccc} 1 & 2 & 3 & 6 & 2 & 4 & 0 & -4 & -2 & -6 & -3 & -2 & -1 \end{array}\right]^{T}$$

Example 4-3: Find the linearised representation of the non-linear function

$$i = \frac{1}{L}\psi \cdot s \tag{4.11}$$

around the base point

$$\psi_0 = \sin \omega_0 t$$

where $s = \cos^2 \omega_0 t$ is a switching function and $L = 0.125$ H.

Substituting the base point and the switching function in (4.11) gives

$$i_0 = \frac{1}{L} \sin \omega_0 t \cos^2 \omega_0 t$$

or, in terms of complex harmonics,

$$\begin{aligned} i_0 &= \frac{1}{0.125}\left(\frac{e^{j\omega_0 t} - e^{-j\omega_0 t}}{2j}\right)\left(\frac{e^{j\omega_0 t} + e^{-j\omega_0 t}}{2}\right)^2 \\ &= \frac{1}{j}\left(e^{j\omega_0 t} - e^{-j\omega_0 t}\right)\left(e^{j2\omega_0 t} + 2 + e^{-j2\omega_0 t}\right) \\ &= -je^{j3\omega_0 t} - je^{j\omega_0 t} + je^{-j\omega_0 t} + je^{-j3\omega_0 t} \end{aligned}$$

This result may be expressed in harmonic coefficients vector form as

$$\mathbf{I}_0 = j\left[\begin{array}{ccccccc} 1 & 0 & 1 & 0 & -1 & 0 & -1 \end{array}\right]^{T}$$

4.3 Convolutions in the Complex Fourier Harmonic Domain

It can be appreciated from the examples above that the mechanics of polynomial evaluation is a straightforward, though cumbersome, process. The number of algebraic operations grows rapidly with the polynomial degree and the harmonic order of the excitation. Clearly, hand calculations are not suitable for most practical harmonic domain calculations and we have to rely instead on numerical procedures that are more suitable for digital computer solutions.

In theory any continuous function defined on a closed, bounded interval can be approximated with an arbitrarily small error by a polynomial function of sufficiently high degree *n*, i.e. the Weierstrass approximation theorem. This is an interesting possibility that opens the door for the solution of many problems in circuit analysis via repeated convolutions. Polynomial evaluation via repeated convolutions is equivalent to the algebraic approach presented in the previous section but it has the advantage of being amenable to efficient computer implementations.

Several harmonic transforms can be used for carrying out convolution operations, namely complex Fourier, real Fourier, Hartley, Walsh, wavelets, etc. In this section the use of the first transform will be described.

4.3.1 Polynomial evaluation

The process of polynomial evaluation via repeated convolutions is illustrated below using complex Fourier series. In order to keep the derivations simple, let us consider a polynomial relation of sufficiently low order, say $n = 3$:

$$u = f(x) = \sum_{q=0}^{3} b_q x(t)^q = b_0 x^0 + b_1 x^1 + b_2 x^2 + b_3 x^3 \tag{4.12}$$

Let us also assume that the variable x contains only fundamental and DC terms:

$$x(t) = \sum_{h=-1}^{1} X_h \mathrm{e}^{\mathrm{j}h\omega_0 t} \tag{4.13}$$

As a first step in the evaluation process of equation (4.12) let us calculate x^2 in the complex Fourier harmonic domain by multiplying equation (4.13) by itself:

$$\begin{aligned} x^2(t) &= \sum_{h=-1}^{1} X_h \mathrm{e}^{\mathrm{j}h\omega_0 t} \sum_{h=-1}^{1} X_h \mathrm{e}^{\mathrm{j}h\omega_0 t} \\ &= \left(X_{-1}\mathrm{e}^{-\mathrm{j}\omega_0 t} + X_0 + X_1 \mathrm{e}^{\mathrm{j}\omega_0 t}\right)\left(X_{-1}\mathrm{e}^{-\mathrm{j}\omega_0 t} + X_0 + X_1 \mathrm{e}^{\mathrm{j}\omega_0 t}\right) \\ &= (X_{-1}X_{-1})\,\mathrm{e}^{-\mathrm{j}2\omega_0 t} + (X_0 X_{-1} + X_{-1}X_0)\,\mathrm{e}^{-\mathrm{j}\omega_0 t} \\ &\quad + (X_1 X_{-1} + X_0 X_0 + X_{-1}X_1) \\ &\quad + (X_0 X_1 + X_1 X_0)\,\mathrm{e}^{\mathrm{j}\omega_0 t} + (X_1 X_1)\,\mathrm{e}^{\mathrm{j}2\omega_0 t} \end{aligned} \tag{4.14}$$

The harmonic domain representation of equation (4.14) is

$$x^2(t) \Rightarrow \mathbf{X}^2 \equiv \mathbf{X} \otimes \mathbf{X} = \begin{bmatrix} 0 \\ X_{-1} \\ X_0 \\ X_1 \\ 0 \end{bmatrix} \otimes \begin{bmatrix} 0 \\ X_{-1} \\ X_0 \\ X_1 \\ 0 \end{bmatrix} = \begin{bmatrix} X_{-1}X_{-1} \\ X_{-1}X_0 + X_0X_{-1} \\ X_1X_{-1} + X_0X_0 + X_{-1}X_1 \\ X_1X_0 + X_0X_1 \\ X_1X_1 \end{bmatrix} \tag{4.15}$$

Since $\mathbf{X}$ is a vector, it is clear that equation (4.15) cannot be directly interpreted in terms of conventional matrix operations but, rather, in terms of convolutions. If $\mathbf{X}$ holds the harmonic coefficients of a discrete-frequency function then $\mathbf{X}^2$ is the convolution of that function by itself. It should be noted that the same result should be obtained by applying equation (2.47). The convolution operation is easily appreciated from the factorised form of equation (4.15),

$$\mathbf{X}^2 = \begin{bmatrix} X_0 & X_{-1} & & & \\ X_1 & X_0 & X_{-1} & & \\ & X_1 & X_0 & X_{-1} & \\ & & X_1 & X_0 & X_{-1} \\ & & & X_1 & X_0 \end{bmatrix} \begin{bmatrix} 0 \\ X_1 \\ X_0 \\ X_1 \\ 0 \end{bmatrix}$$

$$= \begin{bmatrix} X_{-1}X_{-1} \\ X_{-1}X_0 + X_0X_{-1} \\ X_1X_{-1} + X_0X_0 + X_{-1}X_1 \\ X_1X_0 + X_0X_1 \\ X_1X_1 \end{bmatrix} = \begin{bmatrix} X_{-2}^{(2)} \\ X_{-1}^{(2)} \\ X_0^{(2)} \\ X_1^{(2)} \\ X_2^{(2)} \end{bmatrix} \tag{4.16}$$

The convolution of $\mathbf{X}$ by $\mathbf{X}$ is termed self-convolution. The matrix in equation (4.16) is formed with the complex harmonic coefficients of equation (4.13). The matrix is Hermitian and has a Toeplitz structure, i.e. it contains equal elements in all diagonals. The band of side-diagonals is half plus one of the order of the harmonic vector that generates it.

The second step in the evaluation of equation (4.12) is the calculation of the cubic term. In harmonic domain representation,

$$\mathbf{X}^3 = \mathbf{X} \otimes \mathbf{X}^2 = \begin{bmatrix} X_0 & X_{-1} & & & & & \\ X_1 & X_0 & X_{-1} & & & & \\ & X_1 & X_0 & X_{-1} & & & \\ & & X_1 & X_0 & X_{-1} & & \\ & & & X_1 & X_0 & X_{-1} & \\ & & & & X_1 & X_0 & X_{-1} \\ & & & & & X_1 & X_0 \end{bmatrix}$$

$$\times \begin{bmatrix} 0 \\ X_{-1}X_{-1} \\ X_{-1}X_0 + X_0X_{-1} \\ X_1X_{-1} + X_0X_0 + X_{-1}X_1 \\ X_1X_0 + X_0X_1 \\ X_1X_1 \\ 0 \end{bmatrix} = \begin{bmatrix} X_{-3}^{(3)} \\ X_{-2}^{(3)} \\ X_{-1}^{(3)} \\ X_0^{(3)} \\ X_1^{(3)} \\ X_2^{(3)} \\ X_3^{(3)} \end{bmatrix} \tag{4.17}$$

The convolution of $\mathbf{X}$ by $\mathbf{X}^2$ is termed mutual convolution.

The remaining problem that needs addressing is the harmonic domain calculation of $\mathbf{X}^0$ in equation (4.12). It corresponds to a DC component whose magnitude is one, i.e

$$\mathbf{X}^0 = \begin{bmatrix} 0 \\ 0 \\ 0 \\ 1 \\ 0 \\ 0 \\ 0 \end{bmatrix} \tag{4.18}$$

Now the evaluation of the polynomial equation (4.12) in the harmonic domain is carried out:

$$\begin{bmatrix} Y_{-3} \\ Y_{-2} \\ Y_{-1} \\ Y_0 \\ Y_1 \\ Y_2 \\ Y_3 \end{bmatrix} = b_0 \begin{bmatrix} 0 \\ 0 \\ 0 \\ 1 \\ 0 \\ 0 \\ 0 \end{bmatrix} + b_1 \begin{bmatrix} 0 \\ 0 \\ X_{-1} \\ X_0 \\ X_1 \\ 0 \\ 0 \end{bmatrix} + b_2 \begin{bmatrix} 0 \\ X_{-2}^{(2)} \\ X_{-1}^{(2)} \\ X_0^{(2)} \\ X_1^{(2)} \\ X_2^{(2)} \\ 0 \end{bmatrix} + b_3 \begin{bmatrix} X_{-3}^{(3)} \\ X_{-2}^{(3)} \\ X_{-1}^{(3)} \\ X_0^{(3)} \\ X_1^{(3)} \\ X_2^{(3)} \\ X_3^{(3)} \end{bmatrix} \tag{4.19}$$

When implemented in a computer program this method provides a very efficient algorithm for evaluating a general polynomial function of the form

$$p(x, y, z, u) = \sum_{i=0}^{n} b_i x_i^j y_i^k z_i^l u_i^m \tag{4.20}$$

For most practical cases, only a small number of self-convolutions and mutual convolutions will be needed, even for cases when a factor x^q in a polynomial expression has to be evaluated for large values of q. For instance, for q=21 only four self-convolutions and two mutual-convolutions are required.

Example 4-4: The polynomial equation

$$i = f(\psi) = 0.001\psi + 0.0743\psi^3$$

is subjected to a base excitation

$$\psi_0 = \sin\omega_0 t + \cos\omega_0 t = \left(\frac{1}{2} - \mathrm{j}\frac{1}{2}\right)\mathrm{e}^{\mathrm{j}\omega_0 t} + \left(\frac{1}{2} + \mathrm{j}\frac{1}{2}\right)\mathrm{e}^{-\mathrm{j}\omega_0 t} \Rightarrow \Psi_0 = \frac{1}{2}\begin{bmatrix} 0 \\ 1+\mathrm{j} \\ 0 \\ 1-\mathrm{j} \\ 0 \end{bmatrix}$$

The harmonic evaluation is carried out as follows:

$$\Psi^2 = \Psi \otimes \Psi = \frac{1}{2}\begin{bmatrix} 0 \\ 1+\mathrm{j} \\ 0 \\ 1-\mathrm{j} \\ 0 \end{bmatrix} \otimes \frac{1}{2}\begin{bmatrix} 0 \\ 1+\mathrm{j} \\ 0 \\ 1-\mathrm{j} \\ 0 \end{bmatrix}$$

or, in matrix form,

$$\Psi^2 = \frac{1}{2}\begin{bmatrix} 0 & 1+\mathrm{j} & & & \\ 1-\mathrm{j} & 0 & 1+\mathrm{j} & & \\ & 1-\mathrm{j} & 0 & 1+\mathrm{j} & \\ & & 1-\mathrm{j} & 0 & 1+\mathrm{j} \\ & & & 1-\mathrm{j} & 0 \end{bmatrix} \frac{1}{2}\begin{bmatrix} 0 \\ 1+\mathrm{j} \\ 0 \\ 1-\mathrm{j} \\ 0 \end{bmatrix} = \frac{1}{2}\begin{bmatrix} \mathrm{j} \\ 0 \\ 2 \\ 0 \\ -\mathrm{j} \end{bmatrix}$$

Also,

$$\Psi^3 = \Psi \otimes \Psi^2 = \frac{1}{2}\begin{bmatrix} 0 \\ 0 \\ 1+\mathrm{j} \\ 0 \\ 1-\mathrm{j} \\ 0 \\ 0 \end{bmatrix} \otimes \frac{1}{2}\begin{bmatrix} 0 \\ \mathrm{j} \\ 0 \\ 2 \\ 0 \\ -\mathrm{j} \\ 0 \end{bmatrix}$$

or, in matrix form,

$$\Psi^3 = \frac{1}{4}\begin{bmatrix} 0 & 1+\mathrm{j} & & & & & \\ 1-\mathrm{j} & 0 & 1+\mathrm{j} & & & & \\ & 1-\mathrm{j} & 0 & 1+\mathrm{j} & & & \\ & & 1-\mathrm{j} & 0 & 1+\mathrm{j} & & \\ & & & 1-\mathrm{j} & 0 & 1+\mathrm{j} & \\ & & & & 1-\mathrm{j} & 0 & 1+\mathrm{j} \\ & & & & & 1-\mathrm{j} & 0 \end{bmatrix}\begin{bmatrix} 0 \\ \mathrm{j} \\ 0 \\ 2 \\ 0 \\ -\mathrm{j} \\ 0 \end{bmatrix}$$

$$= \frac{1}{4}\begin{bmatrix} -1+\mathrm{j} \\ 0 \\ 3+3\mathrm{j} \\ 0 \\ 3-3\mathrm{j} \\ 0 \\ -1-\mathrm{j} \end{bmatrix}$$

The cubic polynomial equation is now evaluated in the complex Fourier harmonic domain:

$$\begin{bmatrix} I_{-3} \\ I_{-2} \\ I_{-1} \\ I_0 \\ I_1 \\ I_2 \\ I_3 \end{bmatrix} = \frac{0.001}{2}\begin{bmatrix} 0 \\ 0 \\ 1+\mathrm{j} \\ 0 \\ 1-\mathrm{j} \\ 0 \\ 0 \end{bmatrix} + \frac{0.0743}{2}\begin{bmatrix} -1+\mathrm{j} \\ 0 \\ 3+3\mathrm{j} \\ 0 \\ 3-3\mathrm{j} \\ 0 \\ -1-\mathrm{j} \end{bmatrix} = \begin{bmatrix} -0.0186+\mathrm{j}0.0186 \\ 0 \\ 0.0562-\mathrm{j}0.0562 \\ 0 \\ 0.0562-\mathrm{j}0.0562 \\ 0 \\ -0.0186-\mathrm{j}0.0186 \end{bmatrix}$$

or

$$\begin{aligned} i &= (-0.0186+\mathrm{j}0.0186)\mathrm{e}^{-\mathrm{j}3\omega_0 t} + (0.0562+\mathrm{j}0.0562)\mathrm{e}^{-\mathrm{j}\omega_0 t} \\ &\quad +(0.0562-\mathrm{j}0.0562)\mathrm{e}^{\mathrm{j}\omega_0 t} + (-0.0186-\mathrm{j}0.0186)\mathrm{e}^{\mathrm{j}3\omega_0 t} \\ &= 0.1124\,(\sin\omega_0 t + \cos\omega_0 t) + 0.0372\,(\sin 3\omega_0 t - \cos 3\omega_0 t) \end{aligned}$$

Alternatively MATLAB™ can be used to obtain the self– and mutual convolutions shown above:

```
>> Flux=1/2*[0;1+i;0;1-i;0]
Flux =
   0
   0.5000 + 0.5000i
   0
   0.5000 - 0.5000i
   0
>> Flux2=conv(Flux,Flux)
Flux2 =
   0
   0
   0 + 0.5000i
   0
   1.0000
   0
   0 - 0.5000i
   0
   0
>> Flux3=conv(Flux,Flux2)
Flux3 =
   0
   0
   0
  -0.2500 + 0.2500i
   0
   0.7500 + 0.7500i
   0
   0.7500 - 0.7500i
   0
  -0.2500 - 0.2500i
   0
   0
   0
>>
```

4.3.2 Rational polynomial evaluation

As discussed in Section 4.1, non-linear functions may also be described by means of rational polynomials, i.e.

$$z = \frac{\sum_{i=0}^{p} a_i x^i}{\sum_{k=0}^{q} b_k y^k} \tag{4.21}$$

For such cases, the harmonic domain representation takes the following form:

$$\mathbf{Z} = \frac{\sum_{i=0}^{p} a_i \mathbf{X}^i}{\sum_{k=0}^{q} b_k \mathbf{Y}^k} = \frac{\mathbf{V}}{\mathbf{W}} = \frac{\mathbf{V}}{\mathbf{T}_W} = \mathbf{T}_W^{-1}\mathbf{V} \tag{4.22}$$

where

$$\begin{aligned} \mathbf{V} &= a_0\mathbf{X}^0 + a_1\mathbf{X}^1 + \cdots + a_p\mathbf{X}^p \\ \mathbf{W} &= b_0\mathbf{Y}^0 + b_1\mathbf{Y}^1 + \cdots + b_q\mathbf{Y}^q \end{aligned} \tag{4.23}$$

In the complex Fourier harmonic domain $\mathbf{X}$, $\mathbf{Y}$, $\mathbf{Z}$, $\mathbf{V}$ and $\mathbf{W}$ become complex vectors with entries which are complex conjugates around the DC harmonic entry, and $\mathbf{T}_W$ becomes an Hermitian Toeplitz matrix, i.e.

$$\mathbf{X} = \begin{bmatrix} \vdots \\ X_{-2} \\ X_{-1} \\ X_0 \\ X_1 \\ X_2 \\ \vdots \end{bmatrix};\ \mathbf{Y} = \begin{bmatrix} \vdots \\ Y_{-2} \\ Y_{-1} \\ Y_0 \\ Y_1 \\ Y_2 \\ \vdots \end{bmatrix};\ \mathbf{Z} = \begin{bmatrix} \vdots \\ Z_{-2} \\ Z_{-1} \\ Z_0 \\ Z_1 \\ Z_2 \\ \vdots \end{bmatrix};\ \mathbf{W} = \begin{bmatrix} \vdots \\ W_{-2} \\ W_{-1} \\ W_0 \\ W_1 \\ W_2 \\ \vdots \end{bmatrix} \tag{4.24}$$

and

$$\mathbf{T}_W = \begin{bmatrix} W_0 & W_{-1} & W_{-2} & & & & \\ W_1 & W_0 & W_{-1} & W_{-2} & & \ddots & \\ W_2 & W_1 & W_0 & W_{-1} & W_{-2} & & \\ & W_2 & W_1 & W_0 & W_{-1} & W_{-2} & \\ & & W_2 & W_1 & W_0 & W_{-1} & W_{-2} \\ & \ddots & & W_2 & W_1 & W_0 & W_{-1} \\ & & & & W_2 & W_1 & W_0 \end{bmatrix} \tag{4.25}$$

The following MATLAB™ function may be used to obtain the matrix $\mathbf{T}_W$:

```
function Fm=calc_Fm(Fv,h)
% Fv vector with the form Fv=[-h ... -1 0 1 ... h];
% Fm Toeplitz matrix
% h harmonic
Fm=zeros(2*h+1,2*h+1);
  for k=-h:h
     for j=-h:h
       if abs(k-j)<=h
          pos=(h+1)+(k-j);
          x=(h+1)+k;
          y=(h+1)+j;
          Fm(x,y)=Fv(pos);
       end
     end
  end
```

Example 4-5: The rational polynomial

$$z = \frac{x + x^3}{1 + y^2} \tag{4.26}$$

is subjected to a base excitation

$$x_0 = \sin \omega_0 t = \frac{\mathrm{e}^{\mathrm{j}\omega_0 t} - \mathrm{e}^{-\mathrm{j}\omega_0 t}}{2\mathrm{j}} \Rightarrow \mathbf{X} = \frac{1}{2}\begin{bmatrix} 0 \\ \mathrm{j} \\ 0 \\ -\mathrm{j} \\ 0 \end{bmatrix}$$

and

$$y_0 = \cos \omega_0 t = \frac{\mathrm{e}^{\mathrm{j}\omega_0 t} + \mathrm{e}^{-\mathrm{j}\omega_0 t}}{2} \Rightarrow \mathbf{Y} = \frac{1}{2}\begin{bmatrix} 0 \\ 1 \\ 0 \\ 1 \\ 0 \end{bmatrix}$$

The harmonic evaluation is carried out as follows:

$$\mathbf{X}^2 = \mathbf{X} \otimes \mathbf{X} = \frac{1}{2}\begin{bmatrix} 0 \\ \mathrm{j} \\ 0 \\ -\mathrm{j} \\ 0 \end{bmatrix} \otimes \frac{1}{2}\begin{bmatrix} 0 \\ \mathrm{j} \\ 0 \\ -\mathrm{j} \\ 0 \end{bmatrix} = \frac{1}{4}\begin{bmatrix} -1 \\ 0 \\ 2 \\ 0 \\ -1 \end{bmatrix}$$

Also,

$$\mathbf{X}^3 = \mathbf{X} \otimes \mathbf{X}^2 = \frac{1}{2}\begin{bmatrix} 0 \\ 0 \\ \mathrm{j} \\ 0 \\ -\mathrm{j} \\ 0 \\ 0 \end{bmatrix} \otimes \frac{1}{2}\begin{bmatrix} 0 \\ -1 \\ 0 \\ 2 \\ 0 \\ -1 \\ 0 \end{bmatrix} = \frac{\mathrm{j}}{8}\begin{bmatrix} -1 \\ 0 \\ 3 \\ 0 \\ -3 \\ 0 \\ 1 \end{bmatrix}$$

and

$$\mathbf{Y}^2 = \mathbf{Y} \otimes \mathbf{Y} = \frac{1}{2}\begin{bmatrix} 0 \\ 1 \\ 0 \\ 1 \\ 0 \end{bmatrix} \otimes \frac{1}{2}\begin{bmatrix} 0 \\ 1 \\ 0 \\ 1 \\ 0 \end{bmatrix} = \frac{1}{4}\begin{bmatrix} 1 \\ 0 \\ 2 \\ 0 \\ 1 \end{bmatrix}$$

The evaluation of relation (4.26) in the complex Fourier harmonic domain is

$$\mathbf{Z} = \frac{\frac{1}{2}\begin{bmatrix} 0 \\ 0 \\ \mathrm{j} \\ 0 \\ -\mathrm{j} \\ 0 \\ 0 \end{bmatrix} + \frac{\mathrm{j}}{8}\begin{bmatrix} -1 \\ 0 \\ 3 \\ 0 \\ -3 \\ 0 \\ 1 \end{bmatrix}}{\begin{bmatrix} 0 \\ 0 \\ 0 \\ 1 \\ 0 \\ 0 \\ 0 \end{bmatrix} + \frac{1}{4}\begin{bmatrix} 0 \\ 1 \\ 0 \\ 2 \\ 0 \\ 1 \\ 0 \end{bmatrix}} = \frac{\frac{\mathrm{j}}{8}\begin{bmatrix} -1 \\ 0 \\ 7 \\ 0 \\ -7 \\ 0 \\ 1 \end{bmatrix}}{\begin{bmatrix} 0 \\ 1/4 \\ 0 \\ 3/2 \\ 0 \\ 1/4 \\ 0 \end{bmatrix}}$$

and, in matrix form,

$$\mathbf{Z} = \begin{bmatrix} 3/2 & 0 & 1/4 & & & & \\ 0 & 3/2 & 0 & 1/4 & & & \\ 1/4 & 0 & 3/2 & 0 & 1/4 & & \\ & 1/4 & 0 & 3/2 & 0 & 1/4 & \\ & & 1/4 & 0 & 3/2 & 0 & 1/4 \\ & & & 1/4 & 0 & 3/2 & 0 \\ & & & & 1/4 & 0 & 3/2 \end{bmatrix}^{-1} \frac{\mathrm{j}}{8}\begin{bmatrix} -1 \\ 0 \\ 7 \\ 0 \\ -7 \\ 0 \\ 1 \end{bmatrix}$$

$$= \begin{bmatrix} 0.6863 & 0 & -0.1177 & 0 & 0.0202 & 0 & -0.0034 \\ 0 & 0.6863 & 0 & -0.1176 & 0 & 0.0196 & 0 \\ -0.1177 & 0 & 0.7065 & 0 & -0.1211 & 0 & 0.0202 \\ 0 & -0.1176 & 0 & 0.7059 & 0 & -0.1176 & 0 \\ 0.0202 & 0 & -0.1211 & 0 & 0.7065 & 0 & -0.1177 \\ 0 & 0.0196 & 0 & -0.1176 & 0 & 0.6863 & 0 \\ -0.0034 & 0 & 0.0202 & 0 & -0.1177 & 0 & 0.6863 \end{bmatrix}$$

$$\times \frac{\mathrm{j}}{8} \begin{bmatrix} -1 \\ 0 \\ 7 \\ 0 \\ -7 \\ 0 \\ 1 \end{bmatrix} = \begin{bmatrix} -\mathrm{j}0.2069 \\ 0 \\ \mathrm{j}0.7414 \\ 0 \\ -\mathrm{j}0.7414 \\ 0 \\ \mathrm{j}0.2069 \end{bmatrix}$$

It is interesting to notice that the Toeplitz structure of the matrix is lost when the inverse operation is carried out.

The next steps in MATLAB™ can be used for this example:

```
>> V=1/8*[-i;0;7*i;0;-7*i;0;i];
>> W=[0;1/4;0;3/2;0;1/4;0];
>> Tw=calc_Fm(W,3)
Tw =
   1.5000         0 0.2500         0         0         0         0
        0 1.5000         0 0.2500         0         0         0
   0.2500         0 1.5000         0 0.2500         0         0
        0 0.2500         0 1.5000         0 0.2500         0
        0         0 0.2500         0 1.5000         0 0.2500
        0         0         0 0.2500         0 1.5000         0
        0         0         0         0 0.2500         0 1.5000
>> Z=inv(Tw)*V
Z =
   0 - 0.2069i
   0
   0 + 0.7414i
   0
   0 - 0.7414i
   0
   0 + 0.2069i
>>
```

4.3.3 Switching function evaluation

The steady-state response of a wide range of electric circuits consists of a series of conducting and non-conducting periods within one fundamental frequency cycle. The steady-state response, $u(t)$, can be obtained by the multiplication of the switching functions, $s(t)$, and the steady-state excitation, $y(t)$,

$$u(t) = s(t)y(t) \tag{4.27}$$

In some applications it is desirable to conduct the above operation in the frequency domain rather than in the time domain, i.e.

$$\mathbf{U} = \mathbf{S} \otimes \mathbf{Y} \tag{4.28}$$

U and **Y** are harmonic vectors corresponding to the output and input waveforms, respectively. The switching function **S** is also a vector of harmonic coefficients, which may be a function of one or more time-variant control parameters, e.g. conduction angles. For instance, one full period of a switching function $s(t)$ is shown in Figure 4.1.

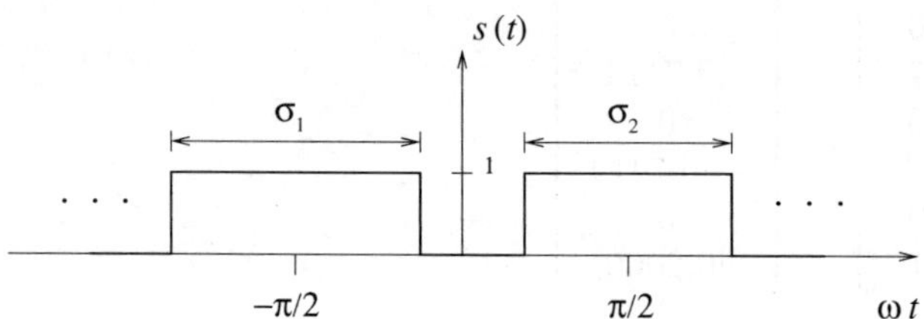

Figure 4.1 A full cycle of the switching function *s*(*t*)

A periodic train of square waves of this form can be represented by the harmonic vector **S**,

$$\mathbf{S} = \begin{bmatrix} S_{-h} & \cdots & S_{-2} & S_{-1} & S_0 & S_1 & S_2 & \cdots & S_h \end{bmatrix}^{\mathrm{T}} \tag{4.29}$$

where the complex Fourier coefficients S_h are given by

$$S_h = \frac{a_h}{2} + \mathrm{j}\frac{b_h}{2} = \frac{1}{\pi}\int_{\pi/2}^{\pi/2} s(t)\mathrm{e}^{-\mathrm{j}h\omega_0 t}\mathrm{d}\omega_0 t \tag{4.30}$$

The Fourier coefficients a_h and b_h are calculated from expressions which take into account the fact that the switching function has a value of one in the following intervals, $\left[-\frac{\pi}{2} - \frac{\sigma_1}{2}, -\frac{\pi}{2} + \frac{\sigma_1}{2}\right]$ and $\left[\frac{\pi}{2} - \frac{\sigma_2}{2}, \frac{\pi}{2} + \frac{\sigma_2}{2}\right]$. Then

$$a_0 = \frac{1}{\pi}\int_{-\frac{\pi}{2}-\frac{\sigma_1}{2}}^{-\frac{\pi}{2}+\frac{\sigma_1}{2}} \mathrm{d}\omega_0 t + \frac{1}{\pi}\int_{\frac{\pi}{2}-\frac{\sigma_2}{2}}^{\frac{\pi}{2}+\frac{\sigma_2}{2}} \mathrm{d}\omega_0 t = \frac{\sigma_1 + \sigma_2}{2\pi} \tag{4.31}$$

$$\begin{aligned} a_h &= \frac{1}{\pi}\int_{-\frac{\pi}{2}-\frac{\sigma_1}{2}}^{-\frac{\pi}{2}+\frac{\sigma_1}{2}} \cos(h\omega_0 t)\mathrm{d}\omega_0 t + \frac{1}{\pi}\int_{\frac{\pi}{2}-\frac{\sigma_2}{2}}^{\frac{\pi}{2}+\frac{\sigma_2}{2}} \cos(h\omega_0 t)\mathrm{d}\omega_0 t \\ &= \frac{2}{h\pi}\left[\sin\left(\frac{h\sigma_1}{2}\right) + \sin\left(\frac{h\sigma_2}{2}\right)\right]\cos\left(\frac{h\pi}{2}\right) \end{aligned} \tag{4.32}$$

$$\begin{aligned} b_n &= \frac{1}{\pi}\int_{-\frac{\pi}{2}-\frac{\sigma_1}{2}}^{-\frac{\pi}{2}+\frac{\sigma_1}{2}} \sin(h\omega_0 t)\mathrm{d}\omega_0 t + \frac{1}{\pi}\int_{\frac{\pi}{2}-\frac{\sigma_2}{2}}^{\frac{\pi}{2}+\frac{\sigma_2}{2}} \sin(h\omega_0 t)\mathrm{d}\omega_0 t \\ &= \frac{2}{h\pi}\left[\sin\left(\frac{h\sigma_2}{2}\right) - \sin\left(\frac{h\sigma_1}{2}\right)\right]\sin\left(\frac{h\pi}{2}\right) \end{aligned} \tag{4.33}$$

Switching functions will be explored more thoroughly in Part III of the book.

4.4 Numeric Evaluation of Non-linear Functions

The harmonic response of a non-linear element can also be obtained by mapping a full period of the input, point by point, upon the non-linear characteristic, and with the use of an FFT or FHT algorithm. The characteristic may be expressed in equation form or by means of a series of points in the x–y plane. In the latter case, the mapping is achieved with the aid of piecewise linear interpolation.

In the case of application to a simple magnetic circuit, the current is obtained as shown in Figure 4.2, i.e. the discretised flux is impressed point by point upon the magnetising characteristic and the corresponding magnetising current is determined. Once a full period of the operating flux, at the fundamental frequency, is obtained in the time domain, it is subdivided into $N = 2^n$ time steps, such that the highest harmonic will be sufficiently well sampled. In the case of magnetic non-linearities, harmonic frequencies beyond the 15th are negligibly small and values of n equal to 9 or 10 are normally adequate. In other applications it becomes necessary to make provision for a larger number of harmonic terms and, accordingly, larger values of n are required.

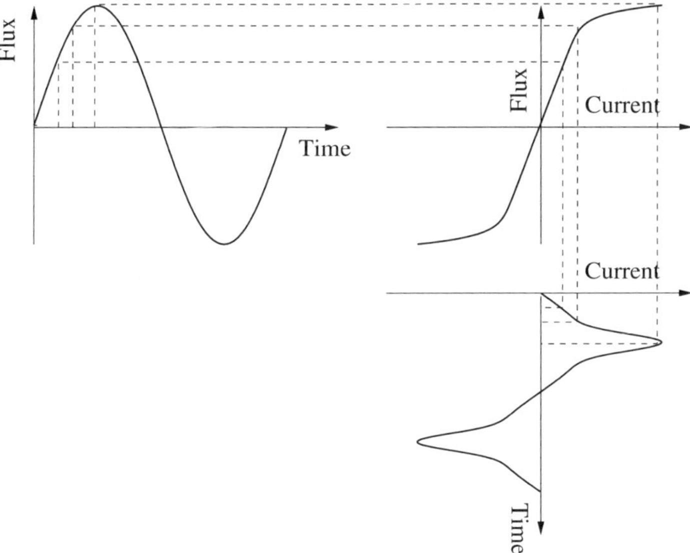

Figure 4.2 Point-by-point derivation of the magnetising characteristic

An FFT or FHT is applied to the magnetising current waveform to extract its harmonic content.

4.5 Analytical Approximation of Saturation

For steady-state harmonic analysis purposes, magnetic saturation is an important non-linear characteristic that is well approximated by an analytical expression. Figure 4.3 shows the top half of a single-valued saturation curve. Several interesting possibilities exist to carry out the fitting and some of them are presented in this section.

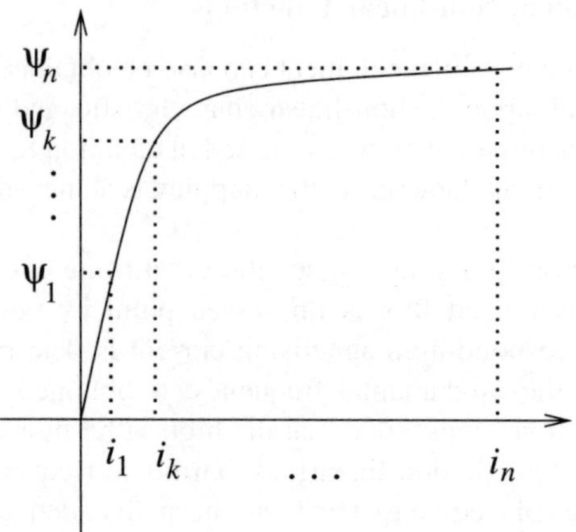

Figure 4.3 Magnetic saturation characteristic

4.5.1 Piecewise linear approximation

This method takes the simple approach of approximating the saturation by using a set of linear segments as shown in Figure 4.4. Very often two or three piecewise linear segments are sufficient to represent well the saturation characteristic. However, in harmonic studies a piecewise linear representation is not necessarily the best option. Selecting this approach would involve by necessity a point-by-point mapping of the non-linear characteristic.

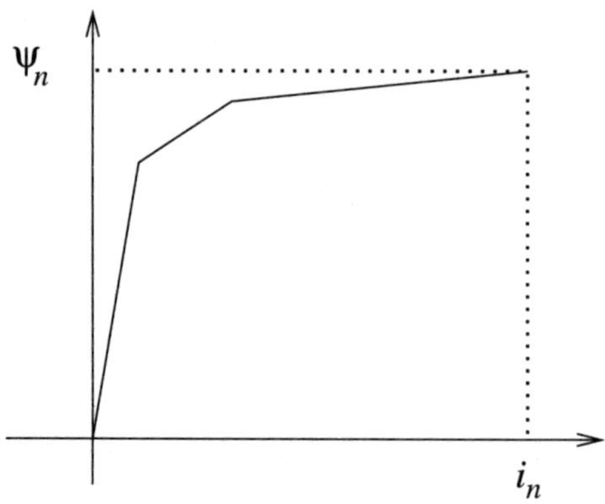

Figure 4.4 Piecewise linear approximation

4.5.2 Rational fraction approximation

The saturation characteristic may also be approximated by rational fractions,

$$i_k = \frac{a_0 + a_1\psi_k + a_2\psi_k^2 + \cdots + a_q\psi_k^q}{1 + b_1\psi_k + b_2\psi_k^2 + \cdots + b_q\psi_k^q}; \qquad k = 1, \ldots, n \tag{4.34}$$

where $(i_1, \psi_1), \ldots, (i_k, \psi_k), \ldots, (i_n, \psi_n)$ are values obtained experimentally and q is the order of the equation. Rewriting (4.34) in the form

$$i_k = a_0 + a_1\psi_k + a_2\psi_k^2 + \cdots + a_q\psi_k^q - b_1 i_k\psi_k - b_2 i_k\psi_k^2 - \cdots - b_q i_k\psi_k^q \tag{4.35}$$

or in matrix form, for all the values n, gives

$$\begin{bmatrix} i_1 \\ \vdots \\ i_k \\ \vdots \\ i_n \end{bmatrix} = \begin{bmatrix} 1 & \psi_1 & \psi_1^2 & \cdots & \psi_1^q & -i_1\psi_1 & -i_1\psi_1^2 & \cdots & -i_1\psi_1^q \\ \vdots & \vdots & \vdots & \vdots & \vdots & \vdots & \vdots & \vdots & \vdots \\ 1 & \psi_k & \psi_k^2 & \cdots & \psi_k^q & -i_k\psi_k & -i_k\psi_k^2 & \cdots & -i_k\psi_k^q \\ \vdots & \vdots & \vdots & \vdots & \vdots & \vdots & \vdots & \vdots & \vdots \\ 1 & \psi_n & \psi_n^2 & \cdots & \psi_n^q & -i_n\psi_n & -i_n\psi_n^2 & \cdots & -i_n\psi_k^q \end{bmatrix} \begin{bmatrix} a_0 \\ a_1 \\ a_2 \\ \vdots \\ a_q \\ b_1 \\ b_2 \\ \vdots \\ b_q \end{bmatrix} \tag{4.36}$$

In compact form,

$$\mathbf{i} = \mathbf{A}\mathbf{x} \tag{4.37}$$

Solving (4.37) for $\mathbf{x}$ may lead to an overdetermined set of equations, depending on the number of points n and the order q selected. In this case the pseudo-inverse of $\mathbf{A}$ can be used to determine $\mathbf{x}$,

$$\mathbf{x} = \mathbf{A}^{+}\mathbf{i} \tag{4.38}$$

where the pseudo-inverse is given as

$$\mathbf{A}^{+} = \left[\mathbf{A}^{\mathrm{T}}\mathbf{A}\mathbf{A}^{\mathrm{T}}\right]^{-1} \tag{4.39}$$

4.5.3 Hyperbolic approximation

The equation of a hyperbola can also be used to model quite accurately the non-linear characteristic in Figure 4.3 by

$$\epsilon\psi_k = (m_1 i_k + b_1 - \psi_k)(m_2 i_k + b_2 - \psi_k) - b_1 b_2 \tag{4.40}$$

where

m_1 = slope of unsaturated region
m_2 = slope of saturated region
b_1 = ordinate to the origin of asymptote to m_1
b_2 = ordinate to the origin of asymptote to m_2
ϵ = correction term

Since b_1 is located at the origin, this term equals zero and (4.40) can be rewritten as

$$(m_1 i_k + \psi_k^2) = (m_1 i_k^2 - i_k \psi_k)\, m_2 + (m_1 i_k - \psi_k)\, b_2 - \epsilon \psi_k \tag{4.41}$$

or, in matrix form, for all the n values,

$$\begin{bmatrix} m_1 i_1 + \psi_1^2 \\ \vdots \\ m_1 i_k + \psi_k^2 \\ \vdots \\ m_1 i_n + \psi_n^2 \end{bmatrix} = \begin{bmatrix} (m_1 i_1^2 - i_1 \psi_1) & (m_1 i_1 - \psi_1) & -\psi_1 \\ \vdots & \vdots & \vdots \\ (m_1 i_k^2 - i_k \psi_k) & (m_1 i_k - \psi_k) & -\psi_k \\ \vdots & \vdots & \vdots \\ (m_1 i_n^2 - i_n \psi_n) & (m_1 i_n - \psi_n) & -\psi_n \end{bmatrix} \begin{bmatrix} m_2 \\ b_2 \\ \epsilon \end{bmatrix} \tag{4.42}$$

m_1 and m_2 may be derived from experimental data, where normally $m_1 \gg m_2$, e.g. $m_1 = 5000$, $m_2 = 0.1$, as shown in Figure 4.5. However, it is more convenient to specify m_1 and to calculate m_2. The correction term ϵ modulates the sharpness of the knee region of the magnetising characteristic. Similarly to the case above, the pseudo-inverse can be used to solve the overdetermined system of equations.

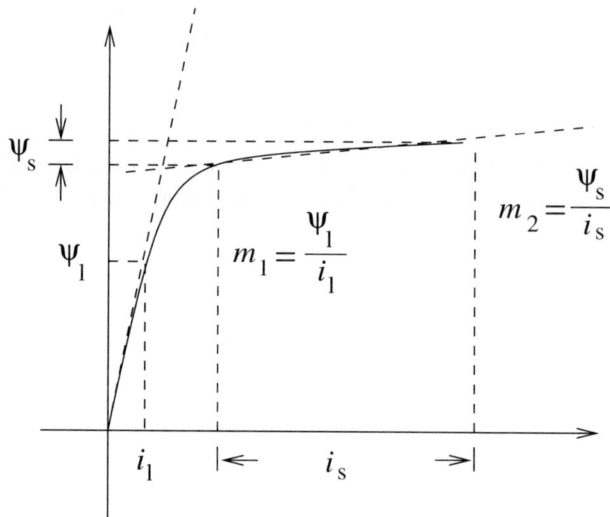

Figure 4.5 Hyperbolic approximation

4.5.4 *Cubic splines interpolation*

The method of cubic splines can also be used to represent the non-linear characteristic in Figure 4.3. In this method the tabulated values

$$a = \psi_0 < \psi_1 < \cdots < \psi_n = b$$

are made to correspond to the cubic functions $i_0 = S(\psi_0),\ i_1 = S(\psi_1),\ \ldots,\ i_n = S(\psi_n)$.

$S(\psi)$ is a cubic spline function defined in the interval $[a, b]$ that should observe the following characteristics:

- $S(\psi)$ is a polynomial of degree 3 in each one of the subintervals

$$[a, \psi_1],\ [\psi_1, \psi_2],\ \ldots,\ [\psi_{n-1}, b]$$

- $S(\psi)$ and its first and second derivatives $S'(\psi)$ and $S''(\psi)$ are continuous functions throughout the range of concern.

A popular expression for the cubic spline function, in the interval $\psi_k \leq \psi \leq \psi_{k+1}$, is expressed as

$$\begin{aligned} S(\psi) &= \frac{(\psi_{k+1} - \psi)^3 M_k + (\psi - \psi_k)^3 M_{k+1}}{6h_k} + \frac{(\psi_{k+1} - \psi)\, i_k + (\psi - \psi_k)\, i_{k+1}}{h_k} \\ &\quad - \frac{h_k}{6}\left[(\psi_{k+1} - \psi) M_k + (\psi - \psi_k) M_{k+1}\right] \end{aligned} \tag{4.43}$$

where $0 \leq k \leq (n-1)$ and $h_k = \psi_{k+1} - \psi_k$.

The unknown values $M_k = S''(\psi_k)$ are calculated from

$$\frac{h_{k-1}}{6} M_{k-1} + \frac{h_k + h_{k-1}}{3} M_k + \frac{h_k}{6} M_{k+1} = \frac{i_{k+1} - i_k}{h_k} - \frac{i_k - i_{k-1}}{h_{k-1}} \tag{4.44}$$

Since this equation gives $n-1$ equations, i.e. for $k = 1, \ldots, n-1$, and $n+1$ unknowns, i.e. $M_0, \ldots, M_n$, the cubic natural interpolating spline is selected at $M_0 = M_n = 0$ in order to have only $n-1$ unknowns.

The system of equations to be solved is

$$\mathbf{D} = \mathbf{AM} \tag{4.45}$$

where

$$\mathbf{D} = \begin{bmatrix} \frac{i_2 - i_1}{h_1} - \frac{i_1 - i_0}{h_0} & \cdots & \frac{i_k - i_{k-1}}{h_{k-1}} - \frac{i_{k-1} - i_{k-2}}{h_{k-2}} & \cdots & \frac{i_n - i_{n-1}}{h_{n-1}} - \frac{i_{n-1} - i_{n-2}}{h_{n-2}} \end{bmatrix}^{\mathrm{T}} \tag{4.46}$$

$$\mathbf{M} = \begin{bmatrix} M_1 & \cdots & M_k & \cdots & M_{n-1} \end{bmatrix}^{\mathrm{T}} \tag{4.47}$$

and

$$\mathbf{A} = \begin{bmatrix} \frac{h_0 + h_1}{3} & \frac{h_1}{6} & 0 & \cdots & 0 \\ \frac{h_1}{6} & \frac{h_1 + h_2}{3} & \frac{h_2}{6} & \ddots & \vdots \\ 0 & \frac{h_2}{6} & \ddots & \ddots & 0 \\ \vdots & \ddots & \ddots & \ddots & \frac{h_{n-2}}{6} \\ 0 & \cdots & 0 & \frac{h_{n-2}}{6} & \frac{h_{n-2} + h_{n-1}}{3} \end{bmatrix} \tag{4.48}$$

4.5.5 *Simple polynomial approximation*

A simple polynomial equation of the form

$$i = a\psi + b\psi^n \tag{4.49}$$

may be made to approximate the non-linear characteristic in Figure 4.3 with reasonably good accuracy. The coefficients a, b and n are derived from the following basic information, corresponding to the saturation magnetising characteristic shown in Figure 4.6:

- Coordinates of the knee, $(\psi_{\text{nom}}, i_{\text{nom}})$.
- Coordinates of the maximum point to be considered, $(\psi_{\text{max}}, i_{\text{max}})$.
- Slope M of the linear part, $a = 1/M$.

A substitution of $(\psi_{\text{max}}, i_{\text{max}})$ in (4.49) leads to the following expression:

$$b = \frac{i_{\text{max}} - a\psi_{\text{max}}}{\psi_{\text{max}}^n} \tag{4.50}$$

Using equation (4.50), an iterative process is started with low, odd values of n, e.g. 3, 5, and the corresponding values of b are obtained. The approximation that best fits the pair b, n and that satisfies (4.49) for $(\psi_{\text{nom}}, i_{\text{nom}})$ is chosen.

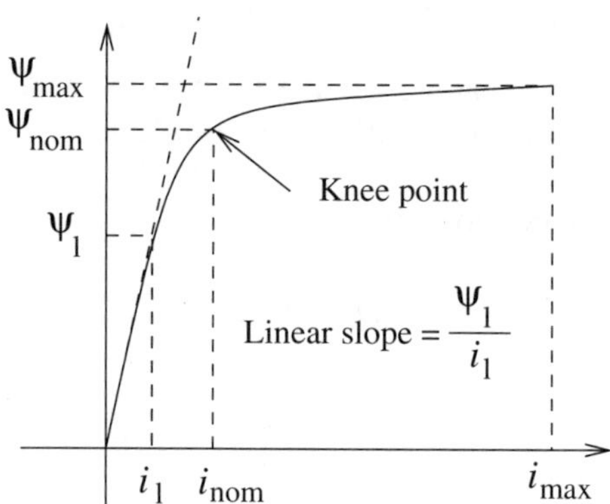

Figure 4.6 Polynomial approximation

Example 4-6: A load, sensitive to poor power quality, will be installed at bus 3 of the power circuit shown in Figure 4.7. This is the one-line diagram of a perfectly balanced three-phase power circuit. Planning studies are conducted to assess the possible adverse harmonic effects introduced by a non-linear load, connected at bus 2, on the sensitive load. Measurements have indicated that the voltage supply at node 1 is sinusoidal. The positive sequence, fundamental impedance values of the two cables and the sensitive load are also shown in Figure 4.7. The non-linear load consists of a saturated magnetic core for which the following positive sequence information exists:

- Linear flux and current values: $\psi_1 = 0.9$ p.u., $i_1 = 0.0009$ p.u.

- Nominal flux and current values: $\psi_{nom} = 1.0$ p.u., $i_{nom} = 0.08$ p.u.
- Maximum flux and current values: $\psi_{max} = 1.1$ p.u., $i_{max} = 0.1$ p.u.

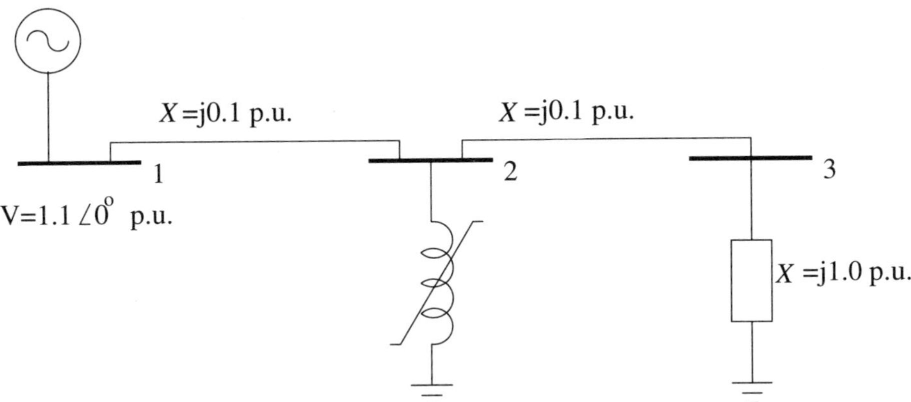

Figure 4.7 Power circuit

(i) Using the information available for the characteristic of the non-linear load, determine the approximated polynomial equation $i = a\psi + b\psi^n$.

The coefficient a is determined first:

$$M = \frac{0.9}{0.0009} = 1000; \; a = \frac{1}{1000} = 0.001$$

Coefficient b is determined by substituting maximum current and flux values, together with the slope value, in the polynomial equation:

$$b = \frac{0.1 - 0.001 \times 1.1}{1.1^n}$$

and solving for b, as a function of n (3,5,7, etc.):

$$b = \frac{0.0989}{1.1^n}$$

$$b = \frac{0.0989}{1.1^3} = 0.0743$$

$b = 0.0743$ is just below 0.08 for $n = 3$ and these values of b and n are selected.

The polynomial equation to be used is

$$i = 0.001\psi + 0.0743\psi^3 \text{ p.u.}$$

(ii) Determine the harmonic currents drawn by the non-linear reactor for the following voltage excitation condition:

$$v = \frac{d\psi}{dt} = \cos \omega_0 t$$

The base operating point

$$v = \frac{d\psi}{dt} = \cos\omega_0 t \implies \psi = \sin\omega_0 t = \frac{e^{j\omega_0 t} - e^{-j\omega_0 t}}{2j}$$

(note that $\omega_0 = 1$ in p.u.) is substituted in the polynomial equation:

$$i = 0.001\left(\frac{e^{j\omega_0 t} - e^{-j\omega_0 t}}{2j}\right) + 0.0743\left(\frac{e^{j\omega_0 t} - e^{-j\omega_0 t}}{2j}\right)^3$$

$$i = j0.0093e^{j3\omega_0 t} - j0.0284e^{j\omega_0 t} + j0.00284e^{-j\omega_0 t} - j0.0093e^{-j3\omega_0 t}$$

The fundamental and harmonic content are,

$$I_{(1)} = -j0.0568\,\text{p.u.}$$

$$I_{(3)} = j0.0186\,\text{p.u.}$$

(iii) Determine the fundamental and harmonic voltages in buses 2 and 3 and the harmonic currents injected into the utility network (bus 1) and into the sensitive load (bus 3).

The nodal admittance matrix representing the power network at the fundamental frequency is

$$\begin{bmatrix} I_1 \\ -I_2 \\ 0 \end{bmatrix}_{(1)} = \begin{bmatrix} -j10 & j10 & 0 \\ j10 & -j20 & j10 \\ 0 & j10 & -j11 \end{bmatrix}_{(1)} \begin{bmatrix} 1.1 \\ V_2 \\ V_3 \end{bmatrix}_{(1)}$$

from which the following equation and result are obtained:

$$\begin{bmatrix} V_2 \\ V_3 \end{bmatrix}_{(1)} = \begin{bmatrix} j0.0917 & j0.0833 \\ j0.0833 & j0.1667 \end{bmatrix}_{(1)} \left\{ \begin{bmatrix} -I_2 \\ 0 \end{bmatrix} - 1.1 \begin{bmatrix} j10 \\ 0 \end{bmatrix} \right\}_{(1)}$$

$$\begin{bmatrix} V_2 \\ V_3 \end{bmatrix}_{(1)} = \begin{bmatrix} j0.0917 & j0.0833 \\ j0.0833 & j0.1667 \end{bmatrix}_{(1)} \begin{bmatrix} j0.0568 - j11 \\ 0 \end{bmatrix}_{(1)} = \begin{bmatrix} 1.0035 \\ 0.9116 \end{bmatrix}_{(1)} \text{p.u.}$$

The nodal admittance matrix representing the power network at the third harmonic frequency is

$$\begin{bmatrix} I_1 \\ -I_2 \\ 0 \end{bmatrix}_{(3)} = \frac{1}{3}\begin{bmatrix} -j10 & j10 & 0 \\ j10 & -j20 & j10 \\ 0 & j10 & -j11 \end{bmatrix}_{(3)} \begin{bmatrix} 0 \\ V_2 \\ V_3 \end{bmatrix}_{(3)}$$

from which the following result is ontained:

$$\begin{bmatrix} V_2 \\ V_3 \end{bmatrix}_{(3)} = 3\begin{bmatrix} j0.0917 & j0.0833 \\ j0.0833 & j0.1667 \end{bmatrix}_{(3)} \begin{bmatrix} -j0.0186 \\ 0 \end{bmatrix}_{(3)} = \begin{bmatrix} 0.0051 \\ 0.0046 \end{bmatrix}_{(3)} \text{p.u.}$$

The third harmonic currents flowing from bus 2 to bus 1 and from bus 2 to bus 3 are

$$\begin{aligned} I_{2-1(3)} &= -j0.017 \implies 1.7\angle -90^\circ\,\% \\ I_{2-3(3)} &= -j0.0017 \implies 0.17\angle -90^\circ\,\% \end{aligned}$$

4.6 Summary

Harmonic evaluation of non-linear functions based on trigonometric operations, or algebraic operations of exponential terms, is a straightforward but cumbersome task. Hence, these methods can only be applied to simple non-linear functions evaluated around a base operating point. Realistic harmonic domain evaluations are carried out using computer calculations. Early techniques relied on point-by-point mapping of the non-linear characteristic in the time domain and the use of FFT and FHT algorithms. These methods have been superseded by direct frequency domain evaluation techniques based on self- and mutual convolutions. Convolutions represent a cleaner way of carrying out these operations, avoiding the errors introduced by numeric interpolation and aliasing due to the FFT and FHT algorithms. Several numeric examples have been presented in full to facilitate the understanding of the theory.

4.7 Bibliography

1. E. Acha, "Modelling of Power System Transformers in the Complex Conjugate Harmonic Space", *PhD Thesis*, University of Canterbury, Christchurch, New Zealand, 1988.
2. J. Arrillaga, B.C. Smith, N.R. Watson, A.R. Wood, *Power System Harmonic Analysis*, John Wiley & Sons, Chichester, 1997.
3. A. Semlyen, N. Rajakovic, "Harmonic Domain Modeling of Laminated Iron Core", *IEEE Transactions on Power Delivery*, Vol. 4, No. 1, January 1989, pp. 382–389.
4. M.L.V Lisboa, "Three-Phase Three-Limb Transformer Models in the Harmonic Domain", *PhD Thesis*, University of Canterbury, Christchurch, New Zealand, 1996.
5. J.J. Rico, "Steady State Modelling of Non-Linear Power Plant Components", *PhD. Thesis*, University of Glasgow, 1997.
6. E. Dick, W. Watson, "Transformer Models for Transient Studies Based on Field Measurements", *IEEE Transactions on Power Apparatus and Systems*, Vol. PAS-100, No. 1, January 1981, pp. 409–419.
7. S.G. Braun, Y.M. Ram, "Structural Parameter Identification in the Frequency Domain: The Use of Overdetermined Systems", *Trans. ASME*, Vol. 109, June 1987, pp. 120–123.
8. C. Mazieres, M. Forquet, "Simulation d'une Bobine a Noyau de Fer par Representation Mathematique du Cycle d'Hysteresis", *Revue Generale de lÉlectricité*, Vol. 77, No. 5, May 1968, pp. 476–481.
9. J. Stoer, R. Bulirsch, *Introduction to Numerical Analysis*, Springer-Verlag, New York and London, 1993.
10. A. Semlyen, E. Acha, J. Arrillaga, "Harmonic Norton Equivalent of the Magnetising Branch of a Transformer", *Proceedings of the IEE*, Part C, Vol. 134, No. 2, March 1987, pp. 162–169.
11. A. Semlyen, E. Acha, J. Arrillaga, "Newton-Type Algorithms for the Harmonic Phasor Analysis of Non-Linear Power Circuits in Periodical Steady-State with Special Reference to Magnetic Non-Linearities", *IEEE Transactions on Power Delivery*, Vol. 3, No. 3, July 1988, pp. 1090–1098.

5

Linearisation in Harmonic Domain

5.1 Introduction

The steady-state operation of most power circuits may be seen as one of fundamental voltage and current with ripple superposition, since the harmonic voltages and currents are only a fraction of the fundamental frequency quantities. This is a key physical characteristic that justifies the use of linearisation of the non-linear equations around a base operating point. The linearisation of each non-linear element of the circuit takes place in the harmonic domain where all the harmonics and cross-coupling between harmonics are represented explicitly.

By way of example, the following magnetising characteristic,

$$i = f(\psi) \tag{5.1}$$

$$v = \dot{\psi} \tag{5.2}$$

when linearised around the base operating point,

$$\psi_{\mathrm{b}} = \Psi_{\mathrm{b,max}} \sin \omega_0 t \Rightarrow v_{\mathrm{b}} = V_{\mathrm{b,max}} \cos \omega_0 t \tag{5.3}$$

is amenable to the following linearised equation:

$$\Delta \mathbf{I} = \mathbf{Y} \Delta \mathbf{V} \tag{5.4}$$

where

$\Delta \mathbf{I}$ is the vector of resulting complex harmonic currents.
$\Delta \mathbf{V}$ is the vector of superimposed complex harmonic voltages.
$\mathbf{Y}$ is a complex harmonic admittance matrix.

The harmonic input–output diagram shown in Figure 5.1 gives a conceptual representation of the linearised equations (5.4).

5.2 Linearisation of a Simple Non-linear Relation

The non-linear relation (5.1) will be used as the starting point to illustrate the basic principles of harmonic domain linearisation.

If the variables i and ψ in (5.1) are periodic, e.g.

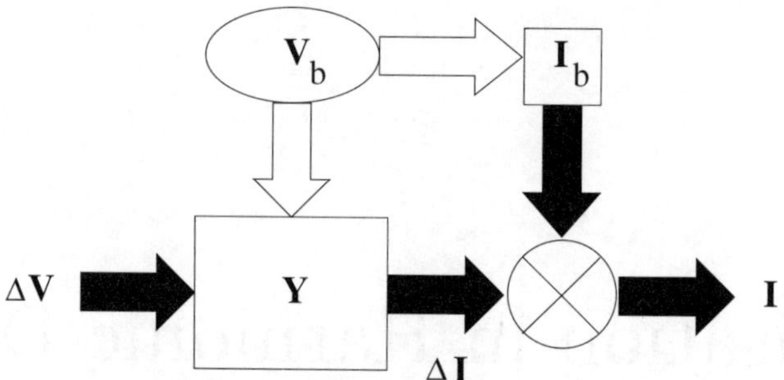

Figure 5.1 Harmonic input–output relationship in a non-linear magnetising characteristic

$$\psi(t) = \sum_{m=-\infty}^{\infty} \Psi_m e^{jm\omega_0 t} \tag{5.5}$$

$$i(t) = \sum_{k=-\infty}^{\infty} I_k e^{jk\omega_0 t} \tag{5.6}$$

and under the assumption that the non-linear function $f(\psi)$ is differentiable, then for small increments of ψ and i around ψ_b and i_b, the following linearised relation exists:

$$\Delta i = \frac{df(\psi_b)}{d\psi} \Delta\psi \tag{5.7}$$

where

$$\Delta\psi(t) = \sum_{m=-\infty}^{\infty} \Delta\Psi_m e^{jm\omega_0 t} \tag{5.8}$$

$$\Delta i(t) = \sum_{k=-\infty}^{\infty} \Delta I_k e^{jk\omega_0 t} \tag{5.9}$$

$$\frac{df(\psi_b)}{d\psi} = \sum_{i=-\infty}^{\infty} \xi_i e^{ji\omega_0 t} \tag{5.10}$$

Alternatively, equation (5.7) can be written in terms of harmonic coefficients. Using the properties of convolutions presented in Section 4.3, the following relation between harmonics coefficients is obtained:

$$\begin{bmatrix} \vdots \\ \Delta I_{-2} \\ \Delta I_{-1} \\ \Delta I_0 \\ \Delta I_1 \\ \Delta I_2 \\ \vdots \end{bmatrix} = \begin{bmatrix} \ddots & & & & & & \\ & \xi_0 & \xi_{-1} & \xi_{-2} & \xi_{-3} & & \\ & \xi_1 & \xi_0 & \xi_{-1} & \xi_{-2} & \xi_{-3} & \\ & \xi_2 & \xi_1 & \xi_0 & \xi_{-1} & \xi_{-2} & \\ & \xi_3 & \xi_2 & \xi_1 & \xi_0 & \xi_{-1} & \\ & & \xi_3 & \xi_2 & \xi_1 & \xi_0 & \\ & & & & & & \ddots \end{bmatrix} \begin{bmatrix} \vdots \\ \Delta\Psi_{-2} \\ \Delta\Psi_{-1} \\ \Delta\Psi_0 \\ \Delta\Psi_1 \\ \Delta\Psi_2 \\ \vdots \end{bmatrix} \tag{5.11}$$

or, in compact form,

$$\Delta \mathbf{I} = \mathbf{F}'\Delta\boldsymbol{\Psi} \tag{5.12}$$

Since $\xi_{-i} = \xi_i^*$, then $\mathbf{F}'$ in equation (5.12) can be written as

$$\mathbf{F}' = \begin{bmatrix} \ddots & & & & & & \\ & \xi_0 & \xi_1^* & \xi_2^* & \xi_3^* & & \\ & \xi_1 & \xi_0 & \xi_1^* & \xi_2^* & \xi_3^* & \\ & \xi_2 & \xi_1 & \xi_0 & \xi_1^* & \xi_2^* & \\ & \xi_3 & \xi_2 & \xi_1 & \xi_0 & \xi_1^* & \\ & & \xi_3 & \xi_2 & \xi_1 & \xi_0 & \\ & & & & & & \ddots \end{bmatrix} \tag{5.13}$$

It can be seen that matrix $\mathbf{F}'$ has the characteristics of a Toeplitz matrix and that it is also a Hermitian matrix due to the fact that $\xi_{-i} = \xi_i^*$. In practical cases, $\mathbf{F}'$ will be of finite dimensions, owing to simplifying assumptions or to truncations during computer evaluation. So, the band of side-diagonals will be of limited width resulting in $\mathbf{F}'$ being a banded Toeplitz matrix.

Example 5-1: To show an application of the theory presented above, the following relation is linearised in the harmonic domain:

$$i = f(\psi) = 0.001\psi + 0.0743\psi^3 \tag{5.14}$$

It should be noted that per unit values will be used in this example, hence $\omega_0 = 1$.

The linearisation takes place around the per unit base operating point,

$$\psi_\mathrm{b} = \sin\omega_0 t \Rightarrow \psi_\mathrm{b} = \frac{\mathrm{e}^{\mathrm{j}\omega_0 t} - \mathrm{e}^{-\mathrm{j}\omega_0 t}}{2\mathrm{j}}$$

Substituting the base operating point into the derivative function of the non-linear characteristic,

$$\frac{\mathrm{d}f(\psi_\mathrm{b})}{\mathrm{d}\psi} = 0.001 + 3(0.0743)\psi_\mathrm{b}^2$$

yields the following results:

$$\begin{aligned} \frac{\mathrm{d}f(\psi_\mathrm{b})}{\mathrm{d}\psi} &= 0.001 + 3(0.0743)\left(\frac{\mathrm{e}^{\mathrm{j}\omega_0 t} - \mathrm{e}^{-\mathrm{j}\omega_0 t}}{2\mathrm{j}}\right)^2 \\ &= -0.0557\mathrm{e}^{-\mathrm{j}2\omega_0 t} + 0.1124 - 0.0557\mathrm{e}^{\mathrm{j}2\omega_0 t} \end{aligned}$$

The vector and matrix of the harmonic inductances are then assembled:

$$\mathbf{F}' = \begin{bmatrix} 0 \\ -0.0557 \\ 0 \\ 0.1124 \\ 0 \\ -0.0557 \\ 0 \end{bmatrix}$$

and according to (5.13):

$$\mathbf{F}' = \begin{bmatrix} 0.1124 & 0 & -0.0557 & & & & \\ 0 & 0.1124 & 0 & -0.0557 & & & \\ -0.0557 & 0 & 0.1124 & 0 & -0.0557 & & \\ & -0.0557 & 0 & 0.1124 & 0 & -0.0557 & \\ & & -0.0557 & 0 & 0.1124 & 0 & -0.0557 \\ & & & -0.0557 & 0 & 0.1124 & 0 \\ & & & & -0.0557 & 0 & 0.1124 \end{bmatrix}$$

Hence, according to (5.12):

$$\Delta\mathbf{I} = \begin{bmatrix} \Delta I_{-3} \\ \Delta I_{-2} \\ \Delta I_{-1} \\ \Delta I_0 \\ \Delta I_1 \\ \Delta I_2 \\ \Delta I_3 \end{bmatrix} \quad \text{and } \Delta\Psi = \begin{bmatrix} \Delta\Psi_{-3} \\ \Delta\Psi_{-2} \\ \Delta\Psi_{-1} \\ \Delta\Psi_0 \\ \Delta\Psi_1 \\ \Delta\Psi_2 \\ \Delta\Psi_3 \end{bmatrix}$$

The function `calc_Fm.m` in MATLAB™, given in Chapter 4, is used to obtain the matrix $\mathbf{F}'$ by using `Fv=[0;-0.0557;0;0.1124;0;-0.0557;0]` and `h=3`:

```
>> Fv=[0;-0.0557;0;0.1124;0;-0.0557;0]; h=3;
>> F=calc_Fm(Fv,h)
F =
  0.1124         0 -0.0557         0         0         0         0
       0    0.1124       0   -0.0557       0         0         0
 -0.0557         0  0.1124         0 -0.0557         0         0
       0   -0.0557       0    0.1124       0   -0.0557         0
       0         0 -0.0557         0  0.1124         0   -0.0557
       0         0       0   -0.0557       0    0.1124         0
       0         0       0         0 -0.0557         0    0.1124
>>
```

5.3 Dynamic Relations

The linearisation of static elements in the harmonic domain has been illustrated in the previous section, and it will now be shown in this section how dynamic elements are handled in the harmonic domain.

Let us consider the basic dynamic relation (5.15),

$$v = \dot{\psi} \tag{5.15}$$

where both variables are periodic, i.e.

$$\psi(t) = \sum_{m=-\infty}^{\infty} \Psi_m e^{jm\omega_0 t} \tag{5.16}$$

$$v(t) = \sum_{n=-\infty}^{\infty} V_n e^{jn\omega_0 t} \tag{5.17}$$

Alternatively, equation (5.15) can be written as

$$\sum_{n=-\infty}^{\infty} V_n e^{jn\omega_0 t} = \frac{d}{dt} \sum_{m=-\infty}^{\infty} \Psi_m e^{jm\omega_0 t} = \sum_{m=-\infty}^{\infty} jm\omega_0 \Psi_m e^{jm\omega_0 t} \tag{5.18}$$

For small increments of ψ and v around ψ_b and v_b, and in terms of harmonic coefficients,

$$\begin{bmatrix} \vdots \\ \Delta V_{-2} \\ \Delta V_{-1} \\ \Delta V_0 \\ \Delta V_1 \\ \Delta V_2 \\ \vdots \end{bmatrix} = \begin{bmatrix} \ddots & & & & & & \\ & -j2\omega_0 & & & & & \\ & & -j\omega_0 & & & & \\ & & & 0 & & & \\ & & & & j\omega_0 & & \\ & & & & & j2\omega_0 & \\ & & & & & & \ddots \end{bmatrix} \begin{bmatrix} \vdots \\ \Delta\Psi_{-2} \\ \Delta\Psi_{-1} \\ \Delta\Psi_0 \\ \Delta\Psi_1 \\ \Delta\Psi_2 \\ \vdots \end{bmatrix} \tag{5.19}$$

or, in compact form,

$$\Delta\mathbf{V} = \mathbf{D}(jm\omega_0)\Delta\Psi \tag{5.20}$$

where matrix $\mathbf{D}(jm\omega_0)$ is called the matrix of differentiation defined by

$$\mathbf{D}(jm\omega_0) = \begin{bmatrix} \ddots & & & & & & \\ & -j2\omega_0 & & & & & \\ & & -j\omega_0 & & & & \\ & & & 0 & & & \\ & & & & j\omega_0 & & \\ & & & & & j2\omega_0 & \\ & & & & & & \ddots \end{bmatrix}$$

It should be noted that the evaluation of dynamic terms is carried out by means of simple algebraic operations.

The relations (5.12) and (5.20), which represent expressions (5.1) and (5.2) in the harmonic domain, are combined together leading to the following harmonic domain expression:

$$\Delta\mathbf{I} = \mathbf{F}'\mathbf{D}^{-1}(jm\omega_0)\Delta\mathbf{V} = \mathbf{Y}\Delta\mathbf{V} \tag{5.21}$$

where $\mathbf{Y} = \mathbf{F}'\mathbf{D}^{-1}(jm\omega_0)$ is a matrix of harmonic admittance and $\omega_0 = 2\pi f_0$.

5.3.1 *Norton and Thévenin circuit equivalents*

The linearisation of equation (5.1) in the harmonic domain is around a point $i_{\rm b}$ and $\psi_{\rm b}$. However, it is more practical to use the voltage as opposed to the flux as the interface variable with the network. Hence the following incremental equations will be used:

$$\begin{aligned} \Delta\mathbf{I} &= \mathbf{I} - \mathbf{I}_{\rm b} \\ \Delta\mathbf{V} &= \mathbf{V} - \mathbf{V}_{\rm b} \end{aligned} \tag{5.22}$$

Substitution of these relations into equation (5.21) produces an alternative expression which yields further insight into harmonic domain techniques:

$$\mathbf{I} = \mathbf{YV} + \mathbf{I}_N \tag{5.23}$$

where

$$\begin{aligned} \mathbf{Y} &= \mathbf{F}'\mathbf{D}^{-1}(jm\omega_0) \\ \mathbf{I}_{\rm N} &= \mathbf{I}_{\rm b} - \mathbf{YV}_{\rm b} \end{aligned} \tag{5.24}$$

The subscript N draws attention to the possibility of interpreting the linearised equation (5.23) as a harmonic Norton equivalent circuit, as shown in Figure 5.2(a). Alternatively, equations

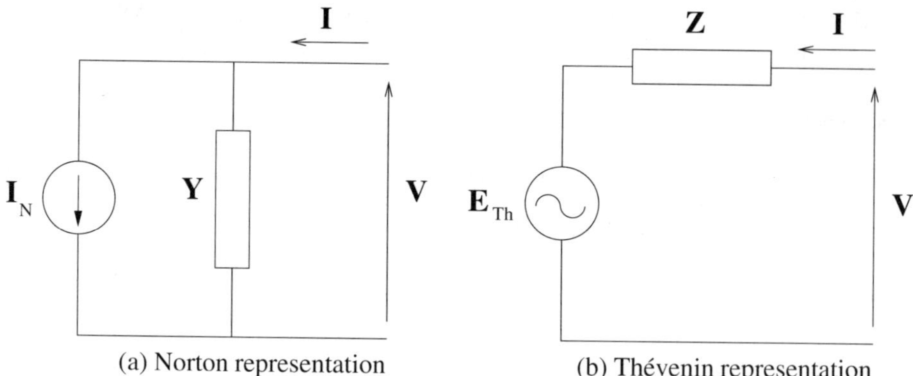

(a) Norton representation (b) Thévenin representation

Figure 5.2 Harmonic equivalents

(5.23) and (5.24) may be written in such a way that they may be interpreted as a harmonic Thévenin equivalent, as shown in Figure 5.2(b), and in equation form,

$$\mathbf{V} = \mathbf{ZI} + \mathbf{E}_{\rm Th} \tag{5.25}$$

where

$$\mathbf{E}_{\rm Th} = \mathbf{V}_{\rm b} - \mathbf{ZI}_{\rm b} \tag{5.26}$$

$$\mathbf{Z} = \mathbf{Y}^{-1} \tag{5.27}$$

Example 5-2: For the non-linear characteristic in Example 5-1, determine the harmonic Norton equivalent using a 1 per unit sinusoidal operating flux.

The evaluation of the non-linear relation yields,

$$\begin{aligned} i_{\mathrm{b}} &= 0.001\left(\frac{\mathrm{e}^{\mathrm{j}\omega_0 t}-\mathrm{e}^{-\mathrm{j}\omega_0 t}}{2\mathrm{j}}\right)+0.0743\left(\frac{\mathrm{e}^{\mathrm{j}\omega_0 t}-\mathrm{e}^{-\mathrm{j}\omega_0 t}}{2\mathrm{j}}\right)^3 \\ &= \frac{\mathrm{j}}{2}\left(-0.0186\mathrm{e}^{-\mathrm{j}3\omega_0 t}+0.0567\mathrm{e}^{-\mathrm{j}\omega_0 t}-0.0567\mathrm{e}^{\mathrm{j}\omega_0 t}+0.0186\mathrm{e}^{\mathrm{j}3\omega_0 t}\right) \end{aligned}$$

This harmonic information is used to assemble the base current vector $\mathbf{I}_{\mathrm{b}}$,

$$\mathbf{I}_{\mathrm{b}} = \begin{bmatrix} -\mathrm{j}0.0093 \\ 0 \\ \mathrm{j}0.0284 \\ 0 \\ -\mathrm{j}0.0284 \\ 0 \\ \mathrm{j}0.0093 \end{bmatrix}$$

which in the time domain is $i_{\mathrm{b}}(t) = 0.0567\sin\omega_0 t - 0.0186\sin 3\omega_0 t$. It should be noted that $\mathbf{I}_b$ may be obtained by convolution, i.e. $\mathbf{I}_{\mathrm{b}} = 0.01 + 0.01\Psi_{\mathrm{b}} + 0.0743\Psi_{\mathrm{b}}^3$.

Also, since the evaluation is around a sinusoidal operating point,

$$v_{\mathrm{b}} = \dot{\psi}_{\mathrm{b}} = \omega_0\cos\omega_0 t \Rightarrow v_{\mathrm{b}} = \frac{\mathrm{e}^{\mathrm{j}\omega_0 t}+\mathrm{e}^{-\mathrm{j}\omega_0 t}}{2} \Rightarrow \mathbf{V}_{\mathrm{b}} = \begin{bmatrix} 0 \\ 0 \\ 0.5 \\ 0 \\ 0.5 \\ 0 \\ 0 \end{bmatrix}$$

It should also be noted that in per unit operations, the coefficient ω_0 in the base voltage v_{b} is 1, i.e. $\omega_0 = 1$.

The inverse of the matrix of differentiation $\mathbf{D}^{-1}(\mathrm{j}m\omega_0)$ is also formed,

$$\mathbf{D}^{-1}(\mathrm{j}m\omega_0) = \begin{bmatrix} \mathrm{j}/3 & & & & & & \\ & \mathrm{j}/2 & & & & & \\ & & \mathrm{j} & & & & \\ & & & \infty & & & \\ & & & & -\mathrm{j} & & \\ & & & & & -\mathrm{j}/2 & \\ & & & & & & -\mathrm{j}/3 \end{bmatrix}$$

where ∞ indicates a very large value.

This matrix pre-multiplies the matrix $\mathbf{F}'$ in Example 5-1 to give the harmonic admittance matrix

$$\mathbf{Y} = \begin{bmatrix} j0.0375 & 0 & -j0.0557 & & & & \\ 0 & j0.0563 & 0 & \infty & & & \\ -j0.0186 & 0 & j0.1124 & \infty & j0.0557 & & \\ & -j0.0278 & 0 & \infty & 0 & j0.0278 & \\ & & -j0.0557 & \infty & -j0.1124 & 0 & j0.0186 \\ & & & \infty & 0 & -j0.0562 & 0 \\ & & & & j0.0557 & 0 & -j0.0375 \end{bmatrix}$$

This is all the information required to form the harmonic Norton current source, $\mathbf{I}_{\mathrm{N}} = \mathbf{I}_{\mathrm{b}} - \mathbf{Y}\mathbf{V}_{\mathrm{b}}$, i.e.

$$\mathbf{I}_{\mathrm{N}} = \begin{bmatrix} j0.0186 \\ 0 \\ -j0.0557 \\ 0 \\ j0.0557 \\ 0 \\ -j0.0186 \end{bmatrix}$$

The linearised representation of the non-linear characteristic in equation (5.14), in the form of a harmonic Norton equivalent, is

$$\mathbf{I} = \mathbf{Y}\mathbf{V} + \mathbf{I}_{\mathrm{N}}$$

where

$$\mathbf{I} = \begin{bmatrix} I_{-3} \\ I_{-2} \\ I_{-1} \\ I_0 \\ I_1 \\ I_2 \\ I_3 \end{bmatrix} \quad \text{and } \mathbf{V} = \begin{bmatrix} V_{-3} \\ V_{-2} \\ V_{-1} \\ V_0 \\ V_1 \\ V_2 \\ V_3 \end{bmatrix}$$

The result of Example 5-2 can be obtained using the following MATLAB™ function for $\omega_0 = 1$. It gives the harmonic Norton equivalent of a non-linear function:

```
function [Y,IN]=calc_Norton(a,b,n,Ve,h,w0)
D          = form_Zm(0,w0,h);
U          = zeros(2*h+1,1);
U(h+1)     = 1;
Phie       = inv(D)*Ve;
Phie_n_1  = Phie;
for k=1:(n-2)
  Phie_n_1=calc_Fm(Phie_n_1,h)*Phie;
end
Phie_n  = calc_Fm(Phie_n_1,h)*Phie;
Ie       = a*Phie+b*Phie_n;
Fe       = a*U+b*n*Phie_n_1;
Y        = calc_Fm(Fe,h)*inv(D);
IN       = Ie-Y*Ve;
```

5.3.2 *Linear relations in dynamic elements*

The response of a linear capacitor is given by the dynamic equation

$$i_C = C\frac{\mathrm{d}v}{\mathrm{d}t} \tag{5.28}$$

This expression has a similar form to equation (5.15) and the result in (5.20) is ready applicable:

$$\mathbf{I}_C = C\mathbf{D}(\mathrm{j}m\omega_0)\mathbf{V} \tag{5.29}$$

This equation represents the harmonic voltages and currents in a capacitive element,

$$\mathbf{I}_C = \mathbf{Y}_C\mathbf{V} \tag{5.30}$$

where $\mathbf{Y}_C$ is an admittance matrix in the harmonic domain given by

$$\mathbf{Y}_C = C\begin{bmatrix} \ddots & & & & & & \\ & -\mathrm{j}2\omega_0 & & & & & \\ & & -\mathrm{j}\omega_0 & & & & \\ & & & 0 & & & \\ & & & & \mathrm{j}\omega_0 & & \\ & & & & & \mathrm{j}2\omega_0 & \\ & & & & & & \ddots \end{bmatrix} \tag{5.31}$$

Similarly, the response of a linear inductor is given by the dynamic equation

$$v = L\frac{\mathrm{d}i_L}{\mathrm{d}t} \tag{5.32}$$

and in the harmonic domain,

$$\mathbf{I}_L = \frac{1}{L}\mathbf{D}^{-1}(\mathrm{j}m\omega_0)\mathbf{V} \tag{5.33}$$

and, in more compact notation,

$$\mathbf{I}_L = \mathbf{Y}_L\mathbf{V} \tag{5.34}$$

where $\mathbf{Y}_L$ is the admittance matrix of a linear inductor in the harmonic domain:

$$\mathbf{Y}_L = \frac{1}{L}\begin{bmatrix} \ddots & & & & & & \\ & \mathrm{j}/2\omega_0 & & & & & \\ & & \mathrm{j}/\omega_0 & & & & \\ & & & \infty & & & \\ & & & & -\mathrm{j}/\omega_0 & & \\ & & & & & -\mathrm{j}/2\omega_0 & \\ & & & & & & \ddots \end{bmatrix} \tag{5.35}$$

5.4 Linear Relations

The linear relation

$$i = f(\psi) = \xi_0 \psi \tag{5.36}$$

is a particular case of equations (5.1), for which the linearised relation is

$$\Delta i = \frac{\mathrm{d}f(\psi)}{\mathrm{d}\psi} \Delta\psi = \xi_0 \Delta\psi \tag{5.37}$$

In the harmonic domain, the constant term ξ_0 corresponds to the DC term and is placed along the main diagonal of the Toeplitz matrix,

$$\begin{bmatrix} \vdots \\ \Delta I_{-2} \\ \Delta I_{-1} \\ \Delta I_0 \\ \Delta I_1 \\ \Delta I_2 \\ \vdots \end{bmatrix} = \begin{bmatrix} \ddots & & & & & & \\ & \xi_0 & & & & & \\ & & \xi_0 & & & & \\ & & & \xi_0 & & & \\ & & & & \xi_0 & & \\ & & & & & \xi_0 & \\ & & & & & & \ddots \end{bmatrix} \begin{bmatrix} \vdots \\ \Delta \Psi_{-2} \\ \Delta \Psi_{-1} \\ \Delta \Psi_0 \\ \Delta \Psi_1 \\ \Delta \Psi_2 \\ \vdots \end{bmatrix} \tag{5.38}$$

The lack of off-diagonal coefficients provides a perfect decoupling between the various harmonic frequencies. No interaction takes place between pre-existing harmonics and no new harmonics are generated. This is the way all frequency-independent linear components are handled in harmonic domain calculations.

This point is well illustrated by looking at the response of a linear resistive element,

$$i_R = \frac{1}{R} v \tag{5.39}$$

This linear relation has the same form as (5.36) and in the harmonic domain it may be expressed as follows:

$$\mathbf{I}_R = \frac{1}{R} \mathbf{U}\mathbf{V} \tag{5.40}$$

or, in more general terms,

$$\mathbf{I}_R = \mathbf{Y}_R \mathbf{V} \tag{5.41}$$

where $\mathbf{U}$ is the identity matrix, and $\mathbf{Y}_R$ is an admittance matrix in the harmonic domain,

$$\mathbf{Y}_R = \frac{1}{R} \begin{bmatrix} \ddots & & & & & & \\ & 1 & & & & & \\ & & 1 & & & & \\ & & & 1 & & & \\ & & & & 1 & & \\ & & & & & 1 & \\ & & & & & & \ddots \end{bmatrix} \tag{5.42}$$

Example 5-3: Apply Kirchhoff's voltage law to the circuit in Figure 5.3 to find a steady-state expression for the current, using harmonic domain techniques.

In the time domain, we have

$$v_s(t) = Ri(t) + L\frac{\mathrm{d}i(t)}{\mathrm{d}t} + \frac{1}{C}\int i(t)\mathrm{d}t$$

and in the harmonic domain

$$\mathbf{V}_s = R\mathbf{UI} + L\mathbf{D}(\mathrm{j}h\omega_0)\mathbf{I} + \frac{1}{C}\mathbf{D}^{-1}(\mathrm{j}h\omega_0)\mathbf{I}$$

From here the current response is derived:

$$\mathbf{I} = \left(R\mathbf{U} + L\mathbf{D}(\mathrm{j}h\omega_0) + \frac{1}{C}\mathbf{D}^{-1}(\mathrm{j}h\omega_0)\right)^{-1}\mathbf{V}_s$$

This expression represent a system of linear equations of the form

$$\mathbf{I} = \mathbf{YV}_s$$

where $\mathbf{Y}$ is the equivalent admittance matrix of the circuit in Figure 5.3.

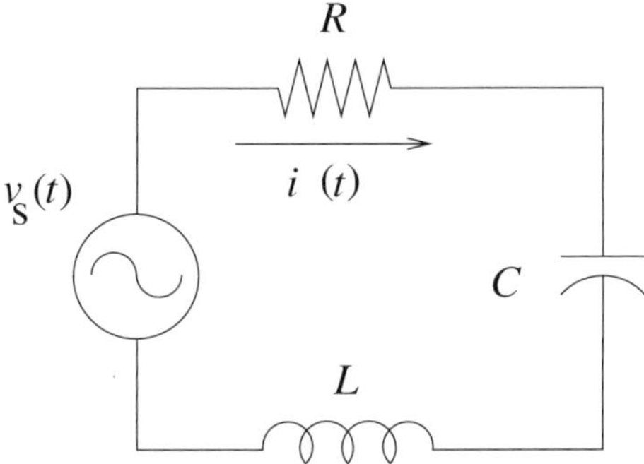

Figure 5.3 Linear series circuit

Example 5-4: Using the theory of harmonic domain linearisation find an expression for the current flowing in the non-linear circuit of Figure 5.4.

The current $i(t)$ is given by the sum of the currents in the three branches,

$$i(t) = i_R(t) + i_L(t) + i_\psi(t)$$

where

$$i_R(t) = \frac{1}{R}v_s(t) \qquad i_L(t) = \frac{1}{L}\int v_s(t)\mathrm{d}t$$

$$i_\psi(t) = f(\psi) \qquad \frac{\mathrm{d}\psi(t)}{\mathrm{d}t} = v_s(t)$$

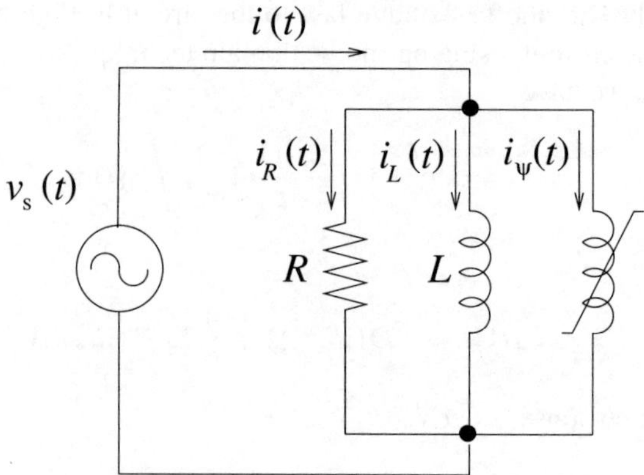

Figure 5.4 Inductor equivalent

Linearising $i_\psi(t)$ around the operation point $\psi_\mathrm{b}(t)$,

$$i_\psi(t) = \frac{\mathrm{d}f(\psi_\mathrm{b})}{\mathrm{d}\psi}\psi(t) + i_\mathrm{N}(t)$$

where

$$i_\mathrm{N}(t) = i_{\psi_\mathrm{b}}(t) - \frac{\mathrm{d}f(\psi_\mathrm{b})}{\mathrm{d}\psi}\psi_\mathrm{b}(t)$$

In the harmonic domain, the expression for the current becomes

$$\mathbf{I} = \mathbf{I}_R + \mathbf{I}_L + \mathbf{I}_\psi = \left(\frac{1}{R}\mathbf{U} + \frac{1}{L}\mathbf{D}^{-1}(\mathrm{j}h\omega_0)\mathbf{F}'\mathbf{D}^{-1}(\mathrm{j}h\omega_0)\right)\mathbf{V}_\mathrm{s} + \mathbf{I}_\mathrm{N}$$

where

$$\mathbf{I}_R = \frac{1}{R}\mathbf{U}\mathbf{V}_\mathrm{s}$$

$$\mathbf{I}_L = \frac{1}{L}\mathbf{D}^{-1}(\mathrm{j}h\omega_0)\mathbf{V}_\mathrm{s}$$

$$\Psi = \mathbf{D}^{-1}(\mathrm{j}h\omega_0)\mathbf{V}_\mathrm{s}$$

$$\mathbf{I}_\psi = \mathbf{F}'\mathbf{D}^{-1}(\mathrm{j}h\omega_0)\mathbf{V}_\mathrm{s} + \mathbf{I}_\mathrm{N}$$

$$\mathbf{I}_\mathrm{N} = \mathbf{I}_{\psi_\mathrm{b}} - \mathbf{F}'\mathbf{D}^{-1}(\mathrm{j}h\omega_0)\mathbf{V}_{\mathrm{s}_\mathrm{b}}$$

In compact form,

$$\mathbf{I} = \mathbf{Y}\mathbf{V}_\mathrm{s} + \mathbf{I}_\mathrm{N}$$

where

$$\mathbf{Y} = \frac{1}{R}\mathbf{U} + \frac{1}{L}\mathbf{D}^{-1}(\mathrm{j}h\omega_0)\mathbf{F}'\mathbf{D}^{-1}(\mathrm{j}h\omega_0)$$

Example 5-5: Linearise, in the harmonic domain, the non-linear relation

$$i = f(\psi) = \psi + \psi^3$$

but this time the linearisation takes place around a non-sinusoidal operating point, i.e.

$$\psi_{\mathrm{b}} = \sin \omega_0 t + \sin 3\omega_0 t \Rightarrow \psi_{\mathrm{b}} = \frac{\mathrm{e}^{\mathrm{j}\omega_0 t} - \mathrm{e}^{-\mathrm{j}\omega_0 t}}{2\mathrm{j}} + \frac{\mathrm{e}^{\mathrm{j}3\omega_0 t} - \mathrm{e}^{-\mathrm{j}3\omega_0 t}}{2\mathrm{j}}$$

Substituting the base operating point in the non-linear characteristic gives

$$\begin{aligned} i_{\mathrm{b}} = \psi_{\mathrm{b}} + \psi_{\mathrm{b}}^3 &= \left(\frac{\mathrm{e}^{\mathrm{j}\omega_0 t} - \mathrm{e}^{-\mathrm{j}\omega_0 t}}{2\mathrm{j}} + \frac{\mathrm{e}^{\mathrm{j}3\omega_0 t} - \mathrm{e}^{-\mathrm{j}3\omega_0 t}}{2\mathrm{j}} \right) \\ &+ \left(\frac{\mathrm{e}^{\mathrm{j}\omega_0 t} - \mathrm{e}^{-\mathrm{j}\omega_0 t}}{2\mathrm{j}} + \frac{\mathrm{e}^{\mathrm{j}3\omega_0 t} - \mathrm{e}^{-\mathrm{j}3\omega_0 t}}{2\mathrm{j}} \right)^3 \end{aligned}$$

After performing some simple but arduous algebra, the following harmonic expression is obtained:

$$\begin{aligned} i_{\mathrm{b}} = & -\mathrm{j}\frac{1}{8}\mathrm{e}^{-\mathrm{j}9\omega_0 t} - \mathrm{j}\frac{3}{8}\mathrm{e}^{-\mathrm{j}7\omega_0 t} + \mathrm{j}\frac{12}{8}\mathrm{e}^{-\mathrm{j}3\omega_0 t} + \mathrm{j}\frac{10}{8}\mathrm{e}^{-\mathrm{j}\omega_0 t} \\ & -\mathrm{j}\frac{10}{8}\mathrm{e}^{\mathrm{j}\omega_0 t} - \mathrm{j}\frac{12}{8}\mathrm{e}^{\mathrm{j}3\omega_0 t} + \mathrm{j}\frac{3}{8}\mathrm{e}^{\mathrm{j}7\omega_0 t} + \mathrm{j}\frac{1}{8}\mathrm{e}^{\mathrm{j}9\omega_0 t} \end{aligned}$$

Also,

$$v_{\mathrm{b}} = \dot{\psi}_{\mathrm{b}} = \cos(\omega_0 t) + 3\cos(3\omega_0 t) \Rightarrow v_{\mathrm{b}} = \frac{\mathrm{e}^{\mathrm{j}\omega_0 t} + \mathrm{e}^{-\mathrm{j}\omega_0 t}}{2} + 3\frac{\mathrm{e}^{\mathrm{j}3\omega_0 t} + \mathrm{e}^{-\mathrm{j}3\omega_0 t}}{2}$$

This gives the relevant harmonic information for assembling the current and voltage vectors $\mathbf{I}_{\mathrm{b}}$ and $\mathbf{V}_{\mathrm{b}}$, respectively, i.e.

$$\begin{aligned} \mathbf{I}_{\mathrm{b}} &= \frac{\mathrm{j}}{8}\begin{bmatrix} -1 & 0 & -3 & 0 & 0 & 0 & 12 & 0 & 10 & 0 & -10 & 0 & -12 & 0 & 0 & 0 & 3 & 0 & 1 \end{bmatrix}^{\mathrm{T}} \\ \mathbf{V}_{\mathrm{b}} &= \frac{1}{2}\begin{bmatrix} 0 & 0 & 0 & 0 & 0 & 0 & 3 & 0 & 1 & 0 & 1 & 0 & 3 & 0 & 0 & 0 & 0 & 0 & 0 \end{bmatrix}^{\mathrm{T}} \end{aligned}$$

Similarly, by substituting the base operating point in the derivative function of the non-linear characteristic,

$$\begin{aligned} f'(\psi_{\mathrm{b}}) &= 1 + 3\psi_{\mathrm{b}}^2 = 1 + 3\left(\frac{\mathrm{e}^{\mathrm{j}\omega_0 t} - \mathrm{e}^{-\mathrm{j}\omega_0 t}}{2\mathrm{j}} + \frac{\mathrm{e}^{\mathrm{j}3\omega_0 t} - \mathrm{e}^{-\mathrm{j}3\omega_0 t}}{2\mathrm{j}} \right)^2 \\ &= -\frac{3}{4}\mathrm{e}^{-\mathrm{j}6\omega_0 t} - \frac{6}{4}\mathrm{e}^{-\mathrm{j}4\omega_0 t} + \frac{3}{4}\mathrm{e}^{-\mathrm{j}2\omega_0 t} + 4 + \frac{3}{4}\mathrm{e}^{\mathrm{j}2\omega_0 t} - \frac{6}{4}\mathrm{e}^{\mathrm{j}4\omega_0 t} - \frac{3}{4}\mathrm{e}^{\mathrm{j}6\omega_0 t} \end{aligned}$$

or, in vector form,

$$\mathbf{F}' = \frac{1}{4}\begin{bmatrix} -3 & 0 & -6 & 0 & 3 & 0 & 16 & 0 & 3 & 0 & -6 & 0 & -3 \end{bmatrix}^{\mathrm{T}}$$

Example 5-6: Figure 5.5 shows an electric circuit and its equivalent in the harmonic domain. Find the values of the mesh currents $i_a(t)$ and $i_b(t)$ in the harmonic domain for the following voltage sources:

(i) $v(t) = 10$ volts,

(ii) $v(t) = 10\cos(\omega_0 t + 45°)$ volts,

(iii) $v(t)$ waveform given in Figure 5.6.

The solution is obtained for the currents when $R_a = 4\,\Omega$, $R_b = 2\,\Omega$, $L = 1\,\text{mH}$ and $C = 100\,\mu\text{F}$.

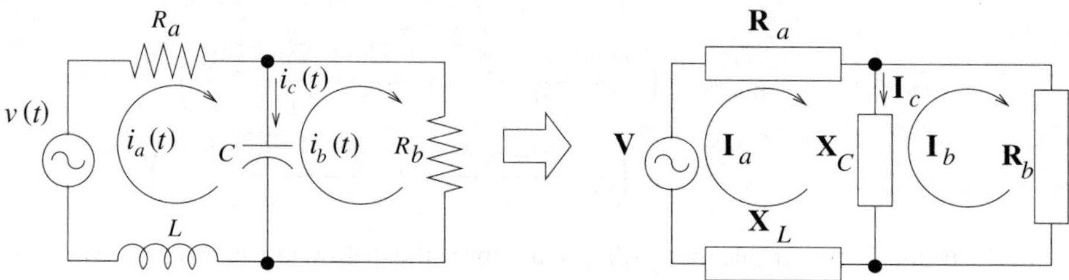

Figure 5.5 Two-mesh electric circuit

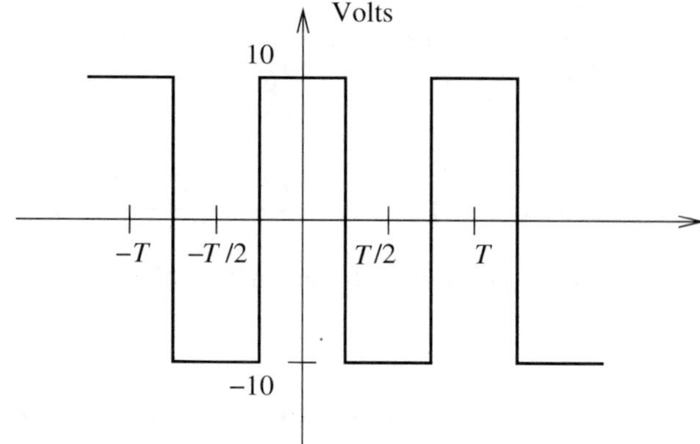

Figure 5.6 Voltage waveform

(i) The source is a DC voltage source with a value of 10 V. Taking into account only the fundamental frequency and the DC term, the elements of the system are

$$\mathbf{V} = \begin{bmatrix} 0 \\ 10 \\ 0 \end{bmatrix}; \qquad \mathbf{R}_b = 2\begin{bmatrix} 1 & 0 & 0 \\ 0 & 1 & 0 \\ 0 & 0 & 1 \end{bmatrix}$$

$$\mathbf{R}_a = 4\begin{bmatrix} 1 & 0 & 0 \\ 0 & 1 & 0 \\ 0 & 0 & 1 \end{bmatrix}; \qquad \mathbf{X}_C = \frac{1}{100 \times 10^{-6}}\begin{bmatrix} \mathrm{j}/\omega_0 & 0 & 0 \\ 0 & \infty & 0 \\ 0 & 0 & -\mathrm{j}/\omega_0 \end{bmatrix}$$

$$\mathbf{X}_L = 1 \times 10^{-3} \begin{bmatrix} -j\omega_0 & 0 & 0 \\ 0 & 0 & 0 \\ 0 & 0 & j\omega_0 \end{bmatrix}$$

At zero frequency, $\mathbf{X}_L$ and $\mathbf{X}_C$ are in short circuit and in open circuit, respectively. Then

$$\mathbf{I}_a = \mathbf{I}_b = (\mathbf{R}_a + \mathbf{R}_b)^{-1}\,\mathbf{V}$$

or

$$\begin{bmatrix} I_{a_{-1}} \\ I_{a_0} \\ I_{a_1} \end{bmatrix} = \left\{ 2 \begin{bmatrix} 1 & 0 & 0 \\ 0 & 1 & 0 \\ 0 & 0 & 1 \end{bmatrix} + 4 \begin{bmatrix} 1 & 0 & 0 \\ 0 & 1 & 0 \\ 0 & 0 & 1 \end{bmatrix} \right\}^{-1} \begin{bmatrix} 0 \\ 10 \\ 0 \end{bmatrix} = \begin{bmatrix} 0 \\ 5/3 \\ 0 \end{bmatrix}$$

and $i_a = i_b = 5/3$ A.

(ii) Using Euler's identity, the voltage source is represented by

$$v(t) = 10\cos(\omega_0 t + 45°) = \left(\frac{5}{\sqrt{2}} - j\frac{5}{\sqrt{2}} \right) e^{-j\omega_0 t} + \left(\frac{5}{\sqrt{2}} + j\frac{5}{\sqrt{2}} \right) e^{j\omega_0 t}$$

Then

$$\mathbf{V} = \begin{bmatrix} 0 \\ \dfrac{5}{\sqrt{2}} - j\dfrac{5}{\sqrt{2}} \\ 0 \\ \dfrac{5}{\sqrt{2}} + j\dfrac{5}{\sqrt{2}} \\ 0 \end{bmatrix}$$

$$\mathbf{R}_b = 2\mathbf{U} \qquad \mathbf{X}_L = 1 \times 10^{-3}\mathbf{D}(jn\omega_0)$$

$$\mathbf{R}_a = 4\mathbf{U} \quad \mathbf{X}_C = \frac{1}{100 \times 10^{-6}}\mathbf{D}^{-1}(jn\omega_0)$$

The mesh equation in matrix form is

$$\begin{bmatrix} \mathbf{I}_a \\ \mathbf{I}_b \end{bmatrix} = \begin{bmatrix} \mathbf{R}_b + \mathbf{X}_C + \mathbf{X}_L & -\mathbf{X}_C \\ -\mathbf{X}_C & \mathbf{X}_C + \mathbf{R}_a \end{bmatrix} \begin{bmatrix} \mathbf{V} \\ 0 \end{bmatrix}$$

Using $f = 50$ Hz, the harmonic domain solution is

$$\mathbf{I}_a = \begin{bmatrix} 0 \\ 0.5768 - j0.6130 \\ 0 \\ 0.5768 + j0.6130 \\ 0 \end{bmatrix}; \ \mathbf{I}_b = \begin{bmatrix} 0 \\ 0.6436 - j0.5321 \\ 0 \\ 0.6436 + j0.5321 \\ 0 \end{bmatrix}$$

and expressed in the time domain,

$$
\begin{aligned}
i_a(t) &= (0.5768 - \mathrm{j}0.6130)\mathrm{e}^{-\mathrm{j}\omega_0 t} + (0.5768 + \mathrm{j}0.6130)\mathrm{e}^{\mathrm{j}\omega_0 t} \\
i_b(t) &= (0.6436 - \mathrm{j}0.5321)\mathrm{e}^{-\mathrm{j}\omega_0 t} + (0.6436 + \mathrm{j}0.5321)\mathrm{e}^{\mathrm{j}\omega_0 t}
\end{aligned}
$$

(iii) The voltage waveform of Figure 5.6 may be represented by the Fourier series

$$
v(t) = \cdots + \frac{20}{5\pi}\mathrm{e}^{-\mathrm{j}5\omega_0 t} - \frac{20}{3\pi}\mathrm{e}^{-\mathrm{j}3\omega_0 t} + \frac{20}{\pi}\mathrm{e}^{-\mathrm{j}\omega_0 t} + \frac{20}{\pi}\mathrm{e}^{\mathrm{j}\omega_0 t} - \frac{20}{3\pi}\mathrm{e}^{\mathrm{j}3\omega_0 t} + \frac{20}{5\pi}\mathrm{e}^{\mathrm{j}5\omega_0 t} + \cdots
$$

Owing to the nature of the voltage waveform, the solution of this problem should include a large number of harmonic terms. It is no longer feasible to carry out hand calculations. Instead, the solution is obtained by computer evaluation using, for instance, MATLAB ™ functions.

The solution below corresponds to the case when $f = 50$ Hz:

$$
\mathbf{V} = \begin{bmatrix} 0 \\ 20/5\pi \\ 0 \\ -20/3\pi \\ 0 \\ 20/\pi \\ 0 \\ 20/\pi \\ 0 \\ -20/3\pi \\ 0 \\ 20/5\pi \\ 0 \end{bmatrix}; \text{ then } \mathbf{I}_a = \begin{bmatrix} 0 \\ 0.2610 - \mathrm{j}0.0124 \\ 0 \\ -0.3839 + \mathrm{j}0.0264 \\ 0 \\ 1.0712 - \mathrm{j}0.0326 \\ 0 \\ 1.0712 + \mathrm{j}0.0326 \\ 0 \\ -0.3839 - \mathrm{j}0.0264 \\ 0 \\ 0.2610 + \mathrm{j}0.0124 \\ 0 \end{bmatrix} \text{ and } \mathbf{I}_b = \begin{bmatrix} 0 \\ 0.1927 + \mathrm{j}0.1087 \\ 0 \\ -0.3448 - \mathrm{j}0.1036 \\ 0 \\ 1.0585 + \mathrm{j}0.1004 \\ 0 \\ 1.0585 - \mathrm{j}0.1004 \\ 0 \\ -0.3448 + \mathrm{j}0.1036 \\ 0 \\ 0.1927 - \mathrm{j}0.1087 \\ 0 \end{bmatrix}
$$

Figure 5.7 shows the current waveform for cases (ii) and (iii). The MATLAB ™ function used to obtain the time representation of a harmonic vector where the FFT is used, as opposed to the complex Fourier series, is:

```
function f=plot_f(F,h,cycle);
% F : vector of harmonic F=[-h ... -1 0 1 ... h]
% cycle : number of cycles
% f : function in time
c=zeros(40,1);
Fa=F(h+1:2*h+1);
Fb=F(1:h);
Fx=[Fa;c;0;c;Fb];
x=size(Fx);
x=x(1);
f=x*real(ifft(Fx));
if cycle>1
  fx=f;
  for k=1:cycle-1
    f=[f;fx];
  end
end
```

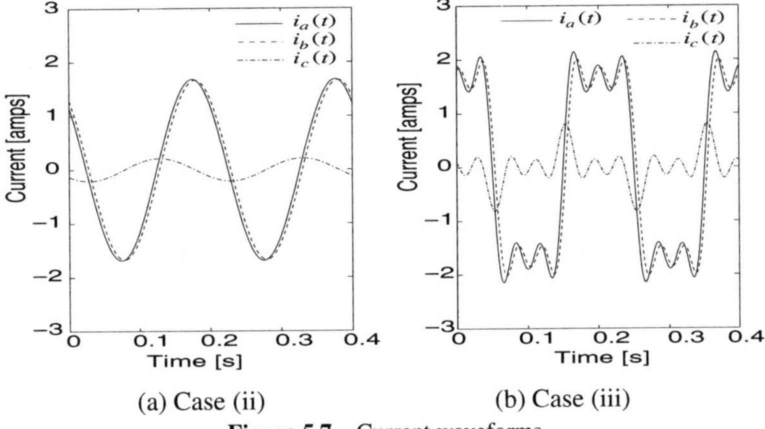

(a) Case (ii) (b) Case (iii)

Figure 5.7 Current waveforms

5.5 Multi-variable Relations

Non-linear differentiable relations which incorporate multiple inputs and multiple outputs are also amenable to treatment in the harmonic domain, i.e.

$$\begin{aligned} i_1 &= f_1(\psi_1, \ldots, \psi_n) \\ &\vdots \\ i_m &= f_m(\psi_1, \ldots, \psi_n) \end{aligned} \tag{5.43}$$

If the linearisation exercise takes place around a base multi-dimensional point

$$\left(\psi_1^0, \ldots, \psi_n^0\right), \left(i_1^0, \ldots, i_m^0\right)$$

then the following expression is obtained:

$$\begin{aligned} \Delta i_1 &= \frac{\partial f_1}{\partial \psi_1}\Delta\psi_1 + \cdots + \frac{\partial f_1}{\partial \psi_n}\Delta\psi_n \\ &\vdots \\ \Delta i_m &= \frac{\partial f_m}{\partial \psi_1}\Delta\psi_1 + \cdots + \frac{\partial f_m}{\partial \psi_n}\Delta\psi_n \end{aligned} \tag{5.44}$$

This equation contains several terms with the same characteristics as those in equation (5.7). By following a similar line of reasoning to that used in Section 5.2, the following multi-dimensional relation is obtained:

$$\begin{bmatrix} \Delta I_1 \\ \vdots \\ \Delta I_m \end{bmatrix} = \begin{bmatrix} \xi_{1,1} & \cdots & \xi_{1,n} \\ \vdots & \ddots & \vdots \\ \xi_{m,1} & \cdots & \xi_{m,n} \end{bmatrix} \begin{bmatrix} \Delta\Psi_1 \\ \vdots \\ \Delta\Psi_n \end{bmatrix} \tag{5.45}$$

Example 5-7: In this example the following multi-dimensional relation is linearised in the harmonic domain:

$$u = f(x, y) = xy^2(1 + y)$$

$$v = g(x,y) = x\left(x^2 - y^2\right)$$

around the base operating point,

$$x_{\mathrm{b}} = \sin\omega_0 t \quad\Rightarrow\quad x_{\mathrm{b}} = \frac{\mathrm{e}^{\mathrm{j}\omega_0 t} - \mathrm{e}^{-\mathrm{j}\omega_0 t}}{2\mathrm{j}}$$

$$y_{\mathrm{b}} = \cos\omega_0 t \quad\Rightarrow\quad y_{\mathrm{b}} = \frac{\mathrm{e}^{\mathrm{j}\omega_0 t} + \mathrm{e}^{-\mathrm{j}\omega_0 t}}{2}$$

and in vector form,

$$\mathbf{X}_{\mathrm{b}} = \mathrm{j}\begin{bmatrix} 0 \\ 0 \\ 0 \\ 1/2 \\ 0 \\ -1/2 \\ 0 \\ 0 \\ 0 \end{bmatrix} ; \quad \mathbf{Y}_{\mathrm{b}} = \begin{bmatrix} 0 \\ 0 \\ 0 \\ 1/2 \\ 0 \\ 1/2 \\ 0 \\ 0 \\ 0 \end{bmatrix}$$

Substituting the base operating point in the non-linear characteristic,

$$\begin{aligned}
u_{\mathrm{b}} &= x_{\mathrm{b}} y_{\mathrm{b}}^2 (1 + y_{\mathrm{b}}) \\
&= \left(\frac{\mathrm{e}^{\mathrm{j}\omega_0 t} - \mathrm{e}^{-\mathrm{j}\omega_0 t}}{2\mathrm{j}}\right)\left(\frac{\mathrm{e}^{\mathrm{j}\omega_0 t} + \mathrm{e}^{-\mathrm{j}\omega_0 t}}{2}\right)^2\left(1 + \frac{\mathrm{e}^{\mathrm{j}\omega_0 t} + \mathrm{e}^{-\mathrm{j}\omega_0 t}}{2}\right) \\
v_{\mathrm{b}} &= x_{\mathrm{b}}\left(x_{\mathrm{b}}^2 - y_{\mathrm{b}}^2\right) \\
&= \left(\frac{\mathrm{e}^{\mathrm{j}\omega_0 t} - \mathrm{e}^{-\mathrm{j}\omega_0 t}}{2\mathrm{j}}\right)\left\{\left(\frac{\mathrm{e}^{\mathrm{j}\omega_0 t} - \mathrm{e}^{-\mathrm{j}\omega_0 t}}{2\mathrm{j}}\right)^2 - \left(\frac{\mathrm{e}^{\mathrm{j}\omega_0 t} + \mathrm{e}^{-\mathrm{j}\omega_0 t}}{2}\right)^2\right\}
\end{aligned}$$

gives rise to the following harmonic expressions:

$$\begin{aligned}
u_{\mathrm{b}} &= \mathrm{j}\frac{1}{16}\mathrm{e}^{-\mathrm{j}4\omega_0 t} + \mathrm{j}\frac{1}{8}\mathrm{e}^{-\mathrm{j}3\omega_0 t} + \mathrm{j}\frac{1}{8}\mathrm{e}^{-\mathrm{j}2\omega_0 t} \\
&\quad + \mathrm{j}\frac{1}{8}\mathrm{e}^{-\mathrm{j}\omega_0 t} - \mathrm{j}\frac{1}{8}\mathrm{e}^{\mathrm{j}\omega_0 t} - \mathrm{j}\frac{1}{8}\mathrm{e}^{\mathrm{j}2\omega_0 t} - \mathrm{j}\frac{1}{8}\mathrm{e}^{\mathrm{j}3\omega_0 t} - \mathrm{j}\frac{1}{16}\mathrm{e}^{\mathrm{j}4\omega_0 t} \\
v_{\mathrm{b}} &= -\mathrm{j}\frac{1}{4}\mathrm{e}^{-\mathrm{j}3\omega_0 t} + \mathrm{j}\frac{1}{4}\mathrm{e}^{-\mathrm{j}\omega_0 t} - \mathrm{j}\frac{1}{4}\mathrm{e}^{\mathrm{j}\omega_0 t} + \mathrm{j}\frac{1}{4}\mathrm{e}^{\mathrm{j}3\omega_0 t}
\end{aligned}$$

This information is used to assemble the harmonic vectors $\mathbf{U}_{\mathrm{b}}$ and $\mathbf{V}_{\mathrm{b}}$,

$$\mathbf{U}_{\rm b} = \frac{\rm j}{8}\begin{bmatrix} 1/2 \\ 1 \\ 1 \\ 1 \\ 0 \\ -1 \\ -1 \\ -1 \\ -1/2 \end{bmatrix} ; \mathbf{V}_{\rm b} = \frac{\rm j}{4}\begin{bmatrix} 0 \\ -1 \\ 0 \\ 1 \\ 0 \\ -1 \\ 0 \\ 1 \\ 0 \end{bmatrix}$$

Similarly, by substituting the base operating point into the derivative functions, the following harmonic information is obtained:

$$\begin{aligned}
\frac{\partial f}{\partial x} &= y^2(1+y) = \left(\frac{{\rm e}^{{\rm j}\omega_0 t} + {\rm e}^{-{\rm j}\omega_0 t}}{2}\right)^2 \left(1 + \frac{{\rm e}^{{\rm j}\omega_0 t} + {\rm e}^{-{\rm j}\omega_0 t}}{2}\right) \\
&= \frac{1}{8}{\rm e}^{-{\rm j}3\omega_0 t} + \frac{1}{4}{\rm e}^{-{\rm j}2\omega_0 t} + \frac{3}{8}{\rm e}^{-{\rm j}\omega_0 t} + \frac{1}{2} + \frac{3}{8}{\rm e}^{{\rm j}\omega_0 t} + \frac{1}{4}{\rm e}^{{\rm j}2\omega_0 t} + \frac{1}{8}{\rm e}^{{\rm j}3\omega_0 t} \\
\frac{\partial f}{\partial y} &= xy(2+3y) = \left(\frac{{\rm e}^{{\rm j}\omega_0 t} - {\rm e}^{-{\rm j}\omega_0 t}}{2{\rm j}}\right)\left(\frac{{\rm e}^{{\rm j}\omega_0 t} + {\rm e}^{-{\rm j}\omega_0 t}}{2}\right)\left(2 + 3\frac{{\rm e}^{{\rm j}\omega_0 t} + {\rm e}^{-{\rm j}\omega_0 t}}{2}\right) \\
&= {\rm j}\frac{3}{8}{\rm e}^{-{\rm j}3\omega_0 t} + {\rm j}\frac{1}{2}{\rm e}^{-{\rm j}2\omega_0 t} + {\rm j}\frac{3}{8}{\rm e}^{-{\rm j}\omega_0 t} - {\rm j}\frac{3}{8}{\rm e}^{{\rm j}\omega_0 t} - {\rm j}\frac{1}{2}{\rm e}^{{\rm j}2\omega_0 t} - {\rm j}\frac{3}{8}{\rm e}^{{\rm j}3\omega_0 t} \\
\frac{\partial g}{\partial x} &= 3x^2 - y^2 = 3\left(\frac{{\rm e}^{{\rm j}\omega_0 t} - {\rm e}^{-{\rm j}\omega_0 t}}{2{\rm j}}\right)^2 - \left(\frac{{\rm e}^{{\rm j}\omega_0 t} + {\rm e}^{-{\rm j}\omega_0 t}}{2}\right)^2 \\
&= -\frac{1}{2}{\rm e}^{-{\rm j}2\omega_0 t} + 2 - \frac{1}{2}{\rm e}^{{\rm j}2\omega_0 t} \\
\frac{\partial g}{\partial y} &= -2xy = -2\left(\frac{{\rm e}^{{\rm j}\omega_0 t} - {\rm e}^{-{\rm j}\omega_0 t}}{2{\rm j}}\right)\left(\frac{{\rm e}^{{\rm j}\omega_0 t} + {\rm e}^{-{\rm j}\omega_0 t}}{2}\right) \\
&= -{\rm j}\frac{1}{2}{\rm e}^{-{\rm j}2\omega_0 t} + {\rm j}\frac{1}{2}{\rm e}^{{\rm j}2\omega_0 t}
\end{aligned}$$

This information is used to assemble the relevant harmonic domain vectors,

$$\mathbf{F}'_X = \begin{bmatrix} 1/8 \\ 1/4 \\ 3/8 \\ 1/2 \\ 3/8 \\ 1/4 \\ 1/8 \end{bmatrix} ; \mathbf{F}'_Y = {\rm j}\begin{bmatrix} 3/8 \\ 1/2 \\ 3/8 \\ 0 \\ -3/8 \\ -1/2 \\ -3/8 \end{bmatrix} ; \mathbf{G}'_X = \begin{bmatrix} 0 \\ -1/2 \\ 0 \\ 2 \\ 0 \\ -1/2 \\ 0 \end{bmatrix} ; \mathbf{G}'_Y = {\rm j}\begin{bmatrix} 0 \\ 0 \\ -1/2 \\ 0 \\ 1/2 \\ 0 \\ 0 \end{bmatrix}$$

Hence, the following multi-dimensional linearised representation is obtained:

$$\begin{bmatrix} \mathbf{U} \\ \mathbf{V} \end{bmatrix} = \begin{bmatrix} \mathbf{F}_{X'} & \mathbf{F}_{Y'} \\ \mathbf{G}_{X'} & \mathbf{G}_{Y'} \end{bmatrix}\begin{bmatrix} \mathbf{X} \\ \mathbf{Y} \end{bmatrix} + \begin{bmatrix} \mathbf{U}_{\rm N} \\ \mathbf{V}_{\rm N} \end{bmatrix}$$

where

$$\mathbf{F}_X{}' = \begin{bmatrix} 1/2 & 3/8 & 1/4 & 1/8 & & & & & \\ 3/8 & 1/2 & 3/8 & 1/4 & 1/8 & & & & \\ 1/4 & 3/8 & 1/2 & 3/8 & 1/4 & 1/8 & & & \\ 1/8 & 1/4 & 3/8 & 1/2 & 3/8 & 1/4 & 1/8 & & \\ & 1/8 & 1/4 & 3/8 & 1/2 & 3/8 & 1/4 & 1/8 & \\ & & 1/8 & 1/4 & 3/8 & 1/2 & 3/8 & 1/4 & 1/8 \\ & & & 1/8 & 1/4 & 3/8 & 1/2 & 3/8 & 1/4 \\ & & & & 1/8 & 1/4 & 3/8 & 1/2 & 3/8 \\ & & & & & 1/8 & 1/4 & 3/8 & 1/2 \end{bmatrix}$$

$$\mathbf{F}_Y{}' = \mathrm{j} \begin{bmatrix} 0 & 3/8 & 1/2 & 3/8 & & & & & \\ -3/8 & 0 & 3/8 & 1/2 & 3/8 & & & & \\ -1/2 & -3/8 & 0 & 3/8 & 1/2 & 3/8 & & & \\ -3/8 & -1/2 & -3/8 & 0 & 3/8 & 1/2 & 3/8 & & \\ & -3/8 & -1/2 & -3/8 & 0 & 3/8 & 1/2 & 3/8 & \\ & & -3/8 & -1/2 & -3/8 & 0 & 3/8 & 1/2 & 3/8 \\ & & & -3/8 & -1/2 & -3/8 & 0 & 3/8 & 1/2 \\ & & & & -3/8 & -1/2 & -3/8 & 0 & 3/8 \\ & & & & & -3/8 & -1/2 & -3/8 & 0 \end{bmatrix}$$

$$\mathbf{G}_X{}' = \begin{bmatrix} 2 & 0 & -1/2 & & & & & & \\ 0 & 2 & 0 & -1/2 & & & & & \\ -1/2 & 0 & 2 & 0 & -1/2 & & & & \\ & -1/2 & 0 & 2 & 0 & -1/2 & & & \\ & & -1/2 & 0 & 2 & 0 & -1/2 & & \\ & & & -1/2 & 0 & 2 & 0 & -1/2 & \\ & & & & -1/2 & 0 & 2 & 0 & -1/2 \\ & & & & & -1/2 & 0 & 2 & 0 \\ & & & & & & -1/2 & 0 & 2 \end{bmatrix}$$

$$\mathbf{G}_Y{}' = \mathrm{j} \begin{bmatrix} 0 & -1/2 & & & & & & & \\ 1/2 & 0 & -1/2 & & & & & & \\ & 1/2 & 0 & -1/2 & & & & & \\ & & 1/2 & 0 & -1/2 & & & & \\ & & & 1/2 & 0 & -1/2 & & & \\ & & & & 1/2 & 0 & -1/2 & & \\ & & & & & 1/2 & 0 & -1/2 & \\ & & & & & & 1/2 & 0 & -1/2 \\ & & & & & & & 1/2 & 0 \end{bmatrix}$$

The multi-dimensional harmonic Norton current vector is calculated by solving the following matrix equation:

$$\begin{bmatrix} \mathbf{U}_{\mathrm{N}} \\ \mathbf{V}_{\mathrm{N}} \end{bmatrix} = \begin{bmatrix} \mathbf{U}_{\mathrm{b}} \\ \mathbf{V}_{\mathrm{b}} \end{bmatrix} - \begin{bmatrix} \mathbf{F}_X{}' & \mathbf{F}_Y{}' \\ \mathbf{G}_X{}' & \mathbf{G}_Y{}' \end{bmatrix} \begin{bmatrix} \mathbf{X}_{\mathrm{b}} \\ \mathbf{Y}_{\mathrm{b}} \end{bmatrix}$$

giving

$$\mathbf{U}_{\mathrm{N}} = \frac{\mathrm{j}}{4}\begin{bmatrix} -3/4 \\ -1 \\ -3/2 \\ -1 \\ 0 \\ 1 \\ 3/2 \\ 1 \\ 3/4 \end{bmatrix} ; \ \mathbf{V}_{\mathrm{N}} = \frac{\mathrm{j}}{4}\begin{bmatrix} 0 \\ 0 \\ 1 \\ -4 \\ 0 \\ 4 \\ -1 \\ 0 \\ 0 \end{bmatrix}$$

5.6 An Application to an Industrial Test Case

Harmonic domain techniques are used in this section to solve a more practical case. It corresponds to an industrial plant with the equivalent circuit shown in Figure 5.8. Harmonic distortion is generated by the induction furnace connected at the secondary side of the transformer where is a capacitor bank to provide power factor correction.

The main parameters of the system are as follow:

Utility system	:	500 kVA, $Z = 10\%$, 13.8 kV/480 V, 60 Hz
Capacitor bank	:	100 kVAR
Linear load	:	112 kVA, PF $= 0.89(-)$
Nonlinear load	:	Electric induction furnance of 40 kVA

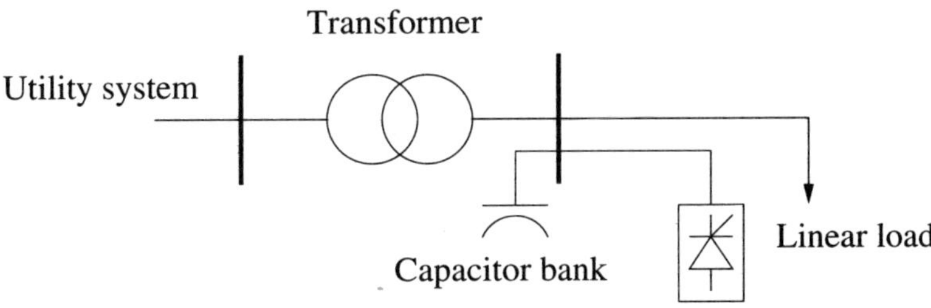

Figure 5.8 Unifilar industrial system

This is an actual installation where a typical harmonic analiser was used to measure the induction furnace current shown in Figure 5.9 and the harmonic content shown in Table 5.1. Most harmonic analysers have digital screens or computer interfaces, which are used to display voltage and current waveforms, harmonic contents, distortion values, powers, RMS values and true power factor. Their internal algorithm is based on the FFT and the sampling process shown in Figure 2.6, for the harmonic content representation, the trigonometric form of the Fourier series is used, i.e. equation (2.16).

The system is operates at rated values, the voltage at fundamental frequency in the capacitor bank is $480 \sin \omega_0 t$ V, and the current $208.34 \cos \omega_0 t$ A.

The capacitor voltage and current in complex form are:

$$V_{\mathrm{s}} = \mathrm{j}240\mathrm{e}^{-\mathrm{j}\omega_0 t} - \mathrm{j}240\mathrm{e}^{\mathrm{j}\omega_0 t} \ \mathrm{V}$$

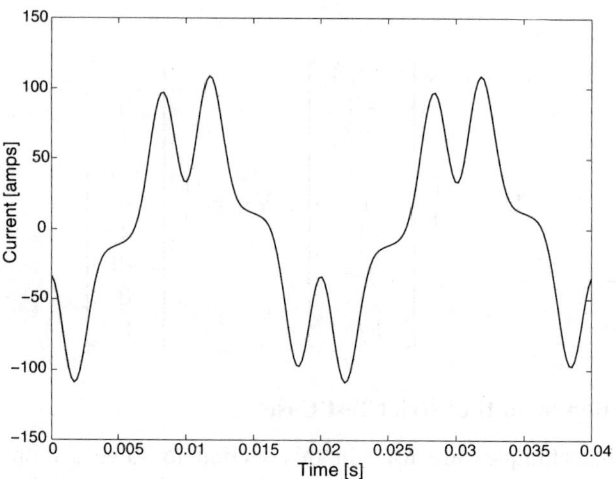

Figure 5.9 Induction furnace current waveform

Table 5.1 Harmonic content of the induction furnace current

Harmonic	Value [A]
1	$79.7247\angle 172.7759°$
3	$5.6958\angle 2.2619°$
5	$30.2500\angle 4.4351°$
7	$9.6932\angle 6.1747°$

and

$$I_{\mathrm{s}} = 104.17\mathrm{e}^{-\mathrm{j}\omega_0 t} + 104.17\mathrm{e}^{\mathrm{j}\omega_0 t} \ \ \mathrm{A}$$

Using a base voltage of 480 V the parameter values referred to the same base are

$$\text{Utility system} \ : \ X_{\mathrm{syst}} = \frac{(\mathrm{kV})^2}{\mathrm{MVA_{CC}}} = \frac{(0.480)^2}{100} = 0.0023\,\Omega$$

$$\text{Transformer} \ : \ X_{\mathrm{trans}} = \frac{(Z\%)(\mathrm{kV})^2}{(100)(\mathrm{MVA})} = \frac{(10)(0.480)^2}{(100)(0.50)} = 0.04608\,\Omega$$

$$\text{Capacitor bank} \ : \ X_{\mathrm{cap}} = \frac{(\mathrm{kV})^2}{\mathrm{MVAR}} = \frac{(0.480)^2}{0.10} = 2.3040\,\Omega$$

The load is represented by a parallel R_L–X_L equivalent obtained from

$$P = (S)(\mathrm{PF}) = (112)(0.89) = 99.68\,\mathrm{kW}$$

$$Q = \sqrt{S^2 - P^2} = \sqrt{112^2 - 99.68^2} = 51.06\,\mathrm{kVAR}$$

$$R_L = \frac{(\text{kV})^2}{\text{MW}} = \frac{(0.480)^2}{0.09968} = 2.3114\,\Omega$$

$$X_L = \frac{(\text{kV})^2}{\text{MVAR}} = \frac{(0.480)^2}{0.05106} = 4.5123\,\Omega$$

The induction furnace is represented by a current injection source i_{f}, and the solution of the system is obtained by superposition of sources effect.

From the point of connection of the non-linear load, the system utility is seen as a short circuit, and the industrial system can be represented by the equivalent circuit of Figure 5.10.

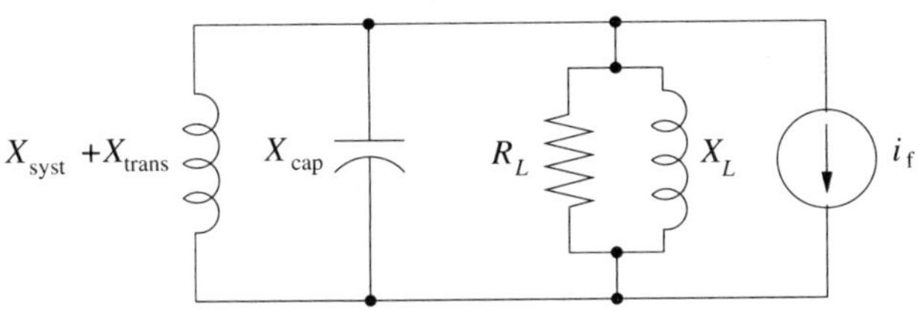

Figure 5.10 Equivalent circuit

The *RLC* equivalent circuit of Figure 5.10 has the following parameters: $R = 2.3114\,\Omega$, $X = 0.04786\,\Omega$ or $L = 0.15236$ mH, $X_C = 2.3040\,\Omega$ or $C = 1381.5\,\mu$F, where the resonant frequency of the *RLC* circuit is:

$$f_{\text{res}} = \frac{1}{2\pi\sqrt{LC}} = 346.90\text{ Hz}$$

The resonant frequency is close to the seventh harmonic at a fundamental frequency of 60 Hz.

In the harmonic domain, the voltage in the capacitor is given by

$$\mathbf{V}_C = -\left(\mathbf{Y}_L + \mathbf{Y}_C + \mathbf{Y}_R\right)^{-1}\mathbf{I}_{\text{f}} + \mathbf{V}_{\text{s}} = \begin{bmatrix} -11.1172 - \text{j}0.1646 \\ 0 \\ -0.9888 + \text{j}7.2942 \\ 0 \\ -0.0184 + \text{j}0.5011 \\ 0 \\ 0.2857 - \text{j}1.9268 \\ 0 \\ 0.2857 + \text{j}1.9268 \\ 0 \\ -0.0184 - \text{j}0.5011 \\ 0 \\ -0.9888 - \text{j}7.2942 \\ 0 \\ -11.1172 + \text{j}0.1646 \end{bmatrix} + \begin{bmatrix} 0 \\ 0 \\ 0 \\ 0 \\ 0 \\ 0 \\ \text{j}240 \\ 0 \\ -\text{j}240 \\ 0 \\ 0 \\ 0 \\ 0 \\ 0 \\ 0 \end{bmatrix}$$

where $\mathbf{I}_{\text{f}}$ is obtained from Table 5.1 and the relation given by equation (2.24):

$$\mathbf{I}_{\mathrm{f}} = \begin{bmatrix} 4.8185 - j0.5213 \\ 0 \\ 15.0797 - j1.1696 \\ 0 \\ 2.8457 - j0.1124 \\ 0 \\ -39.5459 - j5.0127 \\ 0 \\ -39.5459 + j5.0127 \\ 0 \\ 2.8457 + j0.1124 \\ 0 \\ 15.0797 + j1.1696 \\ 0 \\ 4.8185 + j0.5213 \end{bmatrix}$$

The current in the capacitor is given by

$$\mathbf{I}_C = -\mathbf{Y}_C \left(\mathbf{Y}_L + \mathbf{Y}_C + \mathbf{Y}_R\right)^{-1} \mathbf{I}_{\mathrm{f}} + \mathbf{I}_{\mathrm{s}} = \begin{bmatrix} -0.5002 + j33.7762 \\ 0 \\ 15.8294 + j2.1457 \\ 0 \\ 0.6525 + j0.0240 \\ 0 \\ -0.8363 - j0.1240 \\ 0 \\ -0.8363 + j0.1240 \\ 0 \\ 0.6525 - j0.0240 \\ 0 \\ 15.8294 - j2.1457 \\ 0 \\ -0.5002 - j33.7762 \end{bmatrix} + \begin{bmatrix} 0 \\ 0 \\ 0 \\ 0 \\ 0 \\ 0 \\ 104.17 \\ 0 \\ 104.17 \\ 0 \\ 0 \\ 0 \\ 0 \\ 0 \\ 0 \end{bmatrix}$$

These results show that the seventh harmonic current in the capacitor contains the largest value and not the fifth as is the case in the induction furnace current given in Table 5.1. This result makes sense since the resonant frequency at the point of common coupling is close to the seventh harmonic. The voltage and current waveforms in the capacitor bank are shown in Figure 5.11(a) and Figure 5.11(b) gives theirs harmonic content.

The next MATLAB™ function may be used to generate a general linear impedance matrix. This function can also be used to obtain the operational matrix, D=form_Zm(0,ω_0,h), and the identity matrix, U=form_Zm(1,0,h):

```
function Zm=form_Zm(R,X,h);
% R resistance in ohms
% X reactance in ohms (+ inductive, - capacitive)
% h harmonic
Zm=zeros(2*h+1,2*h+1);
```

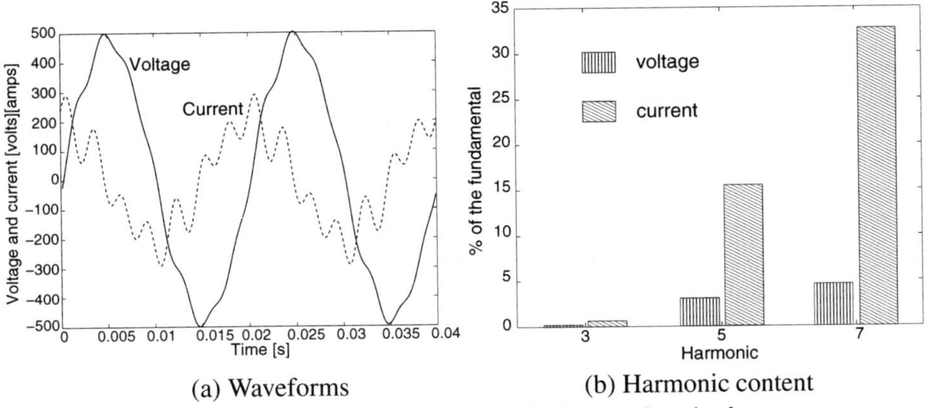

(a) Waveforms (b) Harmonic content

Figure 5.11 Voltage and current in the capacitors bank

```
n=1;
for k=-h:h
  Zm(n,n)=R+i*k^(sign(X))*X;
  n=n+1;
end
Zm(h+1,h+1)=R+1e-9;
```

The next steps in MATLAB™ were used to solve this example:

```
>> h=7;
>> YL=inv(form_Zm(0,0.04786,h));
>> YC=inv(form_Zm(0,-2.3040,h));
>> YR=inv(form_Zm(2.3114,0,h));
>> Vc=-inv(YL+YC+YR)*If+Vs;
>> Ic=-YC*inv(YL+YC+YR)*If+YC*Vs;
>> Vct=plot_f(Vc,h,2);
>> Ict=plot_f(Ic,h,2);
```

5.7 Summary

This chapter has developed from first principles the theory of harmonic domain linearisation. It has been shown that a non-linear function, when linearised around a base operating point, leads to a harmonic Norton equivalent representation. In particular, the admittance matrix of the Norton equivalent contains some very interesting characteristics; it has a Toeplitz matrix structure which, owing to the practical need to consider only a limited number of harmonics, becomes a banded Toeplitz matrix. Moreover, because of the complex conjugate nature of the Fourier series the Toeplitz matrix becomes Hermitian, i.e. $\xi_{-i} = \xi_i^*$. The band width provides a sound indication of function non-linearity. For instance, a linear function will only contain elements along the main diagonal of the matrix. Also, the evaluation of dynamic terms in the harmonic domain has turned out to be a very simple algebraic operation. The theory of linearisation in the harmonic domain is completely general, and applies equally well to static single-variable functions and to static multi-variable functions. The more general case of dynamic functions is also linearised with ease in the harmonic domain. Several hand-worked examples have been presented in detail to aid understanding of the theory.

5.8 Bibliography

1. E. Acha, "Modelling of Power System Transformers in the Complex Conjugate Harmonic Space", *PhD Thesis*, University of Canterbury, Christchurch, New Zealand, 1988.
2. A. Semlyen, E. Acha, J. Arrillaga, "Harmonic Norton Equivalent of the Magnetising Branch of a Transformer", *Proceedings of the IEE*, Part C, Vol. 134, No. 2, March 1987, pp. 162–169.
3. A. Semlyen, E. Acha, J. Arrillaga, "Newton-Type Algorithms for the Harmonic Phasor Analysis of Non-Linear Power Circuits in Periodical Steady-State with Special Reference to Magnetic Non-Linearities", *IEEE Transactions on Power Delivery*, Vol. 3, No. 3, July 1988, pp. 1090–1098.
4. A. Semlyen, N. Rajakovic, "Harmonic Domain Modeling of Laminated Iron Core", *IEEE Transactions on Power Delivery*, Vol. 4, No. 1, January 1989, pp. 382–390.
5. N. Rajakovic, A. Semlyen, "Investigation of the Inrush Phenomenon: A Quasi-Stationary Approach in the Harmonic Domain", *IEEE Transactions on Power Delivery*, Vol. 4, No. 4, October 1989, pp. 2114–2120.
6. E. Acha, A. Semlyen, N. Rajakovic, "A Harmonic Domain Computational Package for Nonlinear Problems and its Application to Electric Arcs", *IEEE Transactions on Power Delivery*, Vol. 5, No. 3, July 1990, pp. 1390–1397.
7. E. Acha, J.J. Rico, "Harmonic Domain Modelling of Non-Linear Power Plant Components", *Proceedings of the IEEE ICHPS VI*, Bologna, Italy, 21–23 September 1994, pp. 206–213.

6

Real Fourier Harmonic Domain

6.1 Introduction

Direct frequency domain evaluations and harmonic domain linearisation using the complex Fourier transform have been presented in Chapters 4 and 5, respectively. This chapter now shows how fully equivalent operations can be carried out using the real Fourier transform. In general, non-linear relations will contain dynamic elements, $\dot{x}$, in addition to static elements, x, e.g.

$$y = f(x, \dot{x}) \tag{6.1}$$

Section 6.2 shows how direct harmonic domain evaluations can be achieved in the real harmonic domain as opposed to the complex Fourier harmonic domain. Sections 6.3 and 6.4 show how the non-linear dynamic relation (6.1) is linearised in the real Fourier harmonic domain. The former section deals with static elements, $y_S = g(x)$, whereas the latter deals with dynamic elements, $y_D = h(\dot{x})$.

6.2 Direct Frequency Domain Evaluations

In practice the harmonic domain evaluation of the static part of equation (6.1) will involve the use of a series of self-convolutions, e.g.

$$x^2 = xx \Rightarrow \mathbf{X} \otimes \mathbf{X} = \mathbf{Y} \tag{6.2}$$

and mutual convolutions, e.g.

$$x^3 = x^2 x \Rightarrow \mathbf{Y} \otimes \mathbf{X} = \mathbf{Z} \tag{6.3}$$

where x is a periodic function and $\mathbf{X}$, $\mathbf{Y}$ and $\mathbf{Z}$ are vectors of harmonic coefficients. $\otimes$ is the convolution symbol.

In the real Fourier harmonic domain the product of two periodic, real variables x and y,

$$x = \sum_{i=0}^{\infty} (X_i^s \sin i\omega_0 t + X_i^c \cos i\omega_0 t) \tag{6.4}$$

$$y = \sum_{h=0}^{\infty} (Y_h^s \sin h\omega_0 t + Y_h^c \cos h\omega_0 t) \tag{6.5}$$

gives rise to another periodic variable, say z,

$$\sum_{k=0}^{\infty} (Z_k^s \sin k\omega_0 t + Z_k^c \cos k\omega_0 t) = \sum_{i=0}^{\infty} (X_i^s \sin i\omega_0 t + X_i^c \cos i\omega_0 t) \times \sum_{h=0}^{\infty} (Y_h^s \sin h\omega_0 t + Y_h^c \cos h\omega_0 t) \tag{6.6}$$

where the coefficients X_0^c, X_i^c and X_i^s correspond to $\frac{1}{2}a_0$, a_i and b_i of the trigonometric Fourier series.

The vector form of equation (6.6) can be obtained by using its harmonic coefficients,

$$\begin{bmatrix} \mathbf{Z}^s \\ \mathbf{Z}^c \end{bmatrix} = \begin{bmatrix} \mathbf{X}^s \\ \mathbf{X}^c \end{bmatrix} \otimes \begin{bmatrix} \mathbf{Y}^s \\ \mathbf{Y}^c \end{bmatrix} \tag{6.7}$$

where

$$\begin{bmatrix} \mathbf{X}^s \\ \mathbf{X}^c \end{bmatrix} = \begin{bmatrix} X_1^s \\ X_2^s \\ X_3^s \\ \vdots \\ - \\ X_0^c \\ X_1^c \\ X_2^c \\ X_3^c \\ \vdots \end{bmatrix} ; \begin{bmatrix} \mathbf{Y}^s \\ \mathbf{Y}^c \end{bmatrix} = \begin{bmatrix} Y_1^s \\ Y_2^s \\ Y_2^s \\ \vdots \\ - \\ Y_0^c \\ Y_1^c \\ Y_2^c \\ Y_3^c \\ \vdots \end{bmatrix} ; \begin{bmatrix} \mathbf{Z}^s \\ \mathbf{Z}^c \end{bmatrix} = \begin{bmatrix} Z_1^s \\ Z_2^s \\ Z_3^s \\ \vdots \\ - \\ Z_0^c \\ Z_1^c \\ Z_2^c \\ Z_3^c \\ \vdots \end{bmatrix} \tag{6.8}$$

Since the terms in equation (6.7) are vectors, it cannot directly be solved in terms of conventional matrix operations. Instead, the matrix form of equation (6.7) provides an alternative for performing actual calculations,

$$\begin{bmatrix} \mathbf{Z}^s \\ \mathbf{Z}^c \end{bmatrix} = \frac{1}{2} \begin{bmatrix} \mathbf{X}^{ss} & \mathbf{X}^{sc} \\ \mathbf{X}^{cs} & \mathbf{X}^{cc} \end{bmatrix} \begin{bmatrix} \mathbf{Y}^s \\ \mathbf{Y}^c \end{bmatrix} \tag{6.9}$$

Notice that the matrix operation in equation (6.9) has replaced the convolution operation in equation (6.7). The elements of matrices $\mathbf{X}^{ss}$, $\mathbf{X}^{sc}$, $\mathbf{X}^{cs}$ and $\mathbf{X}^{cc}$ are identified by taking one term of y at a time in equation (6.6).

Consider, say, Y_h^s:

$$\begin{aligned} &\sum_{i=0}^{\infty} (X_i^s \sin i\omega_0 t + X_i^c \cos i\omega_0 t)\, Y_h^s \sin h\omega_0 t = \\ &\frac{1}{2} \sum_{i=0}^{\infty} [\cos(i-h)\omega_0 t - \cos(i+h)\omega_0 t]\, X_i^s Y_h^s \\ &+ \frac{1}{2} \sum_{i=0}^{\infty} [-\sin(i-h)\omega_0 t + \sin(i+h)\omega_0 t]\, X_i^c Y_h^s \end{aligned} \tag{6.10}$$

and Y_h^c:

$$\sum_{i=0}^{\infty}(X_i^s \sin i\omega_0 t + X_i^c \cos i\omega_0 t)\, Y_h^c \cos h\omega_0 t =$$
$$\frac{1}{2}\sum_{i=0}^{\infty}[\sin(i-h)\omega_0 t - \sin(i+h)\omega_0 t]\, X_i^s Y_h^c$$
$$+\frac{1}{2}\sum_{i=0}^{\infty}[-\cos(i-h)\omega_0 t + \cos(i+h)\omega_0 t]\, X_i^c Y_h^c \tag{6.11}$$

After considering all the terms of the input $\mathbf{Y}^s$ and $\mathbf{Y}^c$, the matrices $\mathbf{X}^{ss}$, $\mathbf{X}^{sc}$, $\mathbf{X}^{cs}$ and $\mathbf{X}^{cc}$ can be readily identified:

$$\mathbf{X}^{ss} = \begin{bmatrix} 2X_0^c - X_2^c & X_1^c - X_3^c & X_2^c - X_4^c & X_3^c - X_5^c & \cdots \\ X_1^c - X_3^c & 2X_0^c - X_4^c & X_1^c - X_5^c & X_2^c - X_6^c & \cdots \\ X_2^c - X_4^c & X_1^c - X_5^c & 2X_0^c - X_6^c & X_1^c - X_7^c & \cdots \\ X_3^c - X_5^c & X_2^c - X_6^c & X_1^c - X_7^c & 2X_0^c - X_8^c & \cdots \\ \vdots & \vdots & \vdots & \vdots & \ddots \end{bmatrix} \tag{6.12}$$

$$\mathbf{X}^{sc} = \begin{bmatrix} 2X_1^s & X_2^s & -X_1^s + X_3^s & -X_2^s + X_4^s & -X_3^s + X_5^s & \cdots \\ 2X_2^s & X_1^s + X_3^s & X_4^s & -X_1^s + X_5^s & -X_2^s + X_6^s & \cdots \\ 2X_3^s & X_2^s + X_4^s & X_1^s + X_5^s & X_6^s & -X_1^s + X_7^s & \cdots \\ 2X_4^s & X_3^s + X_5^s & X_2^s + X_6^s & X_1^s + X_7^s & X_8^s & \cdots \\ \vdots & \vdots & \vdots & \vdots & \vdots & \ddots \end{bmatrix} \tag{6.13}$$

$$\mathbf{X}^{cs} = \begin{bmatrix} X_1^s & X_2^s & X_3^s & X_4^s & \cdots \\ X_2^s & X_1^s + X_3^s & X_2^s + X_4^s & X_3^s + X_5^s & \cdots \\ -X_1^s + X_3^s & X_4^s & X_1^s + X_5^s & X_2^s + X_6^s & \cdots \\ -X_2^s + X_4^s & -X_1^s + X_5^s & X_6^s & X_1^s + X_7^s & \cdots \\ -X_3^s + X_5^s & -X_2^s + X_6^s & -X_1^s + X_7^s & X_8^s & \cdots \\ \vdots & \vdots & \vdots & \vdots & \ddots \end{bmatrix} \tag{6.14}$$

$$\mathbf{X}^{cc} = \begin{bmatrix} 2X_0^c & X_1^c & X_2^c & X_3^c & X_4^c & \cdots \\ 2X_1^c & 2X_0^c + X_2^c & X_1^c + X_3^c & X_2^c + X_4^c & X_3^c + X_5^c & \cdots \\ 2X_2^c & X_1^c + X_3^c & 2X_0^c + X_4^c & X_1^c + X_5^c & X_2^c + X_6^c & \cdots \\ 2X_3^c & X_2^c + X_4^c & X_1^c + X_5^c & 2X_0^c + X_6^c & X_1^c + X_7^c & \cdots \\ 2X_4^c & X_3^c + X_5^c & X_2^c + X_6^c & X_1^c + X_7^c & 2X_0^c + X_8^c & \cdots \\ \vdots & \vdots & \vdots & \vdots & \vdots & \ddots \end{bmatrix} \tag{6.15}$$

Example 6-1: The polynomial equation given below will be evaluated in the real Fourier harmonic domain,

$$i = f(\psi) = 0.001\psi + 0.0743\psi^3 \tag{6.16}$$

around the base excitation,

$$\psi_{\rm b} = \sin \omega_0 t$$

or, in vector form,

$$\Psi_{\rm b} = \begin{bmatrix} 1 \\ 0 \\ - \\ 0 \\ 0 \\ 0 \end{bmatrix}$$

The harmonic evaluation is carried out as follows:

$$\Psi^2 = \Psi \otimes \Psi = \begin{bmatrix} 1 \\ 0 \\ - \\ 0 \\ 0 \\ 0 \end{bmatrix} \otimes \begin{bmatrix} 1 \\ 0 \\ - \\ 0 \\ 0 \\ 0 \end{bmatrix}$$

or, in matrix form, using submatrices $\mathbf{X}^{ss}$, $\mathbf{X}^{sc}$, $\mathbf{X}^{cs}$ and $\mathbf{X}^{cc}$,

$$\Psi^2 = \frac{1}{2} \left[\begin{array}{cc|ccc} 0 & 0 & 2 & 0 & -1 \\ 0 & 0 & 0 & 1 & 0 \\ - & - & - & - & - \\ 1 & 0 & 0 & 0 & 0 \\ 0 & 1 & 0 & 0 & 0 \\ -1 & 0 & 0 & 0 & 0 \end{array} \right] \begin{bmatrix} 1 \\ 0 \\ - \\ 0 \\ 0 \\ 0 \end{bmatrix} = \frac{1}{2} \begin{bmatrix} 0 \\ 0 \\ - \\ 1 \\ 0 \\ -1 \end{bmatrix} \tag{6.17}$$

Also,

$$\Psi^3 = \Psi \otimes \Psi^2 = \begin{bmatrix} 1 \\ 0 \\ 0 \\ - \\ 0 \\ 0 \\ 0 \\ 0 \end{bmatrix} \otimes \frac{1}{2} \begin{bmatrix} 0 \\ 0 \\ 0 \\ - \\ 1 \\ 0 \\ -1 \\ 0 \end{bmatrix}$$

or, in matrix form, using submatrices $\mathbf{X}^{ss}$, $\mathbf{X}^{sc}$, $\mathbf{X}^{cs}$ and $\mathbf{X}^{cc}$,

$$\Psi^3 = \frac{1}{4} \left[\begin{array}{ccc|cccc} 0 & 0 & 0 & 2 & 0 & -1 & 0 \\ 0 & 0 & 0 & 0 & 1 & 0 & -1 \\ 0 & 0 & 0 & 0 & 0 & 1 & 0 \\ - & - & - & - & - & - & - \\ 1 & 0 & 0 & 0 & 0 & 0 & 0 \\ 0 & 1 & 0 & 0 & 0 & 0 & 0 \\ -1 & 0 & 1 & 0 & 0 & 0 & 0 \\ 0 & -1 & 0 & 0 & 0 & 0 & 0 \end{array} \right] \otimes \begin{bmatrix} 0 \\ 0 \\ 0 \\ - \\ 1 \\ 0 \\ -1 \\ 0 \end{bmatrix} = \frac{1}{4} \begin{bmatrix} 3 \\ 0 \\ -1 \\ - \\ 0 \\ 0 \\ 0 \\ 0 \end{bmatrix}$$

The evaluation of the polynomial equation in this domain is

$$\begin{bmatrix} I_1^s \\ I_2^s \\ I_3^s \\ - \\ I_0^c \\ I_1^c \\ I_2^c \\ I_3^c \end{bmatrix} = 0.001 \begin{bmatrix} 1 \\ 0 \\ 0 \\ - \\ 0 \\ 0 \\ 0 \\ 0 \end{bmatrix} + \frac{0.0743}{4} \begin{bmatrix} 3 \\ 0 \\ -1 \\ - \\ 0 \\ 0 \\ 0 \\ 0 \end{bmatrix} = \begin{bmatrix} 0.0567 \\ 0 \\ -0.0186 \\ - \\ 0 \\ 0 \\ 0 \\ 0 \end{bmatrix}$$

or in the time domain

$$i = 0.0567 \sin \omega_0 t - 0.0186 \sin 3\omega_0 t$$

It should be noted that this result is identical to that obtained in Example 5-2 for $i_{\mathrm{b}}(t)$, using the complex Fourier transform.

6.3 Linearisation in the Real Fourier Harmonic Domain

If the variables making up the static non-linear relation

$$y_{\mathrm{s}} = g(x) \tag{6.18}$$

are differentiable around the base operating point $(x_{\mathrm{b}}, y_{\mathrm{b}})$ then the following linearised expression will exist:

$$\Delta y_{\mathrm{s}} = \frac{\partial g}{\partial x} \Delta x \tag{6.19}$$

Furthermore, if the variables in equation (6.19) are periodic they may be expressed in the real Fourier harmonic domain as,

$$\Delta x = \sum_{h=0}^{\infty} (\Delta X_h^s \sin h\omega_0 t + \Delta X_h^c \cos h\omega_0 t) \tag{6.20}$$

$$\Delta y_{\mathrm{s}} = \sum_{k=0}^{\infty} (\Delta Y_k^s \sin k\omega_0 t + \Delta Y_k^c \cos k\omega_0 t) \tag{6.21}$$

$$\frac{\partial g}{\partial x} = \sum_{i=0}^{\infty} (G_i^s \sin i\omega_0 t + G_i^c \cos i\omega_0 t) \tag{6.22}$$

Substituting equations (6.20), (6.21) and (6.22) into (6.19) gives

$$\begin{aligned} \sum_{k=0}^{\infty} (\Delta Y_k^s \sin k\omega_0 t + \Delta Y_k^c \cos k\omega_0 t) &= \sum_{i=0}^{\infty} (G_i^s \sin i\omega_0 t + G_i^c \cos i\omega_0 t) \\ &\quad \times \sum_{h=0}^{\infty} (\Delta X_h^s \sin h\omega_0 t + \Delta X_h^c \cos h\omega_0 t) \end{aligned} \tag{6.23}$$

By following a similar line of reasoning as in Section 6.2, a harmonic matrix expression is determined, which is suitable for carrying out iterative solutions in the real Fourier harmonic domain,

$$\begin{bmatrix} \Delta\mathbf{Y}^s \\ \Delta\mathbf{Y}^c \end{bmatrix} = \begin{bmatrix} \mathbf{G}^s \\ \mathbf{G}^c \end{bmatrix} \otimes \begin{bmatrix} \Delta\mathbf{X}^s \\ \Delta\mathbf{X}^c \end{bmatrix} = \frac{1}{2}\begin{bmatrix} \mathbf{G}^{ss} & \mathbf{G}^{sc} \\ \mathbf{G}^{cs} & \mathbf{G}^{cc} \end{bmatrix}\begin{bmatrix} \Delta\mathbf{X}^s \\ \Delta\mathbf{X}^c \end{bmatrix} \tag{6.24}$$

where

$$\begin{bmatrix} \Delta\mathbf{X}^s \\ \Delta\mathbf{X}^c \end{bmatrix} = \begin{bmatrix} \Delta X_1^s \\ \Delta X_2^s \\ \Delta X_3^s \\ \vdots \\ - \\ \Delta X_0^c \\ \Delta X_1^c \\ \Delta X_2^c \\ \Delta X_3^c \\ \vdots \end{bmatrix}; \quad \begin{bmatrix} \Delta\mathbf{Y}^s \\ \Delta\mathbf{Y}^c \end{bmatrix} = \begin{bmatrix} \Delta Y_1^s \\ \Delta Y_2^s \\ \Delta Y_3^s \\ \vdots \\ - \\ \Delta Y_0^c \\ \Delta Y_1^c \\ \Delta Y_2^c \\ \Delta Y_3^c \\ \vdots \end{bmatrix}; \quad \begin{bmatrix} \mathbf{G}^s \\ \mathbf{G}^c \end{bmatrix} = \begin{bmatrix} G_1^s \\ G_2^s \\ G_3^s \\ \vdots \\ - \\ G_0^c \\ G_1^c \\ G_2^c \\ G_3^c \\ \vdots \end{bmatrix} \tag{6.25}$$

$$\mathbf{G}^{ss} = \begin{bmatrix} 2G_0^c - G_2^c & G_1^c - G_3^c & G_2^c - G_4^c & G_3^c - G_5^c & \cdots \\ G_1^c - G_3^c & 2G_0^c - G_4^c & G_1^c - G_5^c & G_2^c - G_6^c & \cdots \\ G_2^c - G_4^c & G_1^c - G_4^c & 2G_0^c - G_6^c & G_1^c - G_7^c & \cdots \\ G_3^c - G_5^c & G_2^c - G_6^c & G_1^c - G_7^c & 2G_0^c - G_8^c & \cdots \\ \vdots & \vdots & \vdots & \vdots & \ddots \end{bmatrix} \tag{6.26}$$

$$\mathbf{G}^{sc} = \begin{bmatrix} 2G_1^s & 2G_2^s & -G_1^s + G_3^s & -G_2^s + G_4^s & -G_3^s + G_5^s & \cdots \\ 2G_2^s & G_1^s + G_3^s & 2G_4^s & -G_1^s + G_5^s & -G_2^s + G_6^s & \cdots \\ 2G_3^s & G_2^s + G_4^s & G_1^s + G_5^s & 2G_6^s & -G_1^s + G_7^s & \cdots \\ 2G_4^s & G_3^s + G_5^s & G_2^s + G_6^s & G_1^s + G_7^s & 2G_8^s & \cdots \\ \vdots & \vdots & \vdots & \vdots & \vdots & \ddots \end{bmatrix} \tag{6.27}$$

$$\mathbf{G}^{cs} = \begin{bmatrix} G_1^s & G_2^s & G_3^s & G_4^s & \cdots \\ G_2^s & G_1^s + G_3^s & G_2^s + G_4^s & G_3^s + G_5^s & \cdots \\ -G_1^s + G_3^s & G_4^s & G_1^s + G_5^s & G_2^s + G_6^s & \cdots \\ -G_2^s + G_4^s & -G_1^s + G_5^s & G_6^s & G_1^s + G_7^s & \cdots \\ -G_3^s + G_5^s & -G_2^s + G_6^s & -G_1^s + G_7^s & G_8^s & \cdots \\ \vdots & \vdots & \vdots & \vdots & \ddots \end{bmatrix} \tag{6.28}$$

$$\mathbf{G}^{cc} = \begin{bmatrix} 2G_0^c & G_1^c & G_2^c & G_3^c & G_4^c & \cdots \\ 2G_1^c & 2G_0^c + G_2^c & G_1^c + G_3^c & G_2^c + G_4^c & G_3^c + G_5^c & \cdots \\ 2G_2^c & G_1^c + G_3^c & 2G_0^c + G_4^c & G_1^c + G_5^c & G_2^c + G_6^c & \cdots \\ 2G_3^c & G_2^c + G_4^c & G_1^c + G_5^c & 2G_0^c + G_6^c & G_1^c + G_7^c & \cdots \\ 2G_4^c & G_3^c + G_5^c & G_2^c + G_6^c & G_1^c + G_7^c & 2G_0^c + G_8^c & \cdots \\ \vdots & \vdots & \vdots & \vdots & \vdots & \ddots \end{bmatrix} \tag{6.29}$$

Also, incorporating the point of linearisation

$$\Delta\mathbf{X} = \mathbf{X} - \mathbf{X}_{\mathrm{b}}$$
$$\Delta\mathbf{Y} = \mathbf{Y} - \mathbf{Y}_{\mathrm{b}} \tag{6.30}$$

in (6.24) produces an alternative expression which is more amenable to harmonic domain computations:

$$\begin{bmatrix} \mathbf{Y}^s \\ \mathbf{Y}^c \end{bmatrix} = \frac{1}{2}\begin{bmatrix} \mathbf{G}^{ss} & \mathbf{G}^{sc} \\ \mathbf{G}^{cs} & \mathbf{G}^{cc} \end{bmatrix}\begin{bmatrix} \mathbf{X}^s \\ \mathbf{X}^c \end{bmatrix} + \begin{bmatrix} \mathbf{Y}^s_{\mathrm{N}} \\ \mathbf{Y}^c_{\mathrm{N}} \end{bmatrix} \tag{6.31}$$

where

$$\begin{bmatrix} \mathbf{Y}^s_{\mathrm{N}} \\ \mathbf{Y}^c_{\mathrm{N}} \end{bmatrix} = \begin{bmatrix} \mathbf{Y}^s_{\mathrm{b}} \\ \mathbf{Y}^c_{\mathrm{b}} \end{bmatrix} - \frac{1}{2}\begin{bmatrix} \mathbf{G}^{ss} & \mathbf{G}^{sc} \\ \mathbf{G}^{cs} & \mathbf{G}^{cc} \end{bmatrix}\begin{bmatrix} \mathbf{X}^s_{\mathrm{b}} \\ \mathbf{X}^c_{\mathrm{b}} \end{bmatrix} \tag{6.32}$$

The subscript N indicates the possibility of interpreting this equation as a harmonic Norton equivalent. This will be the case when $\mathbf{Y}$ is a vector of harmonic currents, $\mathbf{X}$ is a vector of harmonic voltages and $\mathbf{G}$ is a matrix of harmonic admittance.

From equations (6.24), (6.31) and (6.32) it is apparent that the solution in the real Fourier harmonic domain of a single-variable, non-linear function requires the evaluation of two harmonic vectors, namely $[\mathbf{Y}^s_{\mathrm{b}}\,\mathbf{Y}^c_{\mathrm{b}}]^{\mathrm{T}}$ and $[\mathbf{G}^s_{\mathrm{b}}\,\mathbf{G}^c_{\mathrm{b}}]^{\mathrm{T}}$. It should be noted that $[\mathbf{X}^s_{\mathrm{b}}\,\mathbf{X}^c_{\mathrm{b}}]^{\mathrm{T}}$ is a given input, $[\mathbf{X}^s\,\mathbf{X}^c]^{\mathrm{T}}$ is a newly determined output and $[\mathbf{Y}^s\,\mathbf{Y}^c]^{\mathrm{T}}$ is a vector that allows the interfacing of equation (6.31) with the equation representing the external network. Moreover, if the input vector $[\mathbf{X}^s_{\mathrm{b}}\,\mathbf{X}^c_{\mathrm{b}}]^{\mathrm{T}}$ is far away from the harmonic solution then the output vector $[\mathbf{X}^s\,\mathbf{X}^c]^{\mathrm{T}}$ becomes the new input vector and the harmonic solution is reached by iteration.

6.4 Dynamic Relations

The linearisation of static elements in the real harmonic domain has been dealt with in the previous section, and now it will be shown how dynamic elements are handled in this domain.

Let us consider the basic relation

$$z = \dot{x} \tag{6.33}$$

between the two periodic variables

$$z = \sum_{k=0}^{\infty}(Z^s_k \sin k\omega_0 t + Z^c_k \cos k\omega_0 t) \tag{6.34}$$

$$x = \sum_{i=0}^{\infty}(X^s_i \sin i\omega_0 t + X^c_i \cos i\omega_0 t) \tag{6.35}$$

Equation (6.33) may be written as

$$\begin{aligned} \sum_{k=0}^{\infty}(Z^s_k \sin k\omega_0 t + Z^c_k \cos k\omega_0 t) &= \frac{\mathrm{d}}{\mathrm{d}t}\sum_{i=0}^{\infty}(X^s_i \sin i\omega_0 t + X^c_i \cos i\omega_0 t) \\ &= \sum_{i=0}^{\infty}(i\omega_0 X^s_i \cos i\omega_0 t - i\omega_0 X^c_i \sin i\omega_0 t) \end{aligned} \tag{6.36}$$

Alternatively, in terms of harmonic coefficients,

$$\begin{bmatrix} Z_1^s \\ Z_2^s \\ Z_3^s \\ \vdots \\ - \\ Z_0^c \\ Z_1^c \\ Z_2^c \\ Z_3^c \\ \vdots \end{bmatrix} = \left[\begin{array}{ccccc|ccccc} 0 & \cdots & \cdots & 0 & 0 & -\omega_0 & 0 & \cdots & 0 \\ \vdots & \ddots & & \vdots & \vdots & \ddots & -2\omega_0 & \ddots & \vdots \\ \vdots & & \ddots & \vdots & \vdots & \ddots & \ddots & -3\omega_0 & 0 \\ 0 & \cdots & \cdots & 0 & 0 & \cdots & \cdots & 0 & \ddots \\ - & - & - & - & - & - & - & - & - \\ 0 & \cdots & \cdots & 0 & 0 & \cdots & \cdots & \cdots & 0 \\ \omega_0 & \ddots & & \vdots & \vdots & \ddots & & & \vdots \\ 0 & 2\omega_0 & \ddots & \vdots & \vdots & & \ddots & & \vdots \\ \vdots & \ddots & 3\omega_0 & 0 & \vdots & & & \ddots & \vdots \\ 0 & \cdots & 0 & \ddots & 0 & \cdots & \cdots & \cdots & 0 \end{array}\right] \begin{bmatrix} X_1^s \\ X_2^s \\ X_3^s \\ \vdots \\ - \\ X_0^c \\ X_1^c \\ X_2^c \\ X_3^c \\ \vdots \end{bmatrix} \tag{6.37}$$

This shows that the evaluation of dynamic terms in the real Fourier harmonic domain is carried out by means of simple algebraic operations. Equation (6.37) can be written in compact form as

$$\begin{bmatrix} \mathbf{Z}^s \\ \mathbf{Z}^c \end{bmatrix} = \begin{bmatrix} 0 & \mathbf{D}(-i\omega_0) \\ \mathbf{D}(i\omega_0) & 0 \end{bmatrix} \begin{bmatrix} \mathbf{X}^s \\ \mathbf{X}^c \end{bmatrix} \tag{6.38}$$

and

$$\begin{bmatrix} \mathbf{X}^s \\ \mathbf{X}^c \end{bmatrix} = \begin{bmatrix} 0 & \mathbf{D}\left(\frac{1}{i\omega_0}\right) \\ \mathbf{D}\left(-\frac{1}{i\omega_0}\right) & 0 \end{bmatrix} \begin{bmatrix} \mathbf{Z}^s \\ \mathbf{Z}^c \end{bmatrix} \tag{6.39}$$

6.5 Linear Elements

It should be noted that a linear function is a particular case of the non-linear function in equation (6.18), e.g.

$$f(x) = y = ax \tag{6.40}$$

and that the harmonic content of the function derivative is a DC-like term, i.e.

$$\dot{f} = a \tag{6.41}$$

In many applications it is advantageous to express equation (6.40) using only real coefficients, even though the coefficient a may be complex, i.e. $a = g + \mathrm{j}b$. For instance, the fundamental frequency expression of equation (6.40) is

$$\begin{bmatrix} Y_1^s \\ Y_1^c \end{bmatrix} = \begin{bmatrix} g & b \\ -b & g \end{bmatrix} \begin{bmatrix} X_1^s \\ X_1^c \end{bmatrix} \tag{6.42}$$

For the case when a frequency-dependent linear function is to be expressed in the real Fourier harmonic domain, equation (6.40) is expressed as

$$\begin{bmatrix} Y_1^s \\ Y_2^s \\ Y_3^s \\ \vdots \\ - \\ Y_1^c \\ Y_2^c \\ Y_3^c \\ \vdots \end{bmatrix} = \left[\begin{array}{cccc|ccccc} g_1 & \cdots & \cdots & 0 & 0 & b_1 & 0 & \cdots & 0 \\ \vdots & g_2 & & \vdots & \vdots & \ddots & b_2 & \ddots & \vdots \\ \vdots & & g_3 & \vdots & \vdots & \ddots & \ddots & b_3 & 0 \\ 0 & \cdots & \cdots & \ddots & 0 & \cdots & \cdots & 0 & \ddots \\ - & - & - & - & - & - & - & - & - \\ 0 & \cdots & \cdots & 0 & g_0 & \cdots & \cdots & \cdots & 0 \\ -b_1 & \ddots & & \vdots & \vdots & g_1 & & & \vdots \\ 0 & -b_2 & \ddots & \vdots & \vdots & & g_2 & & \vdots \\ \vdots & \ddots & -b_3 & 0 & \vdots & & & g_3 & \vdots \\ 0 & \cdots & 0 & \ddots & 0 & \cdots & \cdots & \cdots & \ddots \end{array}\right] \begin{bmatrix} X_1^s \\ X_2^s \\ X_3^s \\ \vdots \\ - \\ X_0^c \\ X_1^c \\ X_2^c \\ X_3^c \\ \vdots \end{bmatrix} \tag{6.43}$$

Equation (6.43) can be written in compact form as

$$\begin{bmatrix} \mathbf{Y}^s \\ \mathbf{Y}^c \end{bmatrix} = \begin{bmatrix} \mathbf{D}(g) & \mathbf{D}(b) \\ -\mathbf{D}(b) & \mathbf{D}(g) \end{bmatrix} \begin{bmatrix} \mathbf{X}^s \\ \mathbf{X}^c \end{bmatrix} \tag{6.44}$$

The lack of off-diagonal coefficients in all four submatrices provides a perfect decoupling between the various harmonic frequencies. Similarly to the case of linear elements in the complex domain, no interactions take place between pre-existing harmonics and no new harmonics are generated.

Equations (6.31), (6.38) and (6.44) provide a suitable means to determining the steady-state solution of general non-linear circuits by iteration. Its nodal nature makes it attractive for incorporation into a general harmonic frame of reference where any number of both linear and linearised non-linear components can be represented together to give a unified iterative solution that exhibits quadratic convergence. In this environment all the harmonics, cross-couplings between harmonics, nodes and phases present in the network share a global nodal admittance matrix, which is also a Jacobian matrix.

Example 6-2: The derivative of the polynomial equation (6.16) is given as

$$\frac{\mathrm{d}f}{\mathrm{d}\psi} = 0.001 + 0.2229\psi^2$$

which is also evaluated around the base excitation,

$$\psi_\mathrm{b} = \sin \omega_0 t$$

From Example 6-1,

$$\mathbf{I}_b = \begin{bmatrix} 0.0567 \\ 0 \\ -0.0186 \\ - \\ 0 \\ 0 \\ 0 \\ 0 \end{bmatrix}$$

using per unit notation, $\omega_0 = 1$,

$$\mathbf{V}_b = \left[\begin{array}{ccc|cccc} 0 & 0 & 0 & 0 & -1 & 0 & 0 \\ 0 & 0 & 0 & 0 & 0 & -2 & 0 \\ 0 & 0 & 0 & 0 & 0 & 0 & -3 \\ - & - & - & - & - & - & - \\ 0 & 0 & 0 & 0 & 0 & 0 & 0 \\ 1 & 0 & 0 & 0 & 0 & 0 & 0 \\ 0 & 2 & 0 & 0 & 0 & 0 & 0 \\ 0 & 0 & 3 & 0 & 0 & 0 & 0 \end{array}\right] \begin{bmatrix} 1 \\ 0 \\ 0 \\ - \\ 0 \\ 0 \\ 0 \\ 0 \end{bmatrix} = \begin{bmatrix} 0 \\ 0 \\ 0 \\ - \\ 0 \\ 1 \\ 0 \\ 0 \end{bmatrix}$$

The harmonic evaluation is carried out using the vector solution in equation (6.17),

$$\Psi^2 = \begin{bmatrix} 0 \\ 0 \\ - \\ 0.5 \\ 0 \\ -0.5 \end{bmatrix}$$

The evaluation of the polynomial equation in this domain is

$$\begin{bmatrix} F_1^s \\ F_2^s \\ - \\ F_0^c \\ F_1^c \\ F_2^c \end{bmatrix} = 0.001 \begin{bmatrix} 0 \\ 0 \\ - \\ 1 \\ 0 \\ 0 \end{bmatrix} + 0.2229 \begin{bmatrix} 0 \\ 0 \\ - \\ 0.5 \\ 0 \\ -0.5 \end{bmatrix} = \begin{bmatrix} 0 \\ 0 \\ - \\ 0.1124 \\ 0 \\ -0.1114 \end{bmatrix}$$

According to (6.24) the linearised equation is given as

$$\begin{bmatrix} \Delta\mathbf{I}^s \\ \Delta\mathbf{I}^c \end{bmatrix} = \frac{1}{2} \begin{bmatrix} \mathbf{F}^{ss} & \mathbf{F}^{sc} \\ \mathbf{F}^{cs} & \mathbf{F}^{cc} \end{bmatrix} \begin{bmatrix} \Delta\Psi^s \\ \Delta\Psi^c \end{bmatrix}$$

where

$$\begin{bmatrix} \mathbf{F}^{ss} & \mathbf{F}^{sc} \\ \mathbf{F}^{cs} & \mathbf{F}^{cc} \end{bmatrix} =$$

$$\left[\begin{array}{ccc|cccc} 0.3362 & 0 & -0.1114 & 0 & 0 & 0 & 0 \\ 0 & 0.2248 & 0 & 0 & 0 & 0 & 0 \\ -0.1114 & 0 & 0.2248 & 0 & 0 & 0 & 0 \\ - & - & - & - & - & - & - \\ 0 & 0 & 0 & 0.2248 & 0 & -0.1114 & 0 \\ 0 & 0 & 0 & 0 & 0.1134 & 0 & -0.1114 \\ 0 & 0 & 0 & -0.2228 & 0 & 0.2248 & 0 \\ 0 & 0 & 0 & 0 & -0.1114 & 0 & 0.2248 \end{array}\right]$$

and

$$\left[\begin{array}{c} \Delta\mathbf{I}^s \\ \Delta\mathbf{I}^c \end{array}\right] = \left[\begin{array}{c} \Delta I_1^s \\ \Delta I_2^s \\ \Delta I_3^s \\ - \\ \Delta I_0^c \\ \Delta I_1^c \\ \Delta I_2^c \\ \Delta I_3^c \end{array}\right]; \quad \left[\begin{array}{c} \Delta\mathbf{\Psi}^s \\ \Delta\mathbf{\Psi}^c \end{array}\right] = \left[\begin{array}{c} \Delta\Psi_1^s \\ \Delta\Psi_2^s \\ \Delta\Psi_3^s \\ - \\ \Delta\Psi_0^c \\ \Delta\Psi_1^c \\ \Delta\Psi_2^c \\ \Delta\Psi_3^c \end{array}\right]$$

or, as a function of incremental voltages,

$$\left[\begin{array}{c} \Delta\mathbf{I}^s \\ \Delta\mathbf{I}^c \end{array}\right] = \left[\begin{array}{cc} \mathbf{Y}^{ss} & \mathbf{Y}^{sc} \\ \mathbf{Y}^{cs} & \mathbf{Y}^{cc} \end{array}\right] \left[\begin{array}{c} \Delta\mathbf{V}^s \\ \Delta\mathbf{V}^c \end{array}\right]$$

where

$$\left[\begin{array}{cc} \mathbf{Y}^{ss} & \mathbf{Y}^{sc} \\ \mathbf{Y}^{cs} & \mathbf{Y}^{cc} \end{array}\right] = \frac{1}{2}\left[\begin{array}{cc} \mathbf{F}^{ss} & \mathbf{F}^{sc} \\ \mathbf{F}^{cs} & \mathbf{F}^{cc} \end{array}\right] \left[\begin{array}{cc} 0 & \mathbf{D}(-i\omega_0) \\ \mathbf{D}(i\omega_0) & 0 \end{array}\right]^{-1}$$

and

$$\left[\begin{array}{cc} 0 & \mathbf{D}(-i\omega_0) \\ \mathbf{D}(i\omega_0) & 0 \end{array}\right]^{-1} = \left[\begin{array}{ccc|cccc} 0 & 0 & 0 & 0 & 1 & 0 & 0 \\ 0 & 0 & 0 & 0 & 0 & 1/2 & 0 \\ 0 & 0 & 0 & 0 & 0 & 0 & 1/3 \\ - & - & - & - & - & - & - \\ 0 & 0 & 0 & \infty & 0 & 0 & 0 \\ -1 & 0 & 0 & 0 & 0 & 0 & 0 \\ 0 & -1/2 & 0 & 0 & 0 & 0 & 0 \\ 0 & 0 & -1/3 & 0 & 0 & 0 & 0 \end{array}\right]$$

Also, by incorporating the base operating point $\mathbf{I}_\mathrm{b}$, $\mathbf{V}_\mathrm{b}$, $\mathbf{I}_\mathrm{N}$ is obtained as,

$$\mathbf{I}_\mathrm{N} = \mathbf{I}_\mathrm{b} - \mathbf{Y}\mathbf{V}_\mathrm{b} = \left[\begin{array}{c} -0.1114 \\ 0 \\ 0.0371 \\ - \\ 0 \\ 0 \\ 0 \\ 0 \end{array}\right]$$

where

$$\mathbf{Y} = \begin{bmatrix} \mathbf{Y}^{ss} & \mathbf{Y}^{sc} \\ \mathbf{Y}^{cs} & \mathbf{Y}^{cc} \end{bmatrix}$$

and finally the Norton equivalent is given by

$$\mathbf{I} = \mathbf{YV} + \mathbf{I}_{\mathrm{N}} \tag{6.45}$$

6.6 Summary

An alternative harmonic domain linearisation technique which uses only real algebra has been presented in this chapter. It is based on the use of the real Fourier series as opposed to the complex Fourier series employed in Chapter 5. Moreover, a method that uses only real algebra for carrying out self- and mutual convolutions has also been introduced.

The combined use of both concepts, i.e. evaluation of the non-linear function and its re-linearisation, using the real Fourier harmonic domain leads to more efficient iterative harmonic solutions than those using the complex Fourier harmonic domain. Broadly speaking, the same number of operations is required in either domain but in the former domain the operations are real as opposed to complex. Although not explicitly shown in this chapter, when the non-linearity is linearised in the real Fourier harmonic domain around a base operating point, it also leads to a harmonic Norton equivalent representation. Hence, from the electric circuit viewpoint both linearisations are fully equivalent but, arguably, the mathematical elegance of the complex Fourier harmonic domain is self-evident.

A complete hand-worked example has been presented to aid understanding of the theory.

6.7 Bibliography

1. A. Semlyen, E. Acha, J. Arrillaga, "Harmonic Norton Equivalent of the Magnetising Branch of a Transformer", *Proceedings of the IEE*, Part C, Vol. 134, No. 2, March 1987, pp. 162–169.
2. E. Acha, "Modelling of Power System Transformers in the Complex Conjugate Harmonic Space", *PhD Thesis*, University of Canterbury, Christchurch, New Zealand, 1988.
3. J.J. Rico, "Steady State Modelling of Non-linear Power Plant Components", *PhD Thesis*, University of Glasgow, 1997.

7

Hartley Harmonic Domain

7.1 Introduction

The general non-linear relation used in the previous chapter will also be used in this chapter as the starting point for deriving, from first principles, the method of harmonic domain linearisation in the Hartley domain. The non-linear function contains both dynamic elements, $\dot{x}$, and static elements, x, e.g.

$$y = f(x, \dot{x}) \tag{7.1}$$

Section 7.2 shows that direct harmonic domain evaluations of polynomial representations, switching functions included, can be achieved very efficiently using the Hartley domain. Sections 7.3 and 7.4 show how the non-linear dynamic relation (7.1) is linearised in this harmonic domain. For the sake of presentation, a distinction is made between static elements, $y_S = g(x)$, and dynamic elements, $y_D = h(\dot{x})$. The former is covered in Section 7.2 and the latter in Section 7.3. Also, the representation of linear functions, which may be complex, are shown to be easily accommodated in this real harmonic domain.

7.2 Direct Frequency Domain Evaluations

Similarly to the convolution operations carried out in both the complex Fourier and the real Fourier harmonic domains, self- and mutual convolutions can also be carried out in the Hartley harmonic domain, e.g.

$$\mathbf{X} \otimes \mathbf{X} = \mathbf{Y} \tag{7.2}$$

$$\mathbf{Y} \otimes \mathbf{X} = \mathbf{Z} \tag{7.3}$$

where the vectors $\mathbf{X}$, $\mathbf{Y}$ and $\mathbf{Z}$ are now vectors of Hartley harmonic coefficients corresponding to the periodic variables x, x^2 and x^3, respectively.

In the Hartley harmonic domain the product of two periodic, real variables x and y,

$$x(t) = \sum_{i=-\infty}^{\infty} X_i \operatorname{cas} i\nu_0 t \tag{7.4}$$

$$y(t) = \sum_{h=-\infty}^{\infty} Y_h \operatorname{cas} h\nu_0 t \tag{7.5}$$

gives rise to another periodic, real variable, say z,

$$\sum_{k=-\infty}^{\infty} Z_k \operatorname{cas} k\nu_0 t = \sum_{i=-\infty}^{\infty} X_i \operatorname{cas} i\nu_0 t \sum_{h=-\infty}^{\infty} Y_h \operatorname{cas} h\nu_0 t \tag{7.6}$$

In harmonic domain applications it is desirable to express equation (7.6) in vector form using only the harmonic coefficients,

$$\begin{bmatrix} \vdots \\ Z_{-2} \\ Z_{-1} \\ Z_0 \\ Z_1 \\ Z_2 \\ \vdots \end{bmatrix} = \begin{bmatrix} \vdots \\ X_{-2} \\ X_{-1} \\ X_0 \\ X_1 \\ X_2 \\ \vdots \end{bmatrix} \otimes \begin{bmatrix} \vdots \\ Y_{-2} \\ Y_{-1} \\ Y_0 \\ Y_1 \\ Y_2 \\ \vdots \end{bmatrix} \tag{7.7}$$

This operation is more easily visualised in terms of standard matrix operations. The relevant matrix expression is derived below, using the following Hartley identity,

$$\operatorname{cas}\alpha \operatorname{cas}\beta = \frac{1}{2}\left[\operatorname{cas}(\alpha+\beta) + \operatorname{cas}(\alpha-\beta) + \operatorname{cas}(-\alpha+\beta) - \operatorname{cas}(-\alpha-\beta)\right] \tag{7.8}$$

which, when applied to equation (7.6), leads to the following equation

$$\begin{aligned}\sum_{k=-\infty}^{\infty} Z_k \operatorname{cas} k\nu_0 t \quad &= \quad \frac{1}{2}\sum_{i=-\infty}^{\infty} X_i Y_h \left[\operatorname{cas}(i+h)\nu_0 t + \operatorname{cas}(i-h)\nu_0 t\right. \\ &\quad \left. + \operatorname{cas}(-i+h)\nu_0 t - \operatorname{cas}(-i-h)\nu_0 t\right]\end{aligned} \tag{7.9}$$

In order to derive the matrix expression, we start by considering one term at a time in the series expansion (7.9), say the generic term h,

$$\begin{bmatrix} \vdots \\ Z_{-2} \\ Z_{-1} \\ Z_0 \\ Z_1 \\ Z_2 \\ \vdots \end{bmatrix} = \frac{1}{2}\left\{\begin{bmatrix} \vdots \\ X_{-2+h} \\ X_{-1+h} \\ X_{0+h} \\ X_{1+h} \\ X_{2+h} \\ \vdots \end{bmatrix} + \begin{bmatrix} \vdots \\ X_{-2-h} \\ X_{-1-h} \\ X_{0-h} \\ X_{1-h} \\ X_{2-h} \\ \vdots \end{bmatrix} + \begin{bmatrix} \vdots \\ X_{2+h} \\ X_{1+h} \\ X_{0+h} \\ X_{-1+h} \\ X_{-2+h} \\ \vdots \end{bmatrix} - \begin{bmatrix} \vdots \\ X_{2-h} \\ X_{1-h} \\ X_{0-h} \\ X_{-1-h} \\ X_{-2-h} \\ \vdots \end{bmatrix}\right\} Y_h \tag{7.10}$$

or

$$\begin{bmatrix} \vdots \\ Z_{-2} \\ Z_{-1} \\ Z_0 \\ Z_1 \\ Z_2 \\ \vdots \end{bmatrix} = \frac{1}{2}\left\{\begin{bmatrix} \vdots \\ X_{-2+h}+X_{-2-h} \\ X_{-1+h}+X_{-1-h} \\ X_{0+h}+X_{0-h} \\ X_{1+h}+X_{1-h} \\ X_{2+h}+X_{2-h} \\ \vdots \end{bmatrix} + \begin{bmatrix} \vdots \\ X_{2+h}-X_{2-h} \\ X_{1+h}-X_{1-h} \\ X_{0+h}-X_{0-h} \\ X_{-1+h}-X_{-1-h} \\ X_{-2+h}-X_{-2-h} \\ \vdots \end{bmatrix}\right\} Y_h \tag{7.11}$$

After considering all the terms of the input $\mathbf{Y}$ two matrices become fully identified,

$$\begin{bmatrix} \vdots \\ Z_{-2} \\ Z_{-1} \\ Z_0 \\ Z_1 \\ Z_2 \\ \vdots \end{bmatrix} = \frac{1}{2}\left\{\begin{bmatrix} \ddots & & & & & & \\ & 2X_0 & X_{-1}+X_1 & X_{-2}+X_2 & X_{-3}+X_3 & & \\ & X_1+X_{-1} & 2X_0 & X_{-1}+X_1 & X_{-2}+X_2 & X_{-3}+X_3 & \\ & X_2+X_{-2} & X_1+X_{-1} & 2X_0 & X_{-1}+X_1 & X_{-2}+X_2 & \\ & X_3+X_{-3} & X_2+X_{-2} & X_1+X_{-1} & 2X_0 & X_{-1}+X_1 & \\ & & X_3+X_{-3} & X_2+X_{-2} & X_1+X_{-1} & 2X_0 & \\ & & & & & & \ddots \end{bmatrix}\right.$$

$$\left. + \begin{bmatrix} \ddots & & & & & & \\ & & X_{-3}-X_3 & X_{-2}-X_2 & X_{-1}-X_1 & 0 & \\ & X_{-3}-X_3 & X_{-2}-X_2 & X_{-1}-X_1 & 0 & X_1-X_{-1} & \\ & X_{-2}-X_2 & X_{-1}-X_1 & 0 & X_1-X_{-1} & X_2-X_{-2} & \\ & X_{-1}-X_1 & 0 & X_1-X_{-1} & X_2-X_{-2} & X_3-X_{-3} & \\ & 0 & X_1-X_{-1} & X_2-X_{-2} & X_3-X_{-3} & & \\ & & & & & & \ddots \end{bmatrix}\right\} \begin{bmatrix} \vdots \\ Y_{-2} \\ Y_{-1} \\ Y_0 \\ Y_1 \\ Y_2 \\ \vdots \end{bmatrix} \tag{7.12}$$

or, in compact form,

$$\mathbf{Z} = \frac{1}{2}\left\{\mathbf{X}_{\mathrm{I}} + \mathbf{X}_{\mathrm{II}}\right\}\mathbf{Y} \tag{7.13}$$

Equation (7.12) is used to carry out self- and mutual convolutions in the Hartley harmonic domain.

It should be noted that if $\mathbf{X}$ is an even function, i.e. $X_i = X_{-i}$, then $\mathbf{X}_{\mathrm{II}}$ is zero. And if $\mathbf{X}$ is an odd function, i.e. $X_i = -X_{-i}$, then $\mathbf{X}_{\mathrm{I}}$ is a diagonal matrix with entries $2X_0$.

Example 7-1: The polynomial equation given below will be evaluated in the Hartley harmonic domain,

$$i = f(\psi) = 0.001\psi + 0.0743\psi^3$$

around the base excitation,

$$\psi_{\mathrm{b}} = \sin\omega_0 t = \frac{1}{2}\operatorname{cas}\nu_0 t - \frac{1}{2}\operatorname{cas}(-\nu_0 t)$$

or, in vector form,

$$\Psi_{\mathrm{b}} = \frac{1}{2}\begin{bmatrix} 0 \\ -1 \\ 0 \\ 1 \\ 0 \end{bmatrix}$$

The convolution is carried out as follows:

$$\Psi^2 = \Psi \otimes \Psi = \frac{1}{2}\begin{bmatrix} 0 \\ -1 \\ 0 \\ 1 \\ 0 \end{bmatrix} \otimes \frac{1}{2}\begin{bmatrix} 0 \\ -1 \\ 0 \\ 1 \\ 0 \end{bmatrix}$$

or, using equation (7.12), where only matrix $\mathbf{X}_{\mathrm{II}}$ exists

$$\mathbf{X}_{\mathrm{II}} = \begin{bmatrix} & & & -1 & 0 \\ & & -1 & 0 & 1 \\ & -1 & 0 & 1 & \\ -1 & 0 & 1 & & \\ 0 & 1 & & & \end{bmatrix}$$

then

$$\Psi^2 = \frac{1}{2}\begin{bmatrix} & & & -1 & 0 \\ & & -1 & 0 & 1 \\ & -1 & 0 & 1 & \\ -1 & 0 & 1 & & \\ 0 & 1 & & & \end{bmatrix} \frac{1}{2}\begin{bmatrix} 0 \\ -1 \\ 0 \\ 1 \\ 0 \end{bmatrix} = \frac{1}{2}\begin{bmatrix} -1/2 \\ 0 \\ 1 \\ 0 \\ -1/2 \end{bmatrix}$$

Also,

$$\Psi^3 = \Psi^2 \otimes \Psi = \frac{1}{2}\begin{bmatrix} 0 \\ -1/2 \\ 0 \\ 1 \\ 0 \\ -1/2 \\ 0 \end{bmatrix} \otimes \frac{1}{2}\begin{bmatrix} 0 \\ 0 \\ -1 \\ 0 \\ 1 \\ 0 \\ 0 \end{bmatrix}$$

or, in matrix form, using using equation (7.12), where only matrix $\mathbf{X}_{\mathrm{I}}$ exists then

$$\mathbf{X}_{\mathrm{I}} = \begin{bmatrix} 1 & 0 & -1/2 & & & & \\ 0 & 1 & 0 & -1/2 & & & \\ -1/2 & 0 & 1 & 0 & -1/2 & & \\ & -1/2 & 0 & 1 & 0 & -1/2 & \\ & & -1/2 & 0 & 1 & 0 & -1/2 \\ & & & -1/2 & 0 & 1 & 0 \\ & & & & -1/2 & 0 & 1 \end{bmatrix}$$

and in matrix form,

$$\Psi^3 = \frac{1}{2}\begin{bmatrix} 1 & 0 & -1/2 & & & & \\ 0 & 1 & 0 & -1/2 & & & \\ -1/2 & 0 & 1 & 0 & -1/2 & & \\ & -1/2 & 0 & 1 & 0 & -1/2 & \\ & & -1/2 & 0 & 1 & 0 & -1/2 \\ & & & -1/2 & 0 & 1 & 0 \\ & & & & -1/2 & 0 & 1 \end{bmatrix} \frac{1}{2}\begin{bmatrix} 0 \\ 0 \\ -1 \\ 0 \\ 1 \\ 0 \\ 0 \end{bmatrix} = \frac{1}{4}\begin{bmatrix} 1/2 \\ 0 \\ -3/2 \\ 0 \\ 3/2 \\ 0 \\ -1/2 \end{bmatrix}$$

The evaluation of the cubic polynomial equation in this domain becomes readily available,

$$\begin{bmatrix} I_{-3} \\ I_{-2} \\ I_{-1} \\ I_0 \\ I_1 \\ I_2 \\ I_3 \end{bmatrix} = \frac{0.001}{2}\begin{bmatrix} 0 \\ 0 \\ -1 \\ 0 \\ 1 \\ 0 \\ 0 \end{bmatrix} + \frac{0.0743}{8}\begin{bmatrix} 1 \\ 0 \\ -3 \\ 0 \\ 3 \\ 0 \\ -1 \end{bmatrix} = \frac{1}{2}\begin{bmatrix} 0.0186 \\ 0 \\ -0.0567 \\ 0 \\ 0.0567 \\ 0 \\ -0.0186 \end{bmatrix}$$

The current as a function of time is given as

$$i = \frac{1}{2}0.0186\,\mathrm{cas}(-3\nu_0 t) - \frac{1}{2}0.0567\,\mathrm{cas}(-\nu_0 t) + \frac{1}{2}0.0567\,\mathrm{cas}(\nu_0 t) - \frac{1}{2}0.0186\,\mathrm{cas}(3\nu_0 t)$$

Since $\sin\alpha = \frac{1}{2}\,\mathrm{cas}\,\alpha - \frac{1}{2}\,\mathrm{cas}(-\alpha)$ then

$$i = 0.0567\sin\omega_0 t - 0.0186\sin 3\omega_0 t$$

As expected, this result agrees with the results obtained in Examples 5-2 and 6-1, using the complex Fourier and real Fourier domain, respectively.

7.3 Linearisation in the Hartley Harmonic Domain

The non-linear relation

$$y = f(x) \tag{7.14}$$

is taken to be differentiable around the base operating point $(x_{\mathrm{b}}, y_{\mathrm{b}})$ leading to the following linearised expression:

$$\Delta y = \frac{\mathrm{d}f}{\mathrm{d}x}\Delta x \tag{7.15}$$

Furthermore, if the variables in equation (7.14) are periodic they may be expressed in the Hartley harmonic domain as

$$\Delta x = \sum_{h=-\infty}^{\infty} \Delta X_h \,\mathrm{cas}\, h\nu_0 t \tag{7.16}$$

$$\Delta y = \sum_{k=-\infty}^{\infty} \Delta Y_k \,\mathrm{cas}\, k\nu_0 t \tag{7.17}$$

$$\frac{\mathrm{d}f}{\mathrm{d}x} = \sum_{i=-\infty}^{\infty} \xi_i \operatorname{cas} i\nu_0 t \tag{7.18}$$

Substituting equations (7.16), (7.17) and (7.18) into (7.15) gives

$$\sum_{k=-\infty}^{\infty} \Delta Y_k \operatorname{cas} k\nu_0 t = \sum_{i=-\infty}^{\infty} \xi_i \operatorname{cas} i\nu_0 t \sum_{h=-\infty}^{\infty} \Delta X_h \operatorname{cas} h\nu_0 t \tag{7.19}$$

By following a similar line of reasoning as in Section 7.2, a harmonic matrix expression is determined in the Hartley harmonic domain:

$$\Delta \mathbf{Y} = \mathbf{F}' \otimes \Delta \mathbf{X} = \frac{1}{2} \left(\mathbf{F}'_{\mathrm{I}} + \mathbf{F}'_{\mathrm{I}} \right) \Delta \mathbf{X} \tag{7.20}$$

where

$$\Delta\mathbf{X} = \begin{bmatrix} \vdots \\ \Delta X_{-2} \\ \Delta X_{-1} \\ \Delta X_0 \\ \Delta X_1 \\ \Delta X_2 \\ \vdots \end{bmatrix}; \ \Delta\mathbf{Y} = \begin{bmatrix} \vdots \\ \Delta Y_{-2} \\ \Delta Y_{-1} \\ \Delta Y_0 \\ \Delta Y_1 \\ \Delta Y_2 \\ \vdots \end{bmatrix}; \ \mathbf{F}' = \begin{bmatrix} \vdots \\ \xi_{-2} \\ \xi_{-1} \\ \xi_0 \\ \xi_1 \\ \xi_2 \\ \vdots \end{bmatrix}$$

or, in expanded matrix form,

$$\begin{bmatrix} \vdots \\ \Delta Y_{-2} \\ \Delta Y_{-1} \\ \Delta Y_0 \\ \Delta Y_1 \\ \Delta Y_2 \\ \vdots \end{bmatrix} = \frac{1}{2} \left\{ \begin{bmatrix} \ddots & & & & & & \\ & 2\xi_0 & \xi_{-1}+\xi_1 & \xi_{-2}+\xi_2 & \xi_{-3}+\xi_3 & & \\ & \xi_1+\xi_{-1} & 2\xi_0 & \xi_{-1}+\xi_1 & \xi_{-2}+\xi_2 & \xi_{-3}+\xi_3 & \\ & \xi_2+\xi_{-2} & \xi_1+\xi_{-1} & 2\xi_0 & \xi_{-1}+\xi_1 & \xi_{-2}+\xi_2 & \\ & \xi_3+\xi_{-3} & \xi_2+\xi_{-2} & \xi_1+\xi_{-1} & 2\xi_0 & \xi_{-1}+\xi_1 & \\ & & \xi_3+\xi_{-3} & \xi_2+\xi_{-2} & \xi_1+\xi_{-1} & 2\xi_0 & \\ & & & & & & \ddots \end{bmatrix} \right.$$

$$\left. + \begin{bmatrix} \ddots & & & & & & \\ & & \xi_{-3}-\xi_3 & \xi_{-2}-\xi_2 & \xi_{-1}-\xi_1 & 0 & \\ & \xi_{-3}-\xi_3 & \xi_{-2}-\xi_2 & \xi_{-1}-\xi_1 & 0 & \xi_1-\xi_{-1} & \\ & \xi_{-2}-\xi_2 & \xi_{-1}-\xi_1 & 0 & \xi_1-\xi_{-1} & \xi_2-\xi_{-2} & \\ & \xi_{-1}-\xi_1 & 0 & \xi_1-\xi_{-1} & \xi_2-\xi_{-2} & \xi_3-\xi_{-3} & \\ & 0 & \xi_1-\xi_{-1} & \xi_2-\xi_{-2} & \xi_3-\xi_{-3} & & \\ & & & & & & \ddots \end{bmatrix} \right\} \begin{bmatrix} \vdots \\ \Delta X_{-2} \\ \Delta X_{-1} \\ \Delta X_0 \\ \Delta X_1 \\ \Delta X_2 \\ \vdots \end{bmatrix} \tag{7.21}$$

Also, if the linearisation takes place about $(\mathbf{X}_{\mathrm{b}}, \mathbf{Y}_{\mathrm{b}})$ in the harmonic domain, i.e.

$$\Delta \mathbf{X} = \mathbf{X} - \mathbf{X}_{\mathrm{b}}$$
$$\Delta \mathbf{Y} = \mathbf{Y} - \mathbf{Y}_{\mathrm{b}} \tag{7.22}$$

then substitution of these relations into equation (7.20) produces an alternative expression which is entirely suitable for carrying out an iterative solution in the Hartley harmonic domain,

$$\mathbf{Y} = \frac{1}{2}\left(\mathbf{F}'_{\mathrm{I}} + \mathbf{F}'_{\mathrm{II}}\right)\mathbf{X} + \mathbf{Y}_{\mathrm{N}} \tag{7.23}$$

where

$$\mathbf{Y}_{\mathrm{N}} = \mathbf{Y}_{\mathrm{b}} - \frac{1}{2}\left(\mathbf{F}'_{\mathrm{I}} + \mathbf{F}'_{\mathrm{II}}\right)\mathbf{X}_{\mathrm{b}} \tag{7.24}$$

The subscript N indicates the possibility of interpreting this equation as a harmonic Norton equivalent. This will be the case when $\mathbf{Y}$ is a vector of harmonic currents, $\mathbf{X}$ is a vector of harmonic voltages and $\mathbf{F}_{\mathrm{I}}$ and $\mathbf{F}_{\mathrm{II}}$ are matrices of harmonic admittances.

In this domain, the harmonic solution of a single-variable, non-linear function also requires the evaluation of two harmonic vectors, namely $\mathbf{Y}_{\mathrm{b}}$ and $\mathbf{F}'$. It should be noted that $\mathbf{X}_{\mathrm{b}}$ is a given input, $\mathbf{X}$ is a newly determined output and $\mathbf{Y}$ is a vector that allows the interfacing of equation (7.23) with the equation representing the external network. Moreover, if the input vector $\mathbf{X}_b$ is far away from the harmonic solution then the output vector $\mathbf{X}$ becomes the new input vector and the harmonic solution is reached by iteration.

7.4 Dynamic Relations

The linearisation of static elements in the Hartley harmonic domain has been dealt with in the previous section, and now it will be shown how dynamic elements are handled in this domain.

Let us consider the basic relation

$$z = \dot{x} \tag{7.25}$$

between the two periodic variables

$$z(t) = \sum_{h=-\infty}^{\infty} Z_h \operatorname{cas} h\nu_0 t \tag{7.26}$$

$$x(t) = \sum_{n=-\infty}^{\infty} X_n \operatorname{cas} n\nu_0 t \tag{7.27}$$

Equation (7.25) may be written as

$$\sum_{h=-\infty}^{\infty} Z_h \operatorname{cas} h\nu_0 t = \frac{\mathrm{d}}{\mathrm{d}t} \sum_{n=-\infty}^{\infty} X_n \operatorname{cas} n\nu_0 t = \sum_{n=-\infty}^{\infty} n\nu_0 X_n \operatorname{cas}(-n\nu_0 t) \tag{7.28}$$

Alternatively, in terms of harmonic coefficients,

$$\begin{bmatrix} \vdots \\ Z_{-2} \\ Z_{-1} \\ Z_0 \\ Z_1 \\ Z_2 \\ \vdots \end{bmatrix} = \begin{bmatrix} & & & & & \ddots \\ & & & & -2\nu_0 & \\ & & & -\nu_0 & & \\ & & 0 & & & \\ & \nu_0 & & & & \\ 2\nu_0 & & & & & \\ \ddots & & & & & \end{bmatrix} \begin{bmatrix} \vdots \\ X_{-2} \\ X_{-1} \\ X_0 \\ X_1 \\ X_2 \\ \vdots \end{bmatrix} \quad (7.29)$$

The evaluation of dynamic terms in the Hartley domain is carried out by means of simple algebraic operations. Equation (7.29) can be written in compact form as

$$\mathbf{Z} = \mathbf{D}(n\nu_0)\mathbf{X} \quad (7.30)$$

7.5 Linear Elements

It should be noted that a linear function is a particular case of the non-linear function in equation (7.14), e.g.

$$f(x) = y = ax \quad (7.31)$$

and that the harmonic content of the function derivative is a DC-like term, i.e.

$$\frac{\mathrm{d}f}{\mathrm{d}t} = a \quad (7.32)$$

In many applications it is advantageous to express equation (7.30) using only real coefficients, even though the coefficient a may be complex, i.e. $a = g + \mathrm{j}b$. For instance, the fundamental frequency expression of equation (7.31) is

$$\begin{bmatrix} Y_{-1} \\ Y_1 \end{bmatrix} = \begin{bmatrix} g & -b \\ b & g \end{bmatrix} \begin{bmatrix} X_{-1} \\ X_1 \end{bmatrix} \quad (7.33)$$

For the case when a frequency-dependent linear function is to be expressed in the Hartley harmonic domain, equation (7.31) is expressed as

$$\begin{bmatrix} \vdots \\ Y_{-2} \\ Y_{-1} \\ Y_0 \\ Y_1 \\ Y_2 \\ \vdots \end{bmatrix} = \begin{bmatrix} \ddots & & & & & & \ddots \\ & g_2 & & & & -b_2 & \\ & & g_1 & & -b_1 & & \\ & & & g_0 & & & \\ & & b_1 & & g_1 & & \\ & b_2 & & & & g_2 & \\ \ddots & & & & & & \ddots \end{bmatrix} \begin{bmatrix} \vdots \\ X_{-2} \\ X_{-1} \\ X_0 \\ X_1 \\ X_2 \\ \vdots \end{bmatrix} \quad (7.34)$$

The lack of off-diagonal coefficients in the various submatrices provides a perfect de-coupling between the various harmonic frequencies. No interaction takes place between pre-existing harmonics and no new harmonics are generated.

Equation (7.34) can be written in compact form as

$$\begin{bmatrix} \mathbf{Y}_- \\ \mathbf{Y}_+ \end{bmatrix} = \begin{bmatrix} \mathbf{D}(g) & -\mathbf{D}(b) \\ \mathbf{D}(b) & \mathbf{D}(g) \end{bmatrix} \begin{bmatrix} \mathbf{V}_- \\ \mathbf{V}_+ \end{bmatrix} \tag{7.35}$$

Equations (7.23), (7.24), (7.30) and (7.35) provide a suitable means of determining the steady-state solution of general non-linear circuits by an iteration method in the Hartley harmonic domain. Its nodal nature makes it attractive for incorporation into a general harmonic frame of reference where any number of both linear and linearised non-linear components can be represented together to give a unified iterative solution that exhibits quadratic convergence.

Example 7-2: The polynomial equation given below will be evaluated in the Hartley harmonic domain, using per unit notation,

$$i = f(\psi) = 0.001\psi + 0.0743\psi^3$$

around the base excitation,

$$\psi_{\mathrm{b}} = \sin \omega_0 t = \frac{1}{2} \operatorname{cas} \nu_0 t - \frac{1}{2} \operatorname{cas}(-\nu_0 t)$$

The derivative of the polynomial is given as

$$\frac{\mathrm{d}f}{\mathrm{d}\psi} = 0.001 + 0.2229\psi^2$$

and the base current and voltage are

$$i_{\mathrm{b}} = 0.001\psi_{\mathrm{b}} + 0.0743\psi_{\mathrm{b}}^3$$

$$v_{\mathrm{b}} = \dot{\psi}_{\mathrm{b}} \Rightarrow v_{\mathrm{b}} = \frac{1}{2} \operatorname{cas} \nu_0 t + \frac{1}{2} \operatorname{cas}(-\nu_0 t)$$

where $\nu_0 = 1$ p.u.

In vector form the base excitation is

$$\Psi_{\mathrm{b}} = \frac{1}{2} \begin{bmatrix} 0 \\ -1 \\ 0 \\ 1 \\ 0 \end{bmatrix}$$

and the following convolution operations,

$$\Psi^2 = \frac{1}{2} \begin{bmatrix} & & & -1 & 0 \\ & & -1 & 0 & 1 \\ & -1 & 0 & 1 & \\ -1 & 0 & 1 & & \\ 0 & 1 & & & \end{bmatrix} \frac{1}{2} \begin{bmatrix} 0 \\ -1 \\ 0 \\ 1 \\ 0 \end{bmatrix} = \frac{1}{2} \begin{bmatrix} -1/2 \\ 0 \\ 1 \\ 0 \\ -1/2 \end{bmatrix}$$

$$
\Psi^3 = \frac{1}{2}\begin{bmatrix} 1 & 0 & -1/2 & & & & \\ 0 & 1 & 0 & -1/2 & & & \\ -1/2 & 0 & 1 & 0 & -1/2 & & \\ & -1/2 & 0 & 1 & 0 & -1/2 & \\ & & -1/2 & 0 & 1 & 0 & -1/2 \\ & & & -1/2 & 0 & 1 & 0 \\ & & & & -1/2 & 0 & 1 \end{bmatrix} \frac{1}{2}\begin{bmatrix} 0 \\ 0 \\ -1 \\ 0 \\ 1 \\ 0 \\ 0 \end{bmatrix} = \frac{1}{4}\begin{bmatrix} 1/2 \\ 0 \\ -3/2 \\ 0 \\ 3/2 \\ 0 \\ -1/2 \end{bmatrix}
$$

provide the relevant harmonic information for evaluating both polynomial equations in this domain,

$$
\mathbf{F}' = 0.001\begin{bmatrix} 0 \\ 0 \\ 1 \\ 0 \\ 0 \end{bmatrix} + \frac{0.2229}{2}\begin{bmatrix} -1/2 \\ 0 \\ 1 \\ 0 \\ -1/2 \end{bmatrix} = \frac{1}{2}\begin{bmatrix} -0.1114 \\ 0 \\ 0.2249 \\ 0 \\ -0.1114 \end{bmatrix}
$$

$$
\mathbf{I}_b = \frac{0.001}{2}\begin{bmatrix} 0 \\ 0 \\ -1 \\ 0 \\ 1 \\ 0 \\ 0 \end{bmatrix} + \frac{0.0743}{8}\begin{bmatrix} 1 \\ 0 \\ -3 \\ 0 \\ 3 \\ 0 \\ -1 \end{bmatrix} = \begin{bmatrix} 0.0093 \\ 0 \\ -0.0284 \\ 0 \\ 0.0284 \\ 0 \\ -0.0093 \end{bmatrix}
$$

Also, the base excitation may be expressed in vector form, in per unit notation

$$
\mathbf{V}_b = \begin{bmatrix} 0 \\ 0 \\ 0.5 \\ 0 \\ 0.5 \\ 0 \\ 0 \end{bmatrix}
$$

Owing to the nature of the harmonic information in vector $\mathbf{F}'$, the matrix $\mathbf{F}'_{\mathrm{II}} = 0$, and only matrix $\mathbf{F}'_{\mathrm{I}}$ exists. The linearised expression is

$$
\begin{bmatrix} \Delta I_{-3} \\ \Delta I_{-2} \\ \Delta I_{-1} \\ \Delta I_0 \\ \Delta I_1 \\ \Delta I_2 \\ \Delta I_3 \end{bmatrix} =
$$

$$\frac{1}{2}\begin{bmatrix} 0.2249 & 0 & -0.1114 & & & & \\ 0 & 0.2249 & 0 & -0.1114 & & & \\ -0.1114 & 0 & 0.2249 & 0 & -0.1114 & & \\ & -0.1114 & 0 & 0.2249 & 0 & -0.1114 & \\ & & -0.1114 & 0 & 0.2249 & 0 & -0.1114 \\ & & & -0.1114 & 0 & 0.2249 & 0 \\ & & & & -0.1114 & 0 & 0.2249 \end{bmatrix}\begin{bmatrix} \Delta\Psi_{-3} \\ \Delta\Psi_{-2} \\ \Delta\Psi_{-1} \\ \Delta\Psi_{0} \\ \Delta\Psi_{1} \\ \Delta\Psi_{2} \\ \Delta\Psi_{3} \end{bmatrix}$$

or, in terms of harmonic voltages,

$$\begin{bmatrix} \Delta I_{-3} \\ \Delta I_{-2} \\ \Delta I_{-1} \\ \Delta I_{0} \\ \Delta I_{1} \\ \Delta I_{2} \\ \Delta I_{3} \end{bmatrix} =$$

$$\frac{1}{2}\begin{bmatrix} 0.2249 & 0 & -0.1114 & & & & \\ 0 & 0.2249 & 0 & -0.1114 & & & \\ -0.1114 & 0 & 0.2249 & 0 & -0.1114 & & \\ & -0.1114 & 0 & 0.2249 & 0 & -0.1114 & \\ & & -0.1114 & 0 & 0.2249 & 0 & -0.1114 \\ & & & -0.1114 & 0 & 0.2249 & 0 \\ & & & & -0.1114 & 0 & 0.2249 \end{bmatrix}$$

$$\times \begin{bmatrix} & & & & & & -1/3 \\ & & & & & -1/2 & \\ & & & & -1 & & \\ & & & \infty & & & \\ & & 1 & & & & \\ & 1/2 & & & & & \\ 1/3 & & & & & & \end{bmatrix}\begin{bmatrix} \Delta V_{-3} \\ \Delta V_{-2} \\ \Delta V_{-1} \\ \Delta V_{0} \\ \Delta V_{1} \\ \Delta V_{2} \\ \Delta V_{3} \end{bmatrix}$$

or

$$\begin{bmatrix} \Delta I_{-3} \\ \Delta I_{-2} \\ \Delta I_{-1} \\ \Delta I_{0} \\ \Delta I_{1} \\ \Delta I_{2} \\ \Delta I_{3} \end{bmatrix} =$$

$$\begin{bmatrix} & & & & 0.0557 & 0 & -0.0375 \\ & & & \infty & 0 & -0.0562 & 0 \\ & & -0.0557 & \infty & -0.1124 & 0 & 0.0186 \\ & -0.0278 & 0 & \infty & 0 & 0.0278 & \\ -0.0186 & 0 & 0.1124 & \infty & 0.0557 & & \\ 0 & 0.0562 & 0 & \infty & & & \\ 0.0375 & 0 & -0.0557 & & & & \end{bmatrix}\begin{bmatrix} \Delta V_{-3} \\ \Delta V_{-2} \\ \Delta V_{-1} \\ \Delta V_{0} \\ \Delta V_{1} \\ \Delta V_{2} \\ \Delta V_{3} \end{bmatrix}$$

Also, by incorporating the base vectors $\mathbf{I}_b$ and $\mathbf{V}_b$ into this relationship we obtain the Norton equivalent representation,

$$\begin{bmatrix} I_{-3} \\ I_{-2} \\ I_{-1} \\ I_0 \\ I_1 \\ I_2 \\ I_3 \end{bmatrix} = \begin{bmatrix} & & & & 0.0557 & 0 & -0.0375 \\ & & & \infty & 0 & -0.0562 & 0 \\ & & -0.0557 & \infty & -0.1124 & 0 & 0.0186 \\ & -0.0278 & 0 & \infty & 0 & 0.0278 & \\ -0.0186 & 0 & 0.1124 & \infty & 0.0557 & & \\ 0 & 0.0562 & 0 & \infty & & & \\ 0.0375 & 0 & -0.0557 & & & & \end{bmatrix} \times \begin{bmatrix} V_{-3} \\ V_{-2} \\ V_{-1} \\ V_0 \\ V_1 \\ V_2 \\ V_3 \end{bmatrix} + \begin{bmatrix} -0.0186 \\ 0 \\ 0.0557 \\ 0 \\ -0.0557 \\ 0 \\ 0.0186 \end{bmatrix}$$

7.6 Computational Advantage

The use of the Hartley harmonic domain has several advantages over the complex Fourier harmonic domain, particularly when applied to the solution of large systems. For instance, to calculate $\Delta\mathbf{I}$ from (5.12) requires $4(2h+1)(h+1)$ real multiplications and $(2h+1)^2$ to calculate $\Delta\mathbf{Y}$ from (7.20) since $\Delta I_h = \Delta I^*_{-h}$. Figure 7.1 illustrates this comparison graphically.

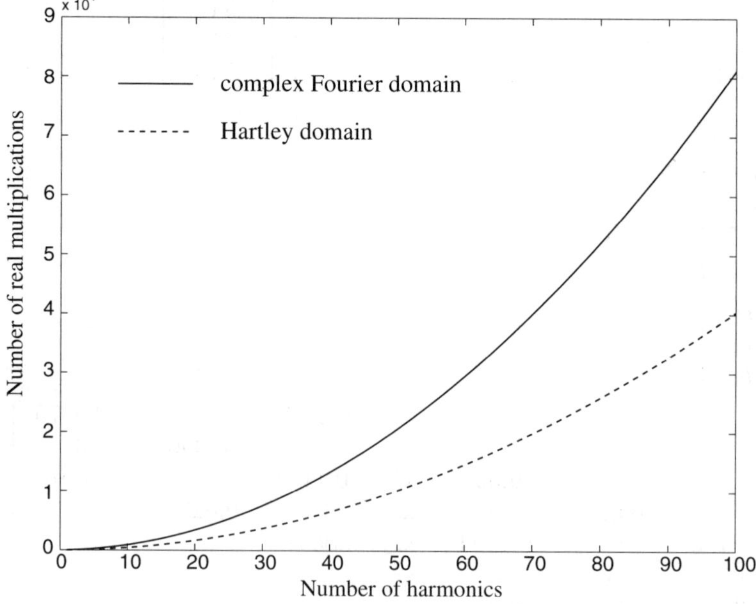

Figure 7.1 Comparison of the number of operations required in complex Fourier and Hartley domains

7.7 General Remarks about the Hartley Harmonic Domain

It should be noted that owing to the real nature of the Hartley transform, matrix $\mathbf{F}'_{\text{I}}$ is real and symmetric. It has a band-diagonal Toeplitz structure and it is represented by the even part of vector $\mathbf{F}'$. Matrix $\mathbf{F}'_{\text{II}}$ is also real but it is asymmetric. It has a reversed band-diagonal structure and it is represented by the odd part of the vector $\mathbf{F}'$. It should also be noted that if the driving function is an odd function then matrix $\mathbf{F}'_{\text{II}}$ will exist but not $\mathbf{F}'_{\text{I}}$. Conversely, if the driving function is an even function then matrix $\mathbf{F}'_{\text{I}}$ will exist but not $\mathbf{F}'_{\text{II}}$. In the more general case, when the driving function contains even and odd parts then both matrices will exist, rendering the overall matrix $\mathbf{F}'$ asymmetric, i.e. $\mathbf{F}' = \frac{1}{2}\{\mathbf{F}'_{\text{I}} + \mathbf{F}'_{\text{II}}\}$. In any event, the admittance matrix $\mathbf{Y} = \frac{1}{2}\{\mathbf{F}'_{\text{I}} + \mathbf{F}'_{\text{II}}\}\mathbf{D}^{-1}(n\nu_0)$ will be asymmetric due to the effect introduced by the dynamic terms, $\mathbf{D}(n\nu_0)$.

7.8 Summary

This chapter has introduced an alternative, harmonic domain linearisation technique, which is based on the use of the Hartley transform. Similarly to the real Fourier harmonic domain, only real algebra is required for conducting all operations. This includes a method for polynomial evaluation using self- and mutual convolutions and the relinearisation process itself. The combined use of both concepts leads to more efficient iterative harmonic solutions than those achieved using the complex Fourier harmonic domain.

Polynomial evaluation using the matrix form of the Hartley harmonic domain requires the same number of operations as those using the Fourier harmonic domain. However, one operation in the Hartley, which is real, will correspond to one complex operation in the complex Fourier, i.e. four real products plus two additions. Arguably, this makes the Hartley domain almost six times more efficient than the complex Fourier domain.

It should be noted that in actual harmonic applications these savings may be quite significant as the evaluation of the harmonic Norton equivalent of each non-linearity involves repeated convolutions and the multiplication of vectors and matrices of relatively large dimensions. Moreover, re-evaluation of the harmonic Norton equivalent is required at every iterative step. Also, the storage requirements of the harmonic Norton equivalent in the Hartley domain are only half of those in the Fourier domain but the property of complex conjugate, in the Fourier domain, can reduce the storage and operation requirements. It is likely, however, that when solving a large-scale network the overall memory storage and computational gains will be less than those indicated above, since the representation of linear elements in the Hartley domain is not much more efficient than in the complex Fourier harmonic domain.

The comparison of the Hartley harmonic domain with the real Fourier harmonic domain is not quite as advantageous for the Hartley as in the analysis above. In the general case, polynomial evaluation using the matrix form of the real Fourier harmonic domain requires a similar number of operations than those required by the Hartley. Note from Example 6-1 the very interesting pattern that the matrix form of the convolution takes in the real Fourier harmonic domain. When the driving function contains no even harmonic terms then only one of the four submatrices is used at a time, making this domain slightly faster that the Hartley. In both cases, all the operations take place in the real plane.

On aspect where the Hartley harmonic domain looks unassailable, notwithstanding its efficiency, is its mathematical elegance. Neither flavours of the Fourier harmonic domain look as attractive as the Hartley domain. Following the same spirit of the previous chapter, a complete hand-worked example has been presented to aid understanding of the theory.

7.9 Bibliography

1. G.T. Heydt, *Electric Power Quality*, Stars in a Circle Publications, Scottsdale, AZ, 1991.
2. G.T. Heydt, "System Analysis Using Hartley Impedance", *IEEE Transactions on Power Delivery*, Vol. 8, No. 2, April 1993, pp. 518–523.
3. K.J. Olejniczak, G.T. Heydt, "Scanning the Special Section on the Hartley Transform", *Proceedings of the IEEE*, Vol. 82, No. 3, March 1994, pp. 372–380.
4. M. Madrigal, "Power System Modelling in the Harmonic Hartley Domain" (in Spanish), *MSc Thesis*, Universidad Autónoma de Nuevo León, Mexico, 1996.
5. J.J. Rico, "Steady State Modelling of Non-Linear Power Plant Components", *PhD Thesis*, University of Glasgow, 1997.
6. E. Acha, J.J. Rico, S. Acha, M. Madrigal, "Harmonic Domain Modelling in Hartley's Domain with Particular Reference to Three Phase Thyristor-Controlled Reactors", *IEEE Transactions on Power Delivery,* Vol. 12, No. 4, October 1997, pp. 1622–1628.
7. J.J. Rico, E. Acha, "Analysis of Linear Time-varying Systems via Hartley Series", *International Journal of Systems Science,* Vol. 29, No. 6, 1998, pp. 541–549.

8

Iterative Solutions of Non-linear Power Plant Components

8.1 Introduction

Harmonic distortion in the power network is on the increase. The tendency of electronics-based plant components and rotating machinery to inject harmonic distortion into the AC network is aggravated by the geometric imbalances present in the network. Magnetic non-linearities acting under saturating conditions and electric arc furnaces may also play an important part in aggravating the problem. Harmonic distortion at the terminals of the power plant component from which the control signals come is likely to upset the control mechanism of the plant component which could give rise to additional harmonic distortion. In these circumstances harmonic instabilities may then occur. Accordingly, harmonic assessment tools with strong convergence characteristics, capable of modelling a wide range of non-linear plant components, will become a valuable resource in both the planning and the operation of future power networks.

In principle at least, the periodic response of a non-linear circuit can be determined by integrating the differential equations that describe the circuit until the transient response dies out, i.e.

$$y = f(x, \dot{x}) \tag{8.1}$$

However, this approach becomes unyielding whenever the transients are governed by time constants that are much larger than the period of the driving force. In such cases it is necessary to carry out numeric integrations for a considerable number of periods before the transient becomes small enough to be ignored. This undesirable characteristic of non-linear circuits has discouraged engineers from using straightforward integration techniques when finding the steady-state response of non-linear circuits. Instead, several efficient solution techniques have been developed to determine the periodic solution of non-linear circuits. In broad terms, they can be classified into harmonic balance methods and shooting methods. Harmonic balance methods solve the non-linear problem to a specified accuracy by employing steady-state, iterative techniques. Shooting methods take the approach of integrating the dynamic equations for one or two full periods and then solving a two-point boundary problem by resorting to iterative techniques.

Modern harmonic balance methods solve the non-linear problem by resorting to Newton-type iterative techniques. They have proved very popular owing to their strong convergence characteristics. They are based on linearisation of the non-linear characteristic around a base operating point, where the linearisation process takes place in the harmonic domain.

For most practical purposes a non-linear relation that can be expanded in a Taylor series is amenable to an iterative solution via the Newton–Raphson method. Higher order terms are neglected in the Taylor series expansion and the resulting linearised equations are used to find a better estimate of the solution. For instance, the expansion of (8.1) in a form suitable for a Newton–Raphson solution is

$$f\left((x,\dot{x})+\Delta(x,\dot{x})\right)=f^{(\mathrm{k})}(x,\dot{x})+\frac{\partial f(x,\dot{x})}{\partial(x,\dot{x})}\Delta(x,\dot{x}) \tag{8.2}$$

or

$$\Delta y=\frac{\partial f(x,\dot{x})}{\partial(x,\dot{x})}\Delta(x,\dot{x}) \tag{8.3}$$

where $\Delta y = f\left((x,\dot{x})+\Delta(x,\dot{x})\right) - f^{(\mathrm{k})}(x,\dot{x})$ and k is an iteration counter.

8.2 Harmonic Domain Newton–Raphson Techniques

If the variables x and y in equation (8.1) are periodic then equation (8.3) can be expressed in the harmonic domain as

$$\Delta\mathbf{Y}=\mathbf{J}\Delta\mathbf{X} \tag{8.4}$$

where $\mathbf{J}$ is a harmonic Jacobian matrix containing first-order partial derivatives of the non-linear function $y = f(x,\dot{x})$ with respect to the harmonic coefficients of x.

By recognising that the linearisation has taken place about $\mathbf{X}^{(0)}$, $\mathbf{Y}^{(0)}$ in the harmonic domain then

$$\begin{aligned}\Delta\mathbf{X}&=\mathbf{X}-\mathbf{X}^{(0)}\\ \Delta\mathbf{Y}&=\mathbf{Y}-\mathbf{Y}^{(0)}\end{aligned} \tag{8.5}$$

Substituting equations (8.5) into equation (8.4) produces an alternative expression which is better suited for carrying out iterative solutions:

$$\mathbf{Y}=\mathbf{J}\mathbf{X}+\mathbf{Y}_{\mathrm{E}} \tag{8.6}$$

where

$$\mathbf{Y}_{\mathrm{E}}=\mathbf{Y}^{(0)}-\mathbf{J}\mathbf{X}^{(0)} \tag{8.7}$$

These equations can be used for interfacing the non-linear component with the external system and for building linearised representations of composite non-linear components. It should be noted that the external system may contain both linear and linearised components. In equations (8.6) and (8.7), $\mathbf{X}^{(0)}$ is a given input, $\mathbf{X}$ is the newly determined output and $\mathbf{Y}$ is a vector that allows the interfacing of equation (8.6) with the external system.

From equations (8.6) and (8.7) it is apparent that the harmonic domain solution of a single-variable, non-linear function requires the evaluation of two vectors. The vector $\mathbf{Y}^{(0)}$ corresponds to the evaluation of the original function $f(\cdot)$ whilst the second vector corresponds to the evaluation of the derivative function $f'(\cdot)$. The harmonic coefficients of $f'(\cdot)$ are used to assemble the Jacobian matrix.

In most practical situations the input vector $\mathbf{X}^{(0)}$ will be far away from the harmonic solution and, hence, the output vector $\mathbf{X}^{(1)}$ is unlikely to be close enough to the actual harmonic solution. However, this output vector $\mathbf{X}^{(1)}$ may still be used as the input vector in a new solution of equations (8.6) and (8.7), leading to an output vector $\mathbf{X}^{(2)}$ which should be closer to the actual solution than $\mathbf{X}^{(1)}$. This iterative pattern is repeated until the absolute difference between two consecutive iterations, say k and k − 1, is smaller than a pre-specified tolerance, i.e. $\left|\mathbf{X}^{(\mathrm{k})} - \mathbf{X}^{(\mathrm{k}-1)}\right| \leq \epsilon$. Harmonic solutions are reached within five to six iterations when the Newton–Raphson method is employed.

Figure 8.1 represents diagrammatically the computational procedures used in harmonic domain calculations. The overall procedure is harmonic domain independent, i.e. Fourier or Hartley harmonic domains can be used. The α, β and γ blocks are alternative routes for harmonic domain evaluations involving non-linearities whilst the δ blocks correspond to the actual harmonic domain Newton–Raphson method.

Some non-linear components, such as electric arcs, fluorescent lamps and magnetic non-linearities, are amenable to a polynomial representation, and route α is preferred for their harmonic evaluation. Other non-linearities, such as power converters and thyristor-controlled reactors, are better represented by means of switching functions and route β is preferred for their harmonic evaluation. Non-linear relations which are not suitably described by either polynomial or switching function approximations can always be represented by a set of (x, y) points. In this situation, route γ represents a very powerful means for performing harmonic domain evaluations.

Harmonic domain evaluations are at the core of the Newton–Raphson algorithm given in Figure 8.1. The polynomial and switching function evaluations in blocks α_1, α_2 and β_1 are achieved very efficiently by means of self- and mutual convolutions. The block β_2 is not executed since the switching vector is evaluated in block β_1, and the derivative function is linearly dependent. The point by point representation of non-linear functions in blocks γ_1 and γ_2 is achieved by means of linear interpolation, the central difference formula and FFT procedures. A full cycle of the fundamental frequency is obtained point-by-point and then an FFT is applied to find the harmonic content of the resultant waveform.

Block δ_1 represents the calculation of the mismatch vector whilst block δ_3 represents the calculation of the Jacobian matrix. The set of linearised equations and the updated state variables are calculated in block δ_4. Sparsity techniques can be used to reduce storage and CPU time requirements in this block. Blocks δ_2, δ_5 and δ_6 are merely control points in the iterative process. The harmonic elements of the first derivative, calculated in block α_2, block β_1 or block γ_2, are used to build the Jacobian matrix.

The general flow chart of Figure 8.1 shows the case when the Jacobian matrix is updated at each iterative step. An alternative approach is to keep the Jacobian constant after the first or some later iterations. It has been found that for weak non-linearities, as in the case of lumped iron cores, this reduces the total execution time, whilst for strong non-linearities, as in the case of laminated iron cores, the execution time will normally be increased. The use of a quasi-Newton–Raphson method based on diagonal relaxation is also a very attractive alternative which retains the Newton's strong convergence whilst reducing the CPU time requirements quite considerably.

These alternative methods and their convergence properties are further addressed in this chapter.

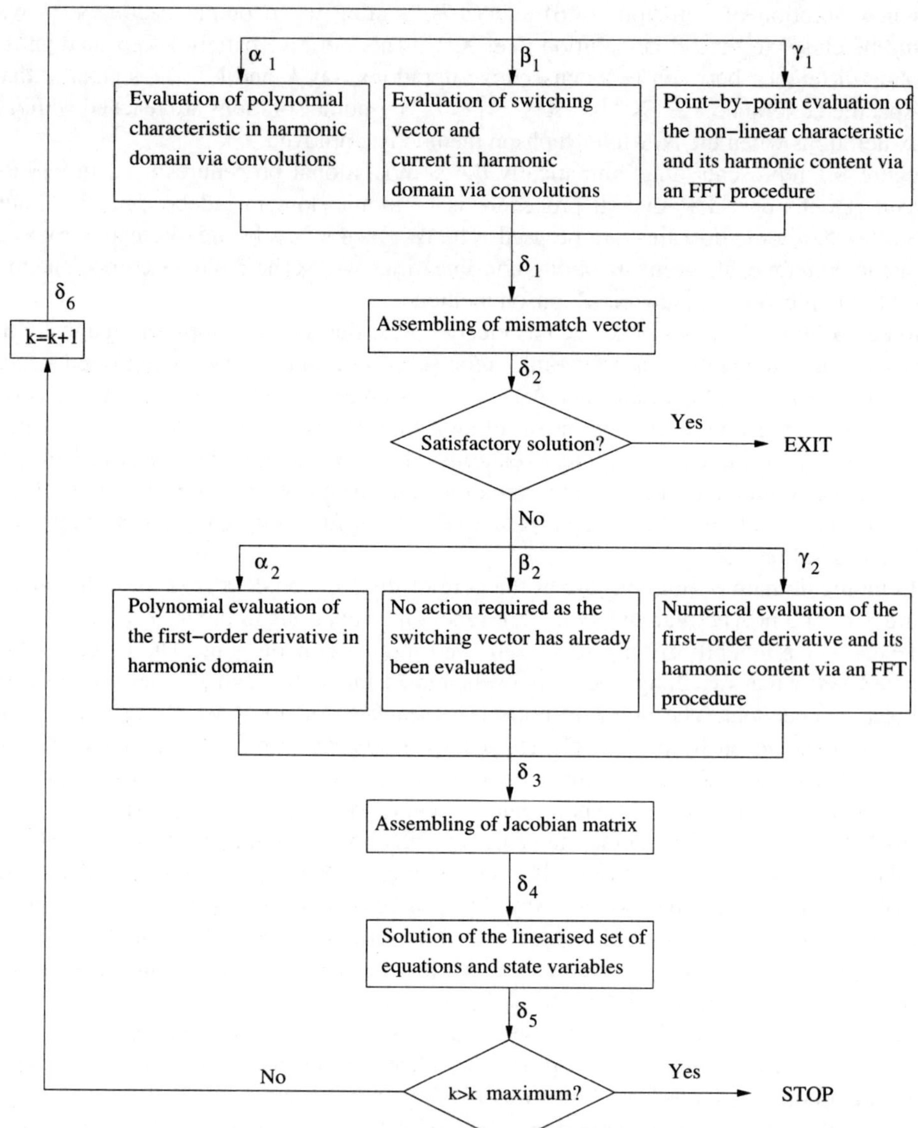

Figure 8.1 Harmonic domain calculations via the Newton–Raphson method

8.2.1 Point-by-point evaluation of the non-linear characteristic

Once the operating voltage and its associated flux has been established, a period is subdivided into $N = 2^n$ time steps, Δt, such that the highest harmonic will be sufficiently well sampled. The discretised flux is impressed upon the experimental characteristic and a corresponding magnetising current determined. This procedure is better appreciated from Figure 8.2.

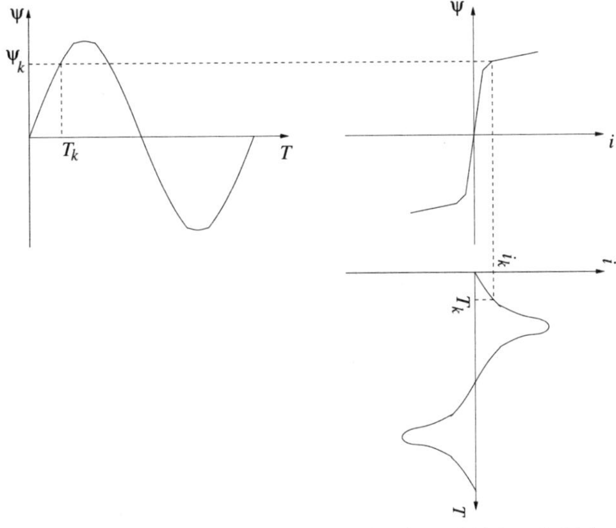

Figure 8.2 Point-by-point determination of $i(t)$ and $\psi(t)$

Provided that the magnetising current characteristic is smooth, its derivative with respect to the magnetising flux can be obtained simply and accurately using the central difference algorithm.

Let t_k be a discrete time in the selected interval and ψ_k and i_k the corresponding values of the magnetising characteristic as shown in Figure 8.2. The derivative for this particular point is given as

$$f'_k = \frac{i_{k+1} - i_{k-1}}{\psi_{k+1} - \psi_{k-1}} \tag{8.8}$$

The following MATLAB™ function is used to obtain the harmonic vector of the derivative f'_k using the point-by-point procedure:

```
function dF=der_point_point(flux,cur,N,h)
% dF : d(flux)/d(cur), i.e. dF=[-h ... -1 0 1 ... h]
% flux : points in the time domain, i.e. x(t)
% cur : point in the time domain, i.e. cur=f(x(t))
% N : number of points
% h : number of harmonics
for t=1:(N-1)
  dflux(t)=(cur(t+1)-cur(t))/(flux(t+1)-flux(t));
end
dflux(N)=dflux(1);
dF=fft(dflux)/N;
dF=fftshift(dF);
```

```
x=round(N/2)-N/2;
if x==0.5
   centre=round(N/2);
else
   centre=round(N/2)+1;
end
dF=dF(centre-h:centre+h);
```

Example 8-1: The point-by-point evaluation procedure is used to derive the harmonic content of the derivative of $i = f(\psi) = 0.001\psi + 0.0743\psi^3$, with respect to the operating point $\psi_b = \sin\omega_0 t$,

$$\mathbf{F}' = \begin{bmatrix} 0 \\ -0.0557 + j0.0017 \\ 0 \\ 0.1124 \\ 0 \\ -0.0557 - j0.0017 \\ 0 \end{bmatrix}$$

This result compares well with that of Example 5-1.

The next MATLAB™ steps are used for this example:

```
>> f=50;
>> w=2*pi*f;
>> t=0:0.0001:(1/f-0.0001);
>> flux=sin(w*t);
>> cur=0.001*flux+0.0743*flux.^3;
>> h=3;
>> size(cur)
ans =
1 200
>> dF=der_point_point(flux,cur,200,h);
>> transpose(dF)
ans =
 0.0000 - 0.0000i
-0.0557 + 0.0017i
-0.0000 + 0.0000i
 0.1124
-0.0000 - 0.0000i
-0.0557 - 0.0017i
 0.0000 + 0.0000i
```

8.3 The Non-linear Inductor in the Harmonic Domain

An electric circuit containing a non-linear inductor will be used to illustrate the convergence properties of the harmonic Newton–Raphson method and derived formulations. Hence, the dynamic equations of the non-linear inductor shown in Figure 8.3 will be linearised in the harmonic domain first.

The dynamic equation of the non-linear inductor is

$$Ri(t) + L\frac{di(t)}{dt} + N\frac{d\psi(t)}{dt} = v(t) \tag{8.9}$$

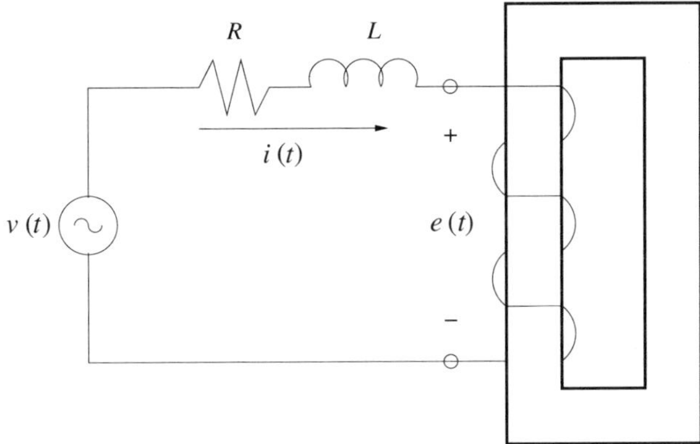

Figure 8.3 Non-linear inductor excited from a periodic voltage source

where

$$N\frac{\mathrm{d}\psi(t)}{\mathrm{d}t} = e(t)$$

Accordingly, the non-linear state equations of the inductor may be written as

$$\dot{\phi} = -(R + \mathrm{j}\omega_0 L)f(\phi) + v \tag{8.10}$$

$$i = f(\phi) \tag{8.11}$$

where equation (8.11) is the magnetising characteristic of the non-linear inductor, N is the number of turns and $\phi = N\psi$ is the total flux linkage.

For harmonic domain calculation purposes the state equations (8.10) and (8.11) must be transformed into algebraic equations. The linearised form of these equation is

$$\mathbf{D}(\mathrm{j}\omega_0 h)\Delta\mathbf{\Phi} = -\mathbf{D}(R + \mathrm{j}\omega_0 hL)\mathbf{F}_{\Phi}\Delta\mathbf{\Phi} + \Delta\mathbf{V} \tag{8.12}$$

$$\Delta\mathbf{I} = \mathbf{F}_{\Phi}\Delta\mathbf{\Phi} \tag{8.13}$$

where $\mathbf{F}_{\Phi}$ is a Toeplitz matrix corresponding to partial derivatives of equation (8.11) with respect to the elements of the harmonic domain vector Φ, $\mathbf{D}(\mathrm{j}\omega_0 h)$ is a diagonal matrix with entries $\mathrm{j}\omega_0 h$ and $\mathbf{D}(R{+}\mathrm{j}\omega_0 hL)$ is a diagonal matrix with entries $R{+}\mathrm{j}\omega_0 hL$. $\Delta\mathbf{V}$, $\Delta\mathbf{\Phi}$ and $\Delta\mathbf{I}$ are vectors of harmonic voltages, fluxes and currents, respectively.

Equations (8.12) and (8.13) can be combined together to give rise to the following input–output relationship:

$$\Delta\mathbf{I} = \mathbf{H}_V\Delta\mathbf{V} \tag{8.14}$$

where $\mathbf{H}_V = \mathbf{F}_{\Phi}\left\{\mathbf{D}(\mathrm{j}\omega_0 h) + \mathbf{D}(R + \mathrm{j}\omega_0 hL)\mathbf{F}_{\Phi}\right\}^{-1}$ is a matrix of harmonic admittances which for the case of an ideal, non-linear inductor reduces to $\mathbf{H}_V = \mathbf{F}_{\Phi}\mathbf{D}^{-1}(\mathrm{j}\omega_0 h)$.

The harmonic Norton equivalent representation of the non-linear inductor shown in Figure 8.4 is realised by incorporating the base operating point $\mathbf{I}^{(0)}$, $\mathbf{V}^{(0)}$ in the linearised equation (8.14),

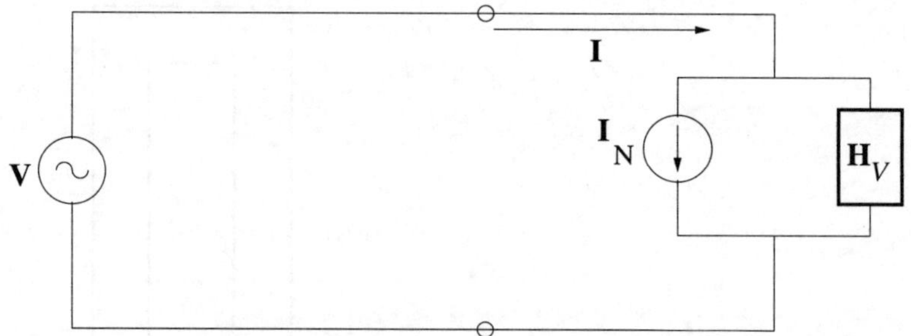

Figure 8.4 Harmonic Norton equivalent of the non-linear inductor

$$\mathbf{I} = \mathbf{H}_V \mathbf{V} + \mathbf{I}_\mathrm{N} \tag{8.15}$$

where $\mathbf{I}_\mathrm{N} = \mathbf{I}^{(0)} - \mathbf{H}_V \mathbf{V}^{(0)}$ is the current source of the harmonic Norton equivalent.

Equation (8.15) possesses several distinct properties. For one, it provides a means for the harmonic solution of the non-linear inductor by iteration. Full evaluation of equation (8.15) at each iterative step produces a solution via the Newton–Raphson method, where the nodal admittance matrix $\mathbf{H}_V$ plays the role of a Jacobian. When close to the solution, the algorithm's rate of convergence is quadratic, i.e. the number of significant figures doubles at each iteration. Furthermore, the equation's structure allows for an efficient interfacing with frequency-dependent transmission line admittances and also with other harmonic Norton equivalents. The interfacing takes place in a harmonic domain frame of reference where all the harmonics, cross-couplings between harmonics, phases and busbars of the network are explicitly represented. The overall nodal admittance matrix of the network plays the role of a global Jacobian matrix.

For harmonic calculation purposes the non-linear inductor of Figure 8.3 is sometimes modelled as the Norton equivalent of an ideal, non-linear inductor in series with a linear admittance, as shown in Figure 8.5.

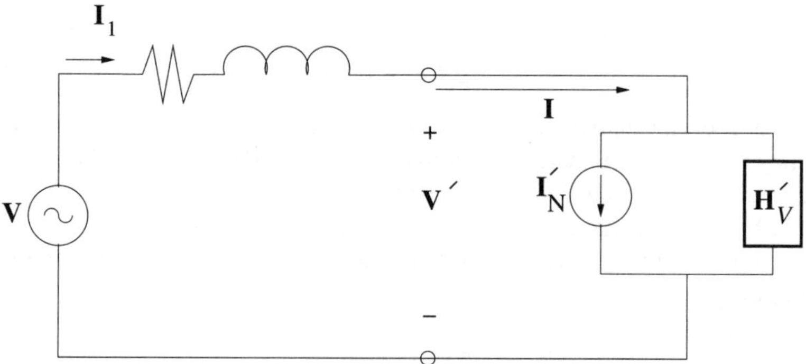

Figure 8.5 Harmonic Norton equivalent of the ideal, non-linear inductor in series with the leakage admittance

This representation allows easy access to the internal (fictitious) inductor's node, a point where the core flux is proportional to the circuit voltage, i.e. $\mathbf{D}^{-1}(\mathrm{j}\omega_0 h)\Phi = \mathbf{V}'$. The following nodal matrix equation represents the circuit of Figure 8.5:

$$\begin{bmatrix} \mathbf{I}_1 \\ -\mathbf{I}'_{\mathrm{N}} \end{bmatrix} = \begin{bmatrix} \mathbf{D}^{-1}(R+\mathrm{j}\omega_0 hL) & -\mathbf{D}^{-1}(R+\mathrm{j}\omega_0 hL) \\ -\mathbf{D}^{-1}(R+\mathrm{j}\omega_0 hL) & \mathbf{D}^{-1}(R+\mathrm{j}\omega_0 hL)+\mathbf{H}'_V \end{bmatrix} \begin{bmatrix} \mathbf{V} \\ \mathbf{V}' \end{bmatrix} \tag{8.16}$$

where $\mathbf{V}$ is a constant, known value and the solution of the harmonic vectors $\mathbf{V}'$ and $\mathbf{I}_1$ is reached by iteration.

8.4 Harmonic Domain Convergence Characteristics

The transmission system of Figure 8.6 is used to illustrate the convergence properties of the Newton-type and Gauss-type methods used in harmonic domain calculations. Figure 8.7 shows its harmonic equivalent circuit and equation (8.17) gives its mathematical description

$$\begin{bmatrix} \mathbf{I}_1 \\ -\mathbf{I}_{\mathrm{N}} \\ 0 \end{bmatrix} = \begin{bmatrix} \mathbf{Y}_{\mathrm{T}} & -\mathbf{Y}_{\mathrm{T}} & 0 \\ -\mathbf{Y}_{\mathrm{T}} & \mathbf{Y}_{\mathrm{T}}+\mathbf{H}_V+\mathbf{Y}_L+\mathbf{Y}_C & -\mathbf{Y}_L \\ 0 & -\mathbf{Y}_L & \mathbf{Y}_L+\mathbf{Y}_C \end{bmatrix} \begin{bmatrix} \mathbf{V}_1 \\ \mathbf{V}_2 \\ \mathbf{V}_3 \end{bmatrix} \tag{8.17}$$

where

$$\begin{aligned} \mathbf{Y}_{\mathrm{T}} &= \mathbf{D}^{-1}(\mathrm{j}h0.0326) \\ \mathbf{Y}_L &= \mathbf{D}^{-1}(0.0265+\mathrm{j}h0.0626) \\ \mathbf{Y}_C &= \mathbf{D}(\mathrm{j}h0.2373) \end{aligned}$$

with $\mathbf{H}_V$ and $\mathbf{I}_{\mathrm{N}}$ representing the elements of the harmonic Norton equivalent of the magnetising branch.

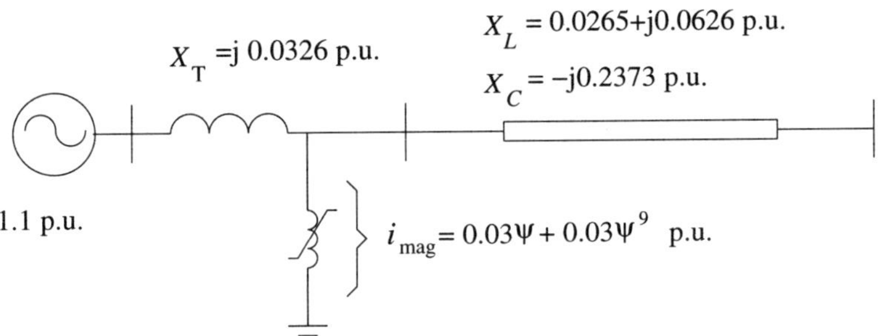

Figure 8.6 Transmission system

Since the Norton equivalent is evaluated at a fix operating point, equation (8.17) must be solved by iteration to update the harmonic Norton equivalent in order to reach the correct solution. The way in which the harmonic Norton equivalent is updated has a bearing on the convergence characteristic of the iterative method.

If a Newton–Raphson algorithm is used to solve (8.17), then the Norton equivalent is updated at each iteration, resulting in a known current source $\mathbf{I}_{\mathrm{N}}$ and a constant admitance matrix $\mathbf{H}_V$ obtained from the past iteration. Since $\mathbf{V}_1$ is a constant voltage source, then Kron's reduction may be used to solve for the unknown variables $\mathbf{I}_1$, $\mathbf{V}_2$ and $\mathbf{V}_3$, i.e.

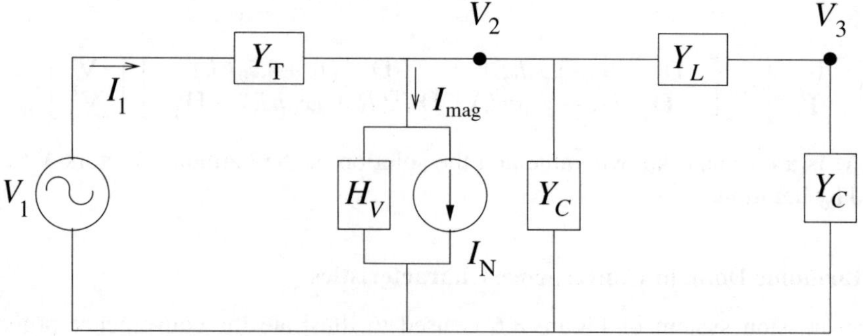

Figure 8.7 Transmission system equivalent in the harmonic domain

$$\begin{bmatrix} \mathbf{V}_1 \\ -\mathbf{I}_{\mathrm{N}} \\ 0 \end{bmatrix} = \begin{bmatrix} \mathbf{Z}_{\mathrm{T}} & \mathbf{U} & 0 \\ -\mathbf{U} & \mathbf{H}_V + \mathbf{Y}_L + \mathbf{Y}_C & -\mathbf{Y}_L \\ 0 & -\mathbf{Y}_L & \mathbf{Y}_L + \mathbf{Y}_C \end{bmatrix} \begin{bmatrix} \mathbf{I}_1 \\ \mathbf{V}_2 \\ \mathbf{V}_3 \end{bmatrix} \tag{8.18}$$

where $\mathbf{U}$ is the identity matrix, and $\mathbf{Z}_{\mathrm{T}} = \mathbf{Y}_{\mathrm{T}}^{-1}$.

The initial values used to evaluate the harmonic Norton equivalent are obtained with no magnetic branch included and $V_1 = 1.1$ p.u. Using these conditions give the following voltages at iteration zero,

$$\mathbf{V}_1^{(0)} = \begin{bmatrix} 0 \\ 0 \\ 0 \\ 0 \\ 0 \\ 0 \\ 0.55 \\ 0 \\ 0.55 \\ 0 \\ 0 \\ 0 \\ 0 \\ 0 \\ 0 \end{bmatrix}; \; \mathbf{V}_2^{(0)} = \begin{bmatrix} 0 \\ 0 \\ 0 \\ 0 \\ 0 \\ 0 \\ 0.8072 + \mathrm{j}0.0328 \\ 0 \\ 0.8072 - \mathrm{j}0.0328 \\ 0 \\ 0 \\ 0 \\ 0 \\ 0 \\ 0 \end{bmatrix}$$

$$\mathbf{V}_3^{(0)} = \begin{bmatrix} 0 \\ 0 \\ 0 \\ 0 \\ 0 \\ 0 \\ 1.0652 + \mathrm{j}0.2062 \\ 0 \\ 1.0652 - \mathrm{j}0.2062 \\ 0 \\ 0 \\ 0 \\ 0 \\ 0 \\ 0 \end{bmatrix} ; \mathbf{I}_1^{(0)} = \begin{bmatrix} 0 \\ 0 \\ 0 \\ 0 \\ 0 \\ 0 \\ 1.0072 - \mathrm{j}7.8906 \\ 0 \\ 1.0072 + \mathrm{j}7.8906 \\ 0 \\ 0 \\ 0 \\ 0 \\ 0 \\ 0 \end{bmatrix}$$

$$\mathbf{F}'^{(0)} = \begin{bmatrix} 0 \\ -0.3804 - \mathrm{j}0.0947 \\ 0 \\ 1.3540 + \mathrm{j}0.2222 \\ 0 \\ -2.7352 - \mathrm{j}0.2229 \\ 0 \\ 3.4603 \\ 0 \\ -2.7352 + \mathrm{j}0.2229 \\ 0 \\ 1.3540 - \mathrm{j}0.2222 \\ 0 \\ -0.3804 + \mathrm{j}0.0947 \\ 0 \end{bmatrix} ; \mathbf{I}_\mathrm{N}^{(0)} = \begin{bmatrix} -0.0790 + \mathrm{j}0.2702 \\ 0 \\ 0.2558 - \mathrm{j}1.2408 \\ 0 \\ -0.3596 + \mathrm{j}2.9342 \\ 0 \\ 0.1802 - \mathrm{j}4.4305 \\ 0 \\ 0.1802 + \mathrm{j}4.4305 \\ 0 \\ -0.3596 - \mathrm{j}2.9342 \\ 0 \\ 0.2558 + \mathrm{j}1.2408 \\ 0 \\ -0.0790 - \mathrm{j}0.2702 \end{bmatrix}$$

Where $\mathbf{F}'$ is the harmonic vector obtained from $\mathrm{d}i_{\mathrm{mag}}/\mathrm{d}\psi$. Substitution of these vectors into the relevant locations of (8.17) leads to the following result at the end of the first iteration:

$$\mathbf{V}_2^{(1)} = \begin{bmatrix} -0.0008 \\ 0 \\ 0.0087 + j0.0005 \\ 0 \\ 0.0345 + j0.0146 \\ 0 \\ 0.7876 + j0.0315 \\ 0 \\ 0.7876 - j0.0315 \\ 0 \\ 0.0345 - j0.0146 \\ 0 \\ 0.0087 - j0.0005 \\ 0 \\ -0.0008 \end{bmatrix} ; \mathbf{V}_3^{(1)} = \begin{bmatrix} 0.0001 \\ 0 \\ -0.0015 + j0.0001 \\ 0 \\ -0.0261 - j0.0042 \\ 0 \\ 1.0395 + j0.2005 \\ 0 \\ 1.0395 - j0.2005 \\ 0 \\ -0.0261 + j0.0042 \\ 0 \\ -0.0015 - j0.0001 \\ 0 \\ 0.0001 \end{bmatrix}$$

Moreover, accurate harmonic current and admittances are derived for the non-linearity using the updated harmonic nodal voltages,

$$\mathbf{F}'^{(1)} = \begin{bmatrix} 0 \\ -0.2483 - j0.0306 \\ 0 \\ 0.9761 + j0.1077 \\ 0 \\ -2.0683 - j0.1254 \\ 0 \\ 2.6610 \\ 0 \\ -2.0683 + j0.1254 \\ 0 \\ 0.9761 - j0.1077 \\ 0 \\ -0.2483 + j0.0306 \\ 0 \end{bmatrix} ; \mathbf{I}_N^{(1)} = \begin{bmatrix} -0.0234 + j0.1669 \\ 0 \\ 0.1072 - j0.8387 \\ 0 \\ -0.1832 + j2.1046 \\ 0 \\ 0.0989 - j3.2634 \\ 0 \\ 0.0989 + j3.2634 \\ 0 \\ -0.1832 - j2.1046 \\ 0 \\ 0.1072 + j0.8387 \\ 0 \\ -0.0234 - j0.1669 \end{bmatrix}$$

At the end of the fourth iteration the solution converges to a voltage tolerance of 10^{-6}. The most relevant harmonic information is presented below and the waveform and harmonic spectrum for the magnetising current $\mathbf{I}_{mag}$ are shown in Figure 8.8.

$$\mathbf{V}_2^{(4)} = \begin{bmatrix} -0.0009 - j0.0001 \\ 0 \\ 0.0092 + j0.0008 \\ 0 \\ 0.0361 + j0.0156 \\ 0 \\ 0.7869 + j0.0314 \\ 0 \\ 0.7869 - j0.0314 \\ 0 \\ 0.0361 - j0.0156 \\ 0 \\ 0.0092 - j0.0008 \\ 0 \\ -0.0009 + j0.0001 \end{bmatrix} ; \mathbf{V}_3^{(4)} = \begin{bmatrix} 0.0001 \\ 0 \\ -0.0016 \\ 0 \\ -0.0274 - j0.0047 \\ 0 \\ 1.0385 + j0.2002 \\ 0 \\ 1.0385 - j0.2002 \\ 0 \\ -0.0274 + j0.0047 \\ 0 \\ -0.0016 \\ 0 \\ 0.0001 \end{bmatrix}$$

$$\mathbf{F}'^{(4)} = \begin{bmatrix} 0 \\ -0.2440 - j0.0280 \\ 0 \\ 0.9630 + j0.1024 \\ 0 \\ -2.0444 - j0.1204 \\ 0 \\ 2.6322 \\ 0 \\ -2.0444 + j0.1204 \\ 0 \\ 0.9630 - j0.1024 \\ 0 \\ -0.2440 + j0.0280 \\ 0 \end{bmatrix} ; \mathbf{I}_1^{(4)} = \begin{bmatrix} -0.0005 + j0.0040 \\ 0 \\ 0.0050 - j0.0567 \\ 0 \\ 0.1600 - j0.3688 \\ 0 \\ 0.9630 - j7.2656 \\ 0 \\ 0.9630 + j7.2656 \\ 0 \\ 0.1600 + j0.3688 \\ 0 \\ 0.0050 + j0.0567 \\ 0 \\ -0.0005 - j0.0040 \end{bmatrix}$$

$$\mathbf{I}_{\mathrm{N}}^{(4)} = \begin{bmatrix} -0.0214 + \mathrm{j}0.1637 \\ 0 \\ 0.1009 - \mathrm{j}0.8254 \\ 0 \\ -0.1749 + \mathrm{j}2.0759 \\ 0 \\ 0.0950 - \mathrm{j}3.2222 \\ 0 \\ 0.0950 + \mathrm{j}3.2222 \\ 0 \\ -0.1749 - \mathrm{j}2.0759 \\ 0 \\ 0.1009 + \mathrm{j}0.8254 \\ 0 \\ -0.0214 - \mathrm{j}0.1637 \end{bmatrix}; \ \mathbf{I}_{\mathrm{mag}}^{(4)} = \begin{bmatrix} 0.0027 - \mathrm{j}0.0205 \\ 0 \\ -0.0126 + \mathrm{j}0.1032 \\ 0 \\ 0.0217 - \mathrm{j}0.2591 \\ 0 \\ -0.0128 + \mathrm{j}0.4264 \\ 0 \\ -0.0128 - \mathrm{j}0.4264 \\ 0 \\ 0.0217 + \mathrm{j}0.2591 \\ 0 \\ -0.0126 - \mathrm{j}0.1032 \\ 0 \\ 0.0027 + \mathrm{j}0.0205 \end{bmatrix}$$

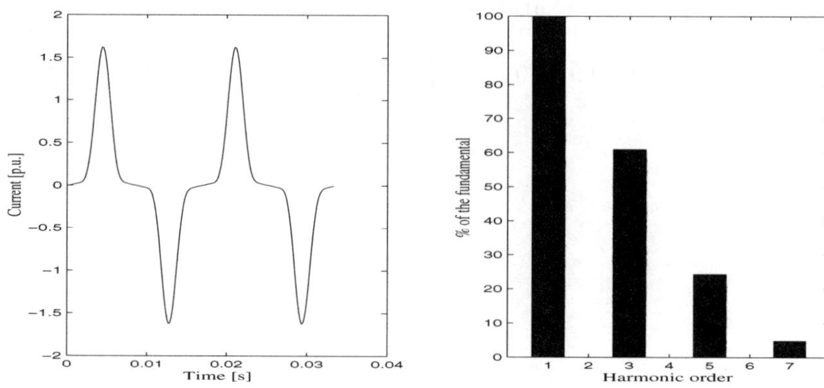

Figure 8.8 Magnetising current

The MATLAB™ program used to solve this examples is given below:

```
clear all
h  = 7;
w0 = 1 % 1 in pu
xT = 0.0326;
R  = 0.0265;
xL = 0.0626;
xC = -0.2373;
a=0.03;
b=0.03;
n=9;
V1=zeros(2*h+1,1);
V1(h)=1.1/2;
V1(h+2)=1.1/2;
cero=zeros(2*h+1,2*h+1);
ZT=form_Zm(0,xT,h);
ZL=form_Zm(R,xL,h);
```

```
ZC=form_Zm(0,xC,h);
U =form_Zm(1,0,h);
YL=inv(ZL);
YC=inv(ZC);
% -- initial conditions, solution with no magnetising branch
Isyst=[V1;V1*0;V1*0];
Ysyst=[ ZT        U      cero
       -U    YC+YL       -YL
     cero       -YL    YL+YC];
Vsyst=inv(Ysyst)*Isyst;
I1=Vsyst(1:2*h+1);
V2=Vsyst(2*h+2:4*h+2);
V3=Vsyst(4*h+3:6*h+3);
% end of initial conditions
error=1;
iter=0;
% -- iterative process
while error>1e-6
  I1e=I1;
  V2e=V2;
  V3e=V3;
  [Hv,IN]=calc_Norton(a,b,n,V2e,h,w0);
  Isyst=[V1;-IN;IN*0];
  Ysyst=[ ZT           U      cero
         -U     Hv+YC+YL       -YL
       cero          -YL    YL+YC];
  Vsyst=inv(Ysyst)*Isyst;
  I1=Vsyst(1:2*h+1);
  V2=Vsyst(2*h+2:4*h+2);
  V3=Vsyst(4*h+3:6*h+3);
  dI1=I1-I1e;
  dV2=V2-V2e;
  dV3=V3-V3e;
  iter=iter+1;
  error=norm([dI1;dV2;dV3]);
end
% -- end of the iterative process
Imag =Hv*V2+IN;
Imagt=plot_f(Imag,h,2);
```

Figure 8.9 shows the magnetising current obtained with a time domain simulation. After the steady state is reached, an FFT is used to obtain the harmonic content of the current, also shown in Figure 8.9. This result agrees very well with that given by the iterative solution in the harmonic domain.

The equivalent circuit used for time domain simulations is shown in Figure 8.10, and the state-space representation is

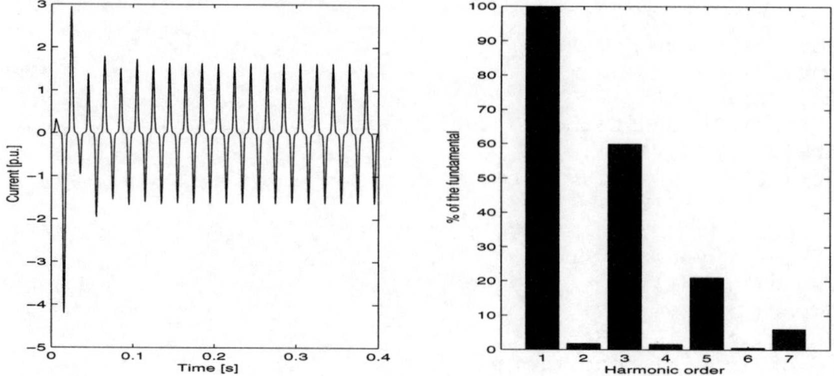

Figure 8.9 Current in the magnetising branch using time domain simulations

$$\begin{bmatrix} di_{\text{aux}}/dt \\ di_1/dt \\ di_3/dt \\ d\psi/dt \\ dv_2/dt \end{bmatrix} = \begin{bmatrix} -\frac{R_1}{L_2} & \frac{1}{L_2C_1} & -\frac{2}{L_2C_1} & -\frac{1}{L_2C_1}\left(a+b\psi^{n-1}\right) & 0 \\ 0 & 0 & 0 & 0 & -\frac{1}{L_1} \\ 1 & 0 & 0 & 0 & 0 \\ 0 & 0 & 0 & 0 & 1 \\ 0 & \frac{1}{C_1} & -\frac{1}{C_1} & -\frac{1}{C_1}\left(a+b\psi^{n-1}\right) & 0 \end{bmatrix}$$

$$\times \begin{bmatrix} i_{\text{aux}} \\ i_1 \\ i_3 \\ \psi \\ v_2 \end{bmatrix} + \begin{bmatrix} 0 \\ \frac{1}{L_1} \\ 0 \\ 0 \\ 0 \end{bmatrix} v_1$$

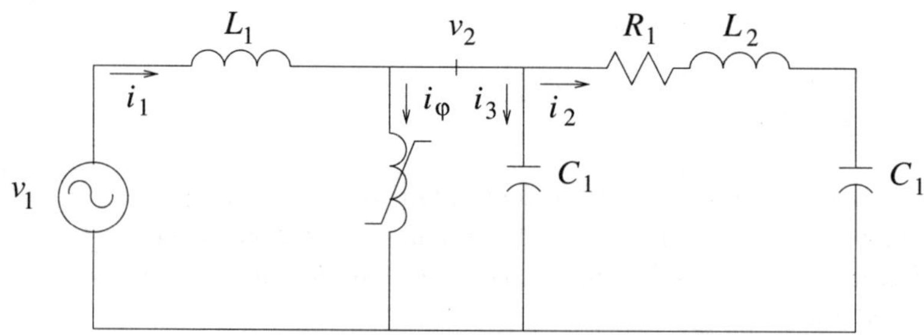

Figure 8.10 Circuit equivalent

The next steps were used to obtain the response of the circuit using the time domain:

```
>> x0=[0 0 0 0 0];
>> [t,x]=ode45('state_space',[0 2*pi*20],x0);
>> Imag=0.03*x(:,4)+0.03*x(:,4).^9;
>> plot(t/(2*pi)/50,Imag)
```

The next function gives the state-space equation used to solve the circuit in the time domain:

```
function dXdt=state_space(t,X);
v1=1.1*cos(t); % w0=1 in p.u.
```

```
L1=0.0326;
L2=0.0626;
R1=0.0265;
C1=1/0.2373;
a=0.03; b=0.03; n=9;
F=X(4);
A=[-R1/L2    1/(L2*C1)    -2/(L2*C1)    -1/(L2*C1)*(a+b*F^(n-1)) 0
    0         0            0             0                        -1/L1
    1         0            0             0                        0
    0         0            0             0                        1
    0         1/C1         -1/C1         -1/C1*(a+b*F^(n-1))      0];
b=[0;1/L1;0;0;0];
U=v1;
dXdt=A*X+b*U;
```

8.4.1 Full Newton–Raphson method

Provided the harmonic Norton equivalent of the magnetising branch of the transformer is recalculated at each iterative step the harmonic solution will be reached in a quadratic convergent fashion, i.e. four iterations regardless of the size of the network and number of harmonics. This performance corresponds to a solution via a harmonic Newton–Raphson method. In harmonic domain applications, one extra iteration may be required owing to the practical need to truncate the Fourier series during the iterative solution. Hence, the solution is obtained very reliably; however, long CPU times are involved even when sparsity techniques are employed since the method requires the inversion, i.e. factorisation, of nodal admittance matrices of large dimensions. The convergence characteristic of this method is shown in Figure 8.11, where maximum voltage mismatches per iteration are plotted for three different voltage excitation conditions, i.e. 1.1, 1.15 and 1.2 p.u.

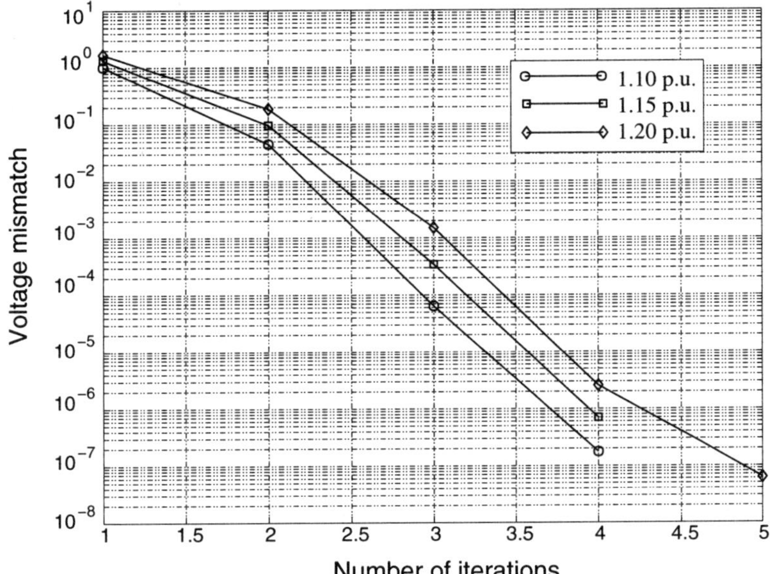

Figure 8.11 Convergence characteristics of full Newton–Raphson method

8.4.2 Single-valued Newton–Raphson method

An alternative iterative scheme, which normally brings CPU time down, consists in evaluating and inverting the nodal admittance matrix in equation (8.17) just once. This action takes place during the first iteration. In subsequent iterations, the admittance matrix of the harmonic Norton equivalent, $\mathbf{H}_V$, is kept constant and the current source vector of the harmonic Norton equivalent, $\mathbf{I}_\mathrm{N}$, is updated from iteration to iteration to reflect the interaction taking place between the various harmonics. This has the advantage that the global nodal admittance matrix of the system remains constant; hence, only one matrix inversion procedure is required. This simplification in the algorithm reduces numeric overheads but at the expense of weakening convergence characteristics, which are linear as opposed to quadratic. This is shown in Figure 8.12 where maximum voltage mismatches per iteration are plotted for three different voltage excitation levels.

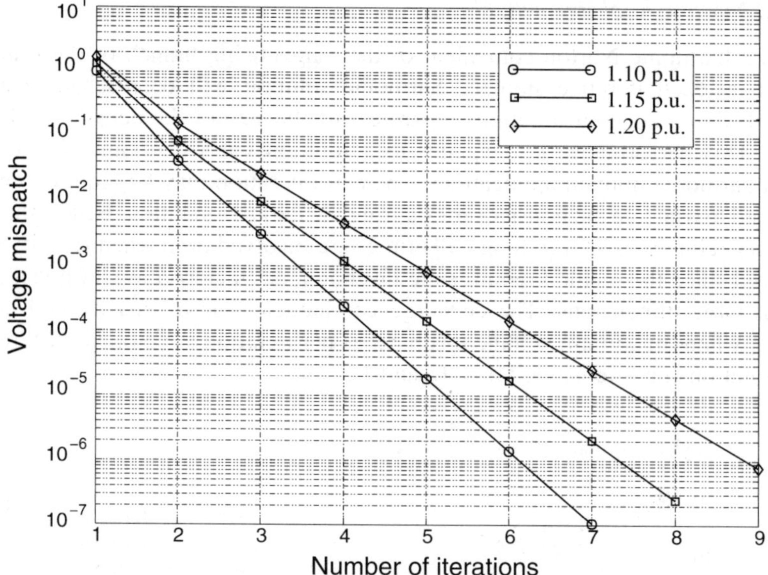

Figure 8.12 Convergence characteristics of single-valued Newton–Raphson method

8.4.3 Gauss–Seidel method

If the admittance matrix of the harmonic Norton equivalent, $\mathbf{H}_V$, is neglected throughout the iterative solution, the magnetising branch of the transformer will then be modelled as a set of harmonic current sources, $\mathbf{I}_0$, as opposed to $\mathbf{I}_\mathrm{N}$.

One harmonic current at a time will be injected into the inverted nodal admittance matrix of equation (8.17), which becomes fully decoupled since $\mathbf{H}_V$ is not present. This is a Gauss–Seidel-type method. The method requires a relatively low number of operations at each iteration but this is at the expense of degrading robustness even further. For circuits with high harmonic distortion this method will converge slowly and it may even fail to converge. Figure 8.13 shows the convergence characteristics of this method.

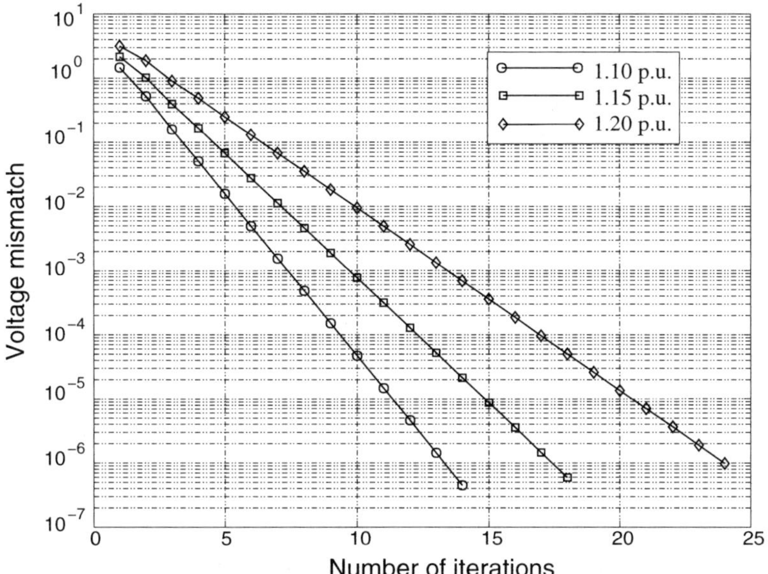

Figure 8.13 Convergence characteristics of Gauss–Seidel method

8.4.4 *Quasi-Newton–Raphson method*

A further variant of the full Newton–Raphson method, which retains most of the Newton's strong convergence characteristics and yet is amenable to a decoupled solution, is presented below. Here, the admittance matrix of the non-linear linearised element, $\mathbf{H}_V$, is separated into a diagonal and an off-diagonal component, say matrices $\mathbf{B}_V$ and $\mathbf{C}_V$, respectively, i.e.

$$\mathbf{H}_V = \mathbf{B}_V + \mathbf{C}_V \tag{8.19}$$

The diagonal elements, $\mathbf{B}_V$, are grouped together with the linear admittance matrix of the transmission network while the off-diagonal elements, $\mathbf{C}_V$, affected by the harmonic voltages obtained in the previous iteration, (k − 1), are grouped together with the vector current source of the harmonic Norton equivalent, $\mathbf{I}_{\mathrm{N}}^{(\mathrm{k})}$. This action decouples, coefficient-wise, the network system equations. The complementary harmonic currents, $\mathbf{C}_V^{(\mathrm{k})}\mathbf{V}^{(\mathrm{k}-1)}$, are calculated using the most up-to-date harmonic voltage information available in order to tighten convergence even further. Figure 8.14 shows the convergence characteristic of this method, which is linear. It should be remarked that the attractiveness of the method is that it does not requires the inversion of large matrices and convergence characteristics are still very strong.

8.5 Summary

Repeated linearisations of a non-linear function in the harmonic domain are amenable to finding the periodic solution of the non-linear function via a harmonic Newton–Raphson technique. The harmonic admittance matrix of the Norton equivalent plays the role of a Jacobian in the iterative harmonic domain solution. The difference between the harmonic currents vector in the present and the previous iteration determines the mismatch vector. A new vector of harmonic voltage increments is determined at each iterative step which, theoretically, should be driven to zero (small value) in a finite number of iterations. From

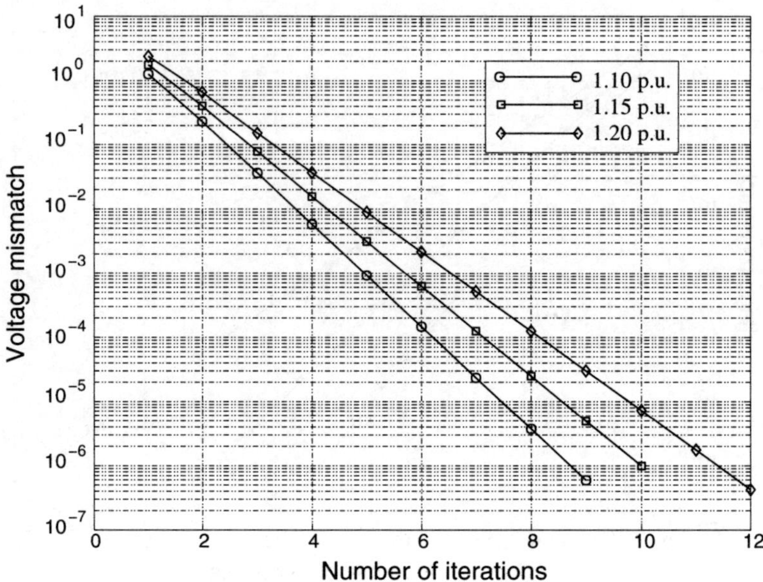

Figure 8.14 Convergence characteristics of the quasi-Newton–Raphson method

a numeric viewpoint, this is the single, most important characteristic of a full Newton–Raphson technique; provided the iterative solution is suitably initialised, it exhibits quadratic convergence.

Several simplifications are possible with respect to the full Newton–Raphson method. For instance, keeping the Jacobian constant after the first or second iteration leads to computing savings per iteration. However, the number of iterations taken for the solution to reach convergence will increase. Neglecting the Jacobian altogether in the solution method is also a possibility. However, the strong convergence properties of the method will be further weakened. In any event, the harmonic solution given by any one of the three methods, provided it converges, should give the same answer. A further variant in this family of methods is the quasi-Newton method. It takes advantage of the best characteristics of the Newton–Raphson and Gauss–Seidel methods, resulting in a method with moderate computing overheads and strong convergence characteristics.

The theory was applied to solve a single-phase test system containing magnetic non-linearities. The harmonic domain solution was compared with the time domain solution, and both results compared rather well with each other.

8.6 Bibliography

1. E. Acha, "Modelling of Power System Transformers in the Complex Conjugate Harmonic Space", *PhD Thesis*, University of Canterbury, Christchurch, New Zealand, 1988.
2. J.J. Rico, "Steady State Modelling of Non-linear Power Plant Components", *PhD Thesis*, University of Glasgow, 1997.
3. A. Semlyen, E. Acha, J. Arrillaga, "Newton-Type Algorithms for the Harmonic Phasor Analysis of Non-Linear Power Circuits in Periodical Steady-State with Special Reference to Magnetic Non-Linearities", *IEEE Transactions on Power Delivery*, Vol. 3, No. 3, July 1988, pp. 1090–1098.

4. E. Acha, J.J. Rico, "Harmonic Domain Modelling of Non-Linear Power Plant Components", *Proceedings of the IEEE ICHPS VI*, Bologna, Italy, 21–23 September 1994, pp. 206–213.
5. M.S. Nakhla, J. Vlach, "A Piecewise Harmonic Balance Technique for Determination of Periodic Response of Nonlinear Systems", *IEEE Transactions on Circuit and Systems*, Vol. CAS-23, No. 2, February 1976, pp. 85–91.
6. L.O. Chua, C.Y. Ng, "Frequency-domain Analysis of Nonlinear Systems", *IEEE Journal of Electronics Circuits and Systems*, Vol. 3, 1979, pp. 165–185.
7. H.W. Dommel, A. Yan, S. Wei, "Harmonics from Transformer Saturation", *IEEE Transactions on Power Systems*, Vol. PWRD-1, No. 2, April 1986, pp. 209–215.
8. E.P. Dick, W. Watson, "Transformer Models for Transients Studies Based on Field Measurements", *IEEE Transactions on Power Apparatus and Systems*, Vol. PAS-100, No. 1, January 1981, pp. 409–419.
9. A. Semlyen, N. Rajakovic, "Harmonic Domain Modeling of Laminated Iron Core", *IEEE Transactions on Power Delivery*, Vol. 4, No. 1, January 1989, pp. 382–390.
10. N. Rajakovic, A. Semlyen, "Investigation of the Inrush Phenomenon: A Quasi-Stationary Approach in the Harmonic Domain", *IEEE Transactions on Power Delivery*, Vol. 4, No. 4, October 1989, pp. 2114–2120.
11. N. Rajakovic, A. Semlyen, "Harmonic Domain of Field Variables Related to Eddy Currents and Hysteresis Losses in Saturated Laminations", *IEEE Transactions on Power Delivery*, Vol. 4, No. 2, April 1989, pp. 1111–1116.

Part II

CONVENTIONAL POWER PLANT EQUIPMENT

9

Synchronous Generator

9.1 Introduction

An electrical power network is a complex system which includes unbalanced transmission lines and cables, dynamic loads, linear elements, electronic power devices, and non-linear elements such as arc furnaces, fluorescent lamps and magnetic cores of saturated transformers. The unbalanced and non-sinusoidal nature of the electric network makes it necessary to model all power plant components in the phase domain (*abc*). This is particularly the case if the main aim is to study the periodic response of the electrical power network.

There is general agreement that a synchronous machine operating under unbalanced conditions can generate harmonics due to an intrinsic frequency conversion mechanism that exists between the stator and rotor circuits. Moreover, if the stator's magnetic circuit becomes saturated then this will constitute an additional source of harmonic distortion. Existing harmonic models of the synchronous machine use an admittance matrix that exhibits cross-coupling between impedances at harmonic frequencies to represent the frequency conversion effect. The saturation effect is included as a Norton-type equivalent in the *abc* frame of reference using the harmonic domain.

In the open literature, harmonic domain models of synchronous machines have been formulated using either complex Fourier series [1,2] or Hartley series [3]. In this chapter the Hartley frame of reference is used to model the synchronous generator since the model that uses complex Fourier series is well described elsewhere [4].

Experimental results obtained with a small synchronous generator in the laboratory are presented towards the end of the chapter, to provide experimental evidence that the synchronous generator does indeed have the ability to generate harmonic distortion.

9.2 Synchronous Machine under Non-sinusoidal Conditions

Synchronous machines represent a source of harmonic currents on two counts: the frequency conversion effect, and the non-linear characteristic effect due to magnetic saturation.

A synchronous generator feeding an unbalanced, three-phase load may experience the flow of a negative sequence current in the rotor, which in turn may induce a third-order harmonic current on the stator winding. This phenomenon is known as the frequency conversion effect.

The saturation of the stator's circuit represents another harmonic source. In this chapter, saturation is modelled using a Norton equivalent in phase coordinates *abc*; the harmonic model is obtained using a Hartley series expansion.

9.2.1 Machine's harmonic admittance matrix in the dq0 axes

For the three-phase winding layout of Figure 9.1, stator and rotor quantities are expressed in the frames of reference attached to their respective physical circuits. The instantaneous voltages of the machine in terms of flux and current are given by the following matrix equation:

$$\mathbf{v} = \mathbf{Ri} + \frac{\mathrm{d}}{\mathrm{d}t}\Phi \tag{9.1}$$

where $\Phi = \mathbf{Li}$, and $\mathbf{R}$ and $\mathbf{L}$ are the machine resistance and inductance matrices. Furthermore, expanding into stator and rotor subsets, equation (9.1) may be written as

$$\begin{bmatrix} \mathbf{v}_\mathrm{s} \\ \mathbf{v}_\mathrm{r} \end{bmatrix} = \begin{bmatrix} \mathbf{R}_\mathrm{s} & \\ & \mathbf{R}_\mathrm{r} \end{bmatrix} \begin{bmatrix} \mathbf{i}_\mathrm{s} \\ \mathbf{i}_\mathrm{r} \end{bmatrix} + \omega_r \begin{bmatrix} \mathbf{G}_\mathrm{ss} & \mathbf{G}_\mathrm{sr} \\ \mathbf{G}_\mathrm{rs} & 0 \end{bmatrix} \begin{bmatrix} \mathbf{i}_\mathrm{s} \\ \mathbf{i}_\mathrm{r} \end{bmatrix} + \begin{bmatrix} \mathbf{L}_\mathrm{ss} & \mathbf{L}_\mathrm{sr} \\ \mathbf{L}_\mathrm{rs} & \mathbf{L}_\mathrm{rr} \end{bmatrix} \begin{bmatrix} p\mathbf{i}_\mathrm{s} \\ p\mathbf{i}_\mathrm{r} \end{bmatrix} \tag{9.2}$$

where $p = \mathrm{d}/\mathrm{d}t$, $\mathbf{G} = \mathrm{d}\mathbf{L}/\mathrm{d}\theta$ and $\omega_\mathrm{r} = \mathrm{d}\theta/\mathrm{d}t$ defines the rotor speed.

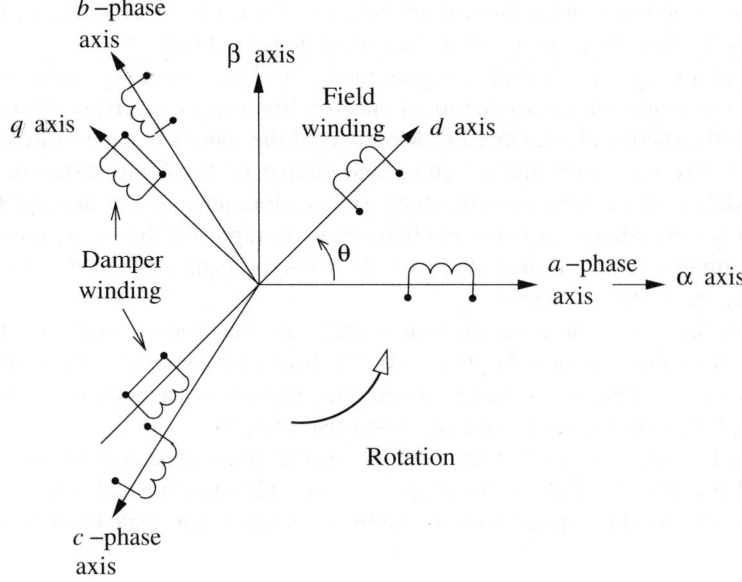

Figure 9.1 Synchronous machine, showing the $dq0$, $\alpha\beta\gamma$ and abc axes

Based on well-known equations [5,6], the form taken by equation (9.2) in the dq axis is given by

$$\begin{bmatrix} v_d \\ v_q \\ v_\mathrm{O} \\ v_\mathrm{f} \\ v_s \\ v_t \end{bmatrix} = \left\{ \begin{bmatrix} R_d & 0 & 0 & 0 & 0 & 0 \\ 0 & R_q & 0 & 0 & 0 & 0 \\ 0 & 0 & R_\mathrm{O} & 0 & 0 & 0 \\ 0 & 0 & 0 & R_\mathrm{f} & 0 & 0 \\ 0 & 0 & 0 & 0 & R_s & 0 \\ 0 & 0 & 0 & 0 & 0 & R_t \end{bmatrix} \right.$$

$$
+p\begin{bmatrix} L_d & 0 & 0 & M_{\rm df} & M_{ds} & 0 \\ 0 & L_q & 0 & 0 & 0 & M_{qt} \\ 0 & 0 & L_0 & 0 & 0 & 0 \\ M_{\rm df} & 0 & 0 & L_f & 0 & 0 \\ M_{ds} & 0 & 0 & 0 & L_s & 0 \\ 0 & M_{qt} & 0 & 0 & 0 & L_t \end{bmatrix}
$$

$$
\left. + \,\omega_{\rm r}\begin{bmatrix} 0 & -L_q & 0 & 0 & 0 & -M_{qt} \\ L_d & 0 & 0 & M_{\rm df} & M_{ds} & 0 \\ 0 & 0 & 0 & 0 & 0 & 0 \\ 0 & 0 & 0 & 0 & 0 & 0 \\ 0 & 0 & 0 & 0 & 0 & 0 \\ 0 & 0 & 0 & 0 & 0 & 0 \end{bmatrix}\right\}\begin{bmatrix} i_d \\ i_q \\ i_0 \\ i_{\rm f} \\ i_s \\ i_t \end{bmatrix} \tag{9.3}
$$

Moreover, since the damping winding circuits are short-circuited, their voltages will be zero, i.e. $v_s = v_t = 0$. For steady-state analysis purposes the rotor speed $\omega_{\rm r}$ is assumed to be constant and equal to the electric speed ω_0. It is also assumed that the field voltage $v_{\rm f}$ has no ripple but only the DC component, so $v_{\rm f} = 0$.

Taking equation (9.3) as the basis for steady-state analysis, we have

$$
\begin{aligned}
\begin{bmatrix} \mathbf{V}_{dq0} \\ 0 \end{bmatrix} &= \left\{\begin{bmatrix} \mathbf{R}_{11} & 0 \\ 0 & \mathbf{R}_{22} \end{bmatrix} + p\begin{bmatrix} \mathbf{L}_{11} & \mathbf{L}_{12} \\ \mathbf{L}_{21} & \mathbf{L}_{22} \end{bmatrix}\right. \\
&\quad \left. + \omega_{\rm r}\begin{bmatrix} \mathbf{J}_{11} & \mathbf{J}_{12} \\ 0 & 0 \end{bmatrix}\right\}\begin{bmatrix} \mathbf{I}_{dq0} \\ \mathbf{I}_{fst} \end{bmatrix}
\end{aligned} \tag{9.4}
$$

In a more compact form,

$$
\begin{bmatrix} \mathbf{V}_{dq0} \\ 0 \end{bmatrix} = \begin{bmatrix} \mathbf{Z}_{11} & \mathbf{Z}_{12} \\ \mathbf{Z}_{21} & \mathbf{Z}_{22} \end{bmatrix}\begin{bmatrix} \mathbf{I}_{dq0} \\ \mathbf{I}_{fst} \end{bmatrix} \tag{9.5}
$$

where

$$
\begin{aligned}
\mathbf{Z}_{11} &= \mathbf{R}_{11} + p\mathbf{L}_{11} + \omega_{\rm r}\mathbf{J}_{11} \\
\mathbf{Z}_{12} &= p\mathbf{L}_{12} + \omega_{\rm r}\mathbf{J}_{12} \\
\mathbf{Z}_{21} &= p\mathbf{L}_{21} \\
\mathbf{Z}_{22} &= \mathbf{R}_{22} + p\mathbf{L}_{22}
\end{aligned} \tag{9.6}
$$

and for a given harmonic term n, $p = \mathrm{j}n\omega_0$ for non-sinusoidal steady-state analysis.

Applying the concept of Hartley impedance [6] to equation (9.6) we have

$$
\mathbf{Z}_{11_n} = \begin{bmatrix} \mathbf{R}_{11} + \omega_{\rm r}\mathbf{J}_{11} & n\omega_0\mathbf{L}_{11} \\ -n\omega_0\mathbf{L}_{11} & \mathbf{R}_{11} + \omega_{\rm r}\mathbf{J}_{11} \end{bmatrix}
$$

$$
\mathbf{Z}_{12_n} = \begin{bmatrix} \omega_{\rm r}\mathbf{J}_{12} & n\omega_0\mathbf{L}_{12} \\ -n\omega_0\mathbf{L}_{12} & \omega_{\rm r}\mathbf{J}_{12} \end{bmatrix}
$$

$$\mathbf{Z}_{21_n} = \begin{bmatrix} 0 & n\omega_0\mathbf{L}_{21} \\ -n\omega_0\mathbf{L}_{21} & 0 \end{bmatrix}$$

$$\mathbf{Z}_{22_n} = \begin{bmatrix} \mathbf{R}_{22} & n\omega_0\mathbf{L}_{22} \\ -n\omega_0\mathbf{L}_{22} & \mathbf{R}_{22} \end{bmatrix}$$

Equation (9.5) may be solved for a given harmonic n for $\mathbf{I}_{dq0}$,

$$\mathbf{I}_{dq0_n} = \mathbf{Y}_{dq0_n}\mathbf{I}_{dq0_n} \tag{9.7}$$

where

$$\mathbf{Y}_{dq0_n} = \left(\mathbf{Z}_{11_n} - \mathbf{Z}_{12_n}\mathbf{Z}_{22_n}^{-1}\mathbf{Z}_{21_n}\right)^{-1} \tag{9.8}$$

and the form $\mathbf{Y}_{dq0_n}$ is given by the Hartley admittance,

$$\mathbf{Y}_{dq0_n} = \begin{bmatrix} \mathbf{G}_n & -\mathbf{B}_n \\ \mathbf{B}_n & \mathbf{G}_n \end{bmatrix} \tag{9.9}$$

where $\mathbf{G}_n$ and $\mathbf{B}_n$ are 3×3 real matrices.

For all harmonics considered in a given analysis, equation (9.10) takes the form

$$\mathbf{I}_{dq0_h} = \mathbf{Y}_{dq0_h}\mathbf{I}_{dq0_h} \tag{9.10}$$

where $\mathbf{Y}_{dq0_h}$ is a Hartley admittance matrix for harmonic analysis.

The following MATLAB™ functions are used to obtain the admittance matrix $\mathbf{Y}_{dq0}$:

```
function Ydq0=calc_Ydq0(data,h)
 hh=h+1;
 Ydq0 =zeros(hh*2+1,hh*2+1);
 [G,B]=calc_GB(0,data);
 Ydq0 =form_A(Ydq0,G+B,hh+1,hh+1);
 for j=1:hh,
  [G,B]=calc_GB(j,data);
  Ydq0=form_A(Ydq0,G,hh+1-j,hh+1-j);
  Ydq0=form_A(Ydq0,G,hh+1+j,hh+1+j);
  Ydq0=form_A(Ydq0,-B,hh+1-j,hh+1+j);
  Ydq0=form_A(Ydq0,B,hh+1+j,hh+1-j);
 end
```

```
function [G,B]=calc_GB(h,data)
% data=[Rd,Rq,Ro,Rf,Rs,Rt,Ld,Lq,Lo,Lf,Ls,Lt,Mdf,Mds,Mqt,Mfs]
Rd=data(1);   Rf=data(4);   Rq=data(2);   Rs=data(5);
Ro=data(3);   Rt=data(6);   Ld=data(7);   Lf=data(10);
Lq=data(8);   Ls=data(11);  Lo=data(9);   Lt=data(12);
Mdf=data(13); Mds=data(14); Mqt=data(15); Mfs=data(16);
wr=1; w=1;
R11=[Rd 0 0 0 Rq 0 0 0 Ro];
R22=[Rf 0 0 0 Rs 0 0 0 Rt];
O=[0 0 0 0 0 0 0 0 0];
L11=[Ld 0 0 0 Lq 0 0 0 Lo];
L12=[Mdf Mds 0 0 0 Mqt 0 0 0];
```

```
L21=[Mdf 0 0 Mds 0 0 0 Mqt 0];
L22=[Lf Mfs 0 Mfs Ls 0 0 0 Lt];
J11=[0 -Lq 0 Ld 0 0 0 0 0];
J12=[0 0 -Mqt Mdf Mds 0 0 0 0 ];
Z11=[R11+wr*J11 h*w*L11 -h*w*L11 R11+wr*J11];
Z12=[wr*J12 h*w*L12 -h*w*L12 wr*J12];
Z21=[ O h*w*L21 -h*w*L21 O];
Z22=[R22 h*w*L22 -h*w*L22 R22];
Y=inv(Z11-Z12*inv(Z22)*Z21);
G=Y(1:3,1:3);
B=Y(4:6,1:3);
```

```
function A=form_A(A,a,i,j)
I=(i-1)*3;
J=(j-1)*3;
A(I+1,J+1)=a(1,1); A(I+1,J+2)=a(1,2); A(I+1,J+3)=a(1,3);
A(I+2,J+1)=a(2,1); A(I+2,J+2)=a(2,2); A(I+2,J+3)=a(2,3);
A(I+3,J+1)=a(3,1); A(I+3,J+2)=a(3,2); A(I+3,J+3)=a(3,3);
```

```
function data=dat_maq1;
Rd=0.05; Rf=0.02; Rq=0.05; Rs=0.01;
Ro=0.05; Rt=0.01; Ld=1.0;  Lf=1.1;
Lq=0.8;  Ls=0.9;  Lo=0.08; Lt=1.0;
Mdf=0.8; Mds=0.8; Mqt=0.6; Mfs=0.8;
data=[Rd,Rq,Ro,Rf,Rs,Rt,Ld,Lq,Lo,Lf,Ls,Lt,Mdf,Mds,Mqt,Mfs];
```

9.2.2 Machine's harmonic admittance matrix in the $\alpha\beta\gamma$ axes

The transformation from the $dq0$ to the $\alpha\beta\gamma$ axes is

$$\begin{bmatrix} v_\alpha \\ v_\beta \\ v_\gamma \end{bmatrix} = \begin{bmatrix} \cos\omega_r t & -\sin\omega_r t & 0 \\ \sin\omega_r t & \cos\omega_r t & 0 \\ 0 & 0 & 1 \end{bmatrix} \begin{bmatrix} v_d \\ v_q \\ v_0 \end{bmatrix} \tag{9.11}$$

Under balanced conditions the voltages v_d, v_q and v_0 are constant, but under unbalanced conditions these voltages become periodic non-sinusoidal functions, which in the Hartley domain are

$$v_d = \sum_{n=-\infty}^{\infty} V_{d_n} \operatorname{cas} n\nu_0 t \tag{9.12}$$

$$v_q = \sum_{m=-\infty}^{\infty} V_{q_m} \operatorname{cas} m\nu_0 t \tag{9.13}$$

$$v_0 = \sum_{i=-\infty}^{\infty} V_{0_i} \operatorname{cas} i\nu_0 t \tag{9.14}$$

Therefore,

$$v_\alpha = \sum_{k=-\infty}^{\infty} V_{\alpha_k} \operatorname{cas} k\nu_0 t \tag{9.15}$$

$$v_\beta = \sum_{j=-\infty}^{\infty} V_{\beta_j} \operatorname{cas} j\nu_0 t \tag{9.16}$$

$$v_\gamma = \sum_{r=-\infty}^{\infty} V_{\gamma_r} \operatorname{cas} r\nu_0 t \tag{9.17}$$

Using $\omega_{\mathrm{r}} = \nu_0$, and substituting (9.12)–(9.14) into (9.11), then after some arduous algebra,

$$\begin{aligned} v_\alpha &= \frac{1}{2}\sum_{n=-\infty}^{\infty} V_{d_n} \operatorname{cas}(n+1)\nu_0 t + \frac{1}{2}\sum_{n=-\infty}^{\infty} V_{d_n} \operatorname{cas}(n-1)\nu_0 t \\ &+ \frac{1}{2}\sum_{m=-\infty}^{\infty} V_{q_m} \operatorname{cas}(-m-1)\nu_0 t - \frac{1}{2}\sum_{m=-\infty}^{\infty} V_{q_m} \operatorname{cas}(-m+1)\nu_0 t \end{aligned} \tag{9.18}$$

$$\begin{aligned} v_\beta &= -\frac{1}{2}\sum_{n=-\infty}^{\infty} V_{d_n} \operatorname{cas}(-n-1)\nu_0 t + \frac{1}{2}\sum_{n=-\infty}^{\infty} V_{d_n} \operatorname{cas}(-n+1)\nu_0 t \\ &+ \frac{1}{2}\sum_{m=-\infty}^{\infty} V_{q_m} \operatorname{cas}(m+1)\nu_0 t + \frac{1}{2}\sum_{m=-\infty}^{\infty} V_{q_m} \operatorname{cas}(m-1)\nu_0 t \end{aligned} \tag{9.19}$$

$$v_\gamma = \sum_{i=-\infty}^{\infty} V_{0_i} \operatorname{cas} i\nu_0 t \tag{9.20}$$

Expanding the series (9.18)–(9.20) and comparing the results term by term, with expressions (9.15)–(9.17) leads to the following harmonic vectors in the Hartley domain:

$$\begin{bmatrix} \vdots \\ V_{\alpha_{-3}} \\ V_{\alpha_{-2}} \\ V_{\alpha_{-1}} \\ V_{\alpha_0} \\ V_{\alpha_1} \\ V_{\alpha_2} \\ V_{\alpha_3} \\ \vdots \end{bmatrix} = \frac{1}{2}\begin{bmatrix} \vdots \\ V_{d_{-4}} \\ V_{d_{-3}} \\ V_{d_{-2}} \\ V_{d_{-1}} \\ V_{d_0} \\ V_{d_1} \\ V_{d_2} \\ \vdots \end{bmatrix} + \frac{1}{2}\begin{bmatrix} \vdots \\ V_{d_{-2}} \\ V_{d_{-1}} \\ V_{d_0} \\ V_{d_1} \\ V_{d_2} \\ V_{d_3} \\ V_{d_4} \\ \vdots \end{bmatrix} + \frac{1}{2}\begin{bmatrix} \vdots \\ V_{q_2} \\ V_{q_1} \\ V_{q_0} \\ V_{q_{-1}} \\ V_{q_{-2}} \\ V_{q_{-3}} \\ V_{q_{-4}} \\ \vdots \end{bmatrix} - \frac{1}{2}\begin{bmatrix} \vdots \\ V_{q_4} \\ V_{q_3} \\ V_{q_2} \\ V_{q_1} \\ V_{q_0} \\ V_{q_{-1}} \\ V_{q_{-2}} \\ \vdots \end{bmatrix} \tag{9.21}$$

$$\begin{bmatrix} \vdots \\ V_{\beta_{-3}} \\ V_{\beta_{-2}} \\ V_{\beta_{-1}} \\ V_{\beta_0} \\ V_{\beta_1} \\ V_{\beta_2} \\ V_{\beta_3} \\ \vdots \end{bmatrix} = -\frac{1}{2}\begin{bmatrix} \vdots \\ V_{d_2} \\ V_{d_1} \\ V_{d_0} \\ V_{d_{-1}} \\ V_{d_{-2}} \\ V_{d_{-3}} \\ V_{d_{-4}} \\ \vdots \end{bmatrix} + \frac{1}{2}\begin{bmatrix} \vdots \\ V_{d_4} \\ V_{d_3} \\ V_{d_2} \\ V_{d_1} \\ V_{d_0} \\ V_{d_{-1}} \\ V_{d_{-2}} \\ \vdots \end{bmatrix} + \frac{1}{2}\begin{bmatrix} \vdots \\ V_{q_4} \\ V_{q_3} \\ V_{q_2} \\ V_{q_1} \\ V_{q_0} \\ V_{q_{-1}} \\ V_{q_{-2}} \\ \vdots \end{bmatrix} + \frac{1}{2}\begin{bmatrix} \vdots \\ V_{q_{-2}} \\ V_{q_{-1}} \\ V_{q_0} \\ V_{q_1} \\ V_{q_2} \\ V_{q_3} \\ V_{q_4} \\ \vdots \end{bmatrix} \tag{9.22}$$

$$\begin{bmatrix} \vdots \\ V_{\gamma_{-3}} \\ V_{\gamma_{-2}} \\ V_{\gamma_{-1}} \\ V_{\gamma_0} \\ V_{\gamma_1} \\ V_{\gamma_2} \\ V_{\gamma_3} \\ \vdots \end{bmatrix} = \begin{bmatrix} \vdots \\ V_{0_{-3}} \\ V_{0_{-2}} \\ V_{0_{-1}} \\ V_{0_0} \\ V_{0_1} \\ V_{0_2} \\ V_{0_3} \\ \vdots \end{bmatrix} \tag{9.23}$$

From (9.21)–(9.23) a general transformation matrix is obtained, and summarised by

$$\begin{bmatrix} \vdots \\ \mathbf{V}_{\alpha\beta\gamma_{-3}} \\ \mathbf{V}_{\alpha\beta\gamma_{-2}} \\ \mathbf{V}_{\alpha\beta\gamma_{-1}} \\ \mathbf{V}_{\alpha\beta\gamma_0} \\ \mathbf{V}_{\alpha\beta\gamma_1} \\ \mathbf{V}_{\alpha\beta\gamma_2} \\ \mathbf{V}_{\alpha\beta\gamma_3} \\ \vdots \end{bmatrix} \begin{bmatrix} \ddots & & & & & & & & \\ & \mathbf{T} & \mathbf{U}' & & & & \mathbf{M} & & \\ & \mathbf{U}' & \mathbf{T} & \mathbf{U}' & & \mathbf{M} & & \mathbf{M}^{\mathrm{T}} & \\ & & \mathbf{U}' & \mathbf{T} & \mathbf{N} & & \mathbf{M}^{\mathrm{T}} & & \\ & & & \mathbf{N} & \mathbf{T} & \mathbf{N}^{\mathrm{T}} & & & \\ & & \mathbf{M} & & \mathbf{N}^{\mathrm{T}} & \mathbf{T} & \mathbf{U}' & & \\ & \mathbf{M} & & \mathbf{M}^{\mathrm{T}} & & \mathbf{U}' & \mathbf{T} & \mathbf{U}' & \\ & & \mathbf{M}^{\mathrm{T}} & & & & \mathbf{U}' & \mathbf{T} & \\ & & & & & & & & \ddots \end{bmatrix} \begin{bmatrix} \vdots \\ \mathbf{V}_{dq0_{-3}} \\ \mathbf{V}_{dq0_{-2}} \\ \mathbf{V}_{dq0_{-1}} \\ \mathbf{V}_{dq0_0} \\ \mathbf{V}_{dq0_1} \\ \mathbf{V}_{dq0_2} \\ \mathbf{V}_{dq0_3} \\ \vdots \end{bmatrix} \tag{9.24}$$

where

$$\mathbf{V}_{\alpha\beta\gamma_i} = \begin{bmatrix} V_{\alpha_i} \\ V_{\beta_i} \\ V_{\gamma_i} \end{bmatrix}; \ \mathbf{V}_{dq0_i} = \begin{bmatrix} V_{d_i} \\ V_{q_i} \\ V_{0_i} \end{bmatrix}$$

$$\mathbf{U}' = \begin{bmatrix} 1 & 0 & 0 \\ 0 & 1 & 0 \\ 0 & 0 & 0 \end{bmatrix}; \ \mathbf{M} = \begin{bmatrix} 0 & 1 & 0 \\ -1 & 0 & 0 \\ 0 & 0 & 0 \end{bmatrix}; \ \mathbf{T} = \begin{bmatrix} 0 & 0 & 0 \\ 0 & 0 & 0 \\ 0 & 0 & 2 \end{bmatrix}; \ \mathbf{N} = \begin{bmatrix} 1 & 1 & 0 \\ -1 & 1 & 0 \\ 0 & 0 & 0 \end{bmatrix}$$

In compact notation,

$$\mathbf{V}_{\alpha\beta\gamma_h} = \mathbf{C}_h \mathbf{V}_{dq0_h} \tag{9.25}$$

The transformation matrix $\mathbf{C}_h$ is singular, and in order to remove the singularity, either the first or the last row and column of the matrix are removed. Moreover, it can be shown that

$$\mathbf{C}_h^{\mathrm{T}}\mathbf{C}_h = \begin{bmatrix} \mathbf{X} & & & & \mathbf{Y}^{\mathrm{T}} \\ & \mathbf{U} & & & \\ & & \mathbf{U} & & \\ & & & \mathbf{U} & \\ \mathbf{Y} & & & & \mathbf{X} \end{bmatrix}$$

where

$$\mathbf{X} = \begin{bmatrix} 1/2 & 0 & 0 \\ 0 & 1/2 & 0 \\ 0 & 0 & 1 \end{bmatrix}; \ \mathbf{U} = \begin{bmatrix} 1 & 0 & 0 \\ 0 & 1 & 0 \\ 0 & 0 & 1 \end{bmatrix}; \ \mathbf{Y} = \frac{1}{2}\mathbf{M}$$

Pre-multiplying (9.25) by $\mathbf{C}^{\mathrm{T}}$, using $h+1$ harmonics,

$$\mathbf{C}_{h+1}^{\mathrm{T}}\mathbf{V}_{\alpha\beta\gamma_{h+1}} = \mathbf{C}_{h+1}^{\mathrm{T}}\mathbf{C}_{h+1}\mathbf{V}_{dq0_{h+1}}$$

and removing the contribution of the $(h+1)$th harmonic term from this equation,

$$\mathbf{V}_{dq0_h} = \mathbf{C}_h^{\mathrm{T}}\mathbf{V}_{\alpha\beta\gamma_h} \tag{9.26}$$

If the machines feeds a balanced load then the v_d and v_q voltages do not have harmonics, only the DC coefficient, and equation (9.24) simplifies to the following terms:

$$\begin{aligned} V_{\alpha_{-1}} &= \frac{1}{2}(V_{d_0} + V_{q_0}) & V_{\beta_{-1}} &= \frac{1}{2}(-V_{d_0} + V_{q_0}) \\ V_{\alpha_1} &= \frac{1}{2}(V_{d_0} - V_{q_0}) & V_{\beta_1} &= \frac{1}{2}(V_{d_0} + V_{q_0}) \end{aligned}$$

and in terms of the cas function,

$$\begin{aligned} v_\alpha &= \frac{1}{2}(V_{d_0} + V_{q_0})\operatorname{cas}(-\nu_0 t) + \frac{1}{2}(V_{d_0} - V_{q_0})\operatorname{cas}\nu_0 t = V_{d_0}\cos\omega_0 t - V_{q_0}\sin\omega_0 t \\ v_\beta &= \frac{1}{2}(-V_{d_0} + V_{q_0})\operatorname{cas}(-\nu_0 t) + \frac{1}{2}(V_{d_0} + V_{q_0})\operatorname{cas}\nu_0 t = V_{d_0}\sin\omega_0 t + V_{q_0}\cos\omega_0 t \end{aligned}$$

This is an expected result for a synchronous machine feeding a balanced load, and the simplification has been used to show the generality of the theory.

We now derive a similar equation to (9.26) but for the currents,

$$\mathbf{I}_{dq0_h} = \mathbf{C}_h^{\mathrm{T}}\mathbf{I}_{\alpha\beta\gamma_h} \tag{9.27}$$

Using (9.26) and (9.27) in (9.7) leads to

$$\mathbf{I}_{\alpha\beta\gamma_h} = \mathbf{Y}_{\alpha\beta\gamma_h}\mathbf{V}_{\alpha\beta\gamma_h} \tag{9.28}$$

where

$$\mathbf{Y}_{\alpha\beta\gamma_h} = \mathbf{C}_h\mathbf{Y}_{dq0_h}\mathbf{C}_h^{\mathrm{T}} \tag{9.29}$$

This matrix is termed the Hartley harmonic admittance in the $\alpha\beta\gamma$ axes.

The MATLAB™ function given below is used to obtain the admittance matrix $\mathbf{Y}_{\alpha\beta\gamma_h}$:

```
function Yabg=calc_Yabg(Ydq0,h)
hh=h+1;
% Form matrix C
C=zeros(hh*2+1,hh*2+1);
U=[1 0 0;0 1 0;0 0 0];
M=[0 1 0;-1 0 0;0 0 0];
N=U+M;
D=[0 0 0;0 0 0;0 0 2];
C=form_A(C,D,1,1);
C=form_A(C,D,hh*2+1,hh*2+1);
C=form_A(C,D,hh+1,hh+1);
for j=1:hh-1,
 C=form_A(C,D,hh+1+j,hh+1+j);
 C=form_A(C,D,hh+1-j,hh+1-j);
 C=form_A(C,U,hh+1+j+1,hh+1+j);
 C=form_A(C,U,hh+1-j,hh+1-j-1);
 C=form_A(C,U,hh+1+j,hh+1+j+1);
 C=form_A(C,U,hh+1-j-1,hh+1-j);
 C=form_A(C,M',hh+1-j,hh+1+j+1);
 C=form_A(C,M',hh+1+j+1,hh+1-j);
 C=form_A(C,M,hh+1+j,hh+1-j-1);
 C=form_A(C,M,hh+1-j-1,hh+1+j);
end
C=form_A(C,N,hh+1,hh+1-1);
C=form_A(C,N,hh+1-1,hh+1);
C=form_A(C,N',hh+1,hh+1+1);
C=form_A(C,N',hh+1+1,hh+1);
C=C/2;
% Form matrix Yabg
Yabg=C*Ydq0*(C');
Yabg=Yabg(4:hh*2*3,4:hh*2*3);
```

9.2.3 *Machine's harmonic admittance matrix in abc axes*

The transformation from $\alpha\beta\gamma$ to *abc* axes is

$$\begin{aligned}\mathbf{V}_{\alpha\beta\gamma} &= \mathbf{T}_t\mathbf{V}_{abc}\\ \mathbf{I}_{\alpha\beta\gamma} &= \mathbf{T}_t\mathbf{I}_{abc}\end{aligned} \tag{9.30}$$

where

$$\mathbf{T}_t = \sqrt{2/3}\begin{bmatrix} 1 & -1/2 & -1/2 \\ 0 & \sqrt{3}/2 & -\sqrt{3}/2 \\ 1/\sqrt{2} & 1/\sqrt{2} & 1/\sqrt{2} \end{bmatrix}$$

and $\mathbf{T}_t^{-1} = \mathbf{T}_t^{\mathrm{T}}$.

For h harmonics, the transformation matrix becomes

$$\mathbf{T}_{t_h} = \begin{bmatrix} \ddots & & & & \\ & \mathbf{T}_t & & 0 & \\ & & \mathbf{T}_t & & \\ & 0 & & \mathbf{T}_t & \\ & & & & \ddots \end{bmatrix} \tag{9.31}$$

A substitution of (9.31) in (9.30) for h harmonics gives

$$\mathbf{I}_{abc_h} = \mathbf{Y}_{abc_h}\mathbf{V}_{abc_h}$$

where

$$\mathbf{Y}_{abc_h} = \mathbf{T}_{t_h}^{\mathrm{T}}\mathbf{Y}_{\alpha\beta\gamma_h}\mathbf{T}_{t_h}$$

represents a Hartley harmonic admittance in the *abc* frame of reference.

The following MATLAB™ function is used to obtain the admittance matrix $\mathbf{Y}_{abc_h}$:

```
function Yabc=calc_Yabc(Yabg,h,connection)
% connection : 0 for star, 1 for delta
% Form matrix T
a=1/2; b=sqrt(3)/2; c=1/sqrt(2); d=sqrt(2/3);
Ti=d*[1 -a -a;0 b -b;c c c ];
T=zeros(2*h+1,2*h+1);
for j=1:2*h+1,
 T=form_A(T,Ti,j,j);
end
% Form matrix Yabc
Yabc=(T')*Yabg*T;
% Change order of matrix Yabc
for j=1:3,
 c=0;
 for k=j:2*h+j,
  Y(k+(2*h)*(j-1),1:(2*h+1)*3)=Yabc(j+c,1:(2*h+1)*3);
  c=c+3;
 end
end
for j=1:3,
 c=0;
 for k=j:2*h+j,
  Yy(1:(2*h+1)*3,k+(2*h)*(j-1))=Y(1:(2*h+1)*3,j+c);
  c=c+3;
 end
end
Yabc=Yy;
% For delta connection
if connection==1,
 U=zeros(2*h+1,2*h+1);
 for j=1:2*h+1,
  U(j,j)=1;
 end
```

```
  Q=[U -U 0*U;0*U U -U;-U 0*U U];
  Yabc=Q*Yy*Q';
end
```

The three Hartley admittance matrices defined above have very interesting patterns. The pattern of $\mathbf{Y}_{\alpha\beta\gamma_h}$ suggests that the harmonic voltage of order i, $V_{\alpha\beta\gamma_i}$, can generate harmonic currents of order $I_{\alpha\beta\gamma_i}$ and $I_{\alpha\beta\gamma_{i\pm1}}$, with the same process being present in $\mathbf{Y}_{abc_h}$. This result illustrates the frequency conversion effect present in synchronous machines working under unbalanced conditions. From $\mathbf{Y}_{\alpha\beta\gamma_h}$, it is seen that the γ equivalent (zero sequence) does not contribute to the frequency conversion process. It is important to notice that $\mathbf{Y}_{abc_h}$ must be ordered differently before it can be combined with the admittance matrix of the external circuit. The original and modified patterns for $\mathbf{Y}_{abc_h}$ are shown in the following diagrams:

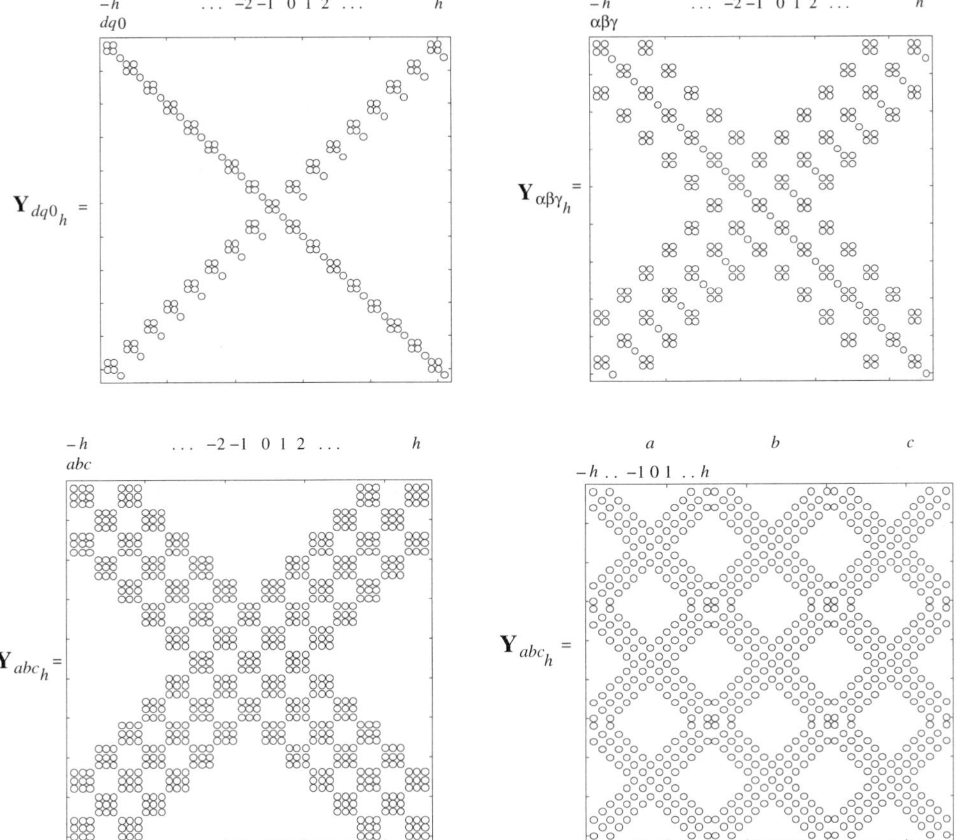

$\mathbf{Y}_{abc_h}$ gives the harmonic model of the synchronous machine, including the frequency conversion effect.

9.3 Saturation in the Synchronous Machine

Magnetic saturation in the synchronous machine is well modelled in phase coordinates by means of a Norton-type linearised equivalent, similar to the single-phase magnetic core.

Synchronous machine saturation may be represented by a non-linear equation of the form

$$i = f(\psi) \tag{9.32}$$

where a dynamic equation that represents the flux is

$$v = \dot{\psi} \tag{9.33}$$

It should be remarked that the current i, voltage v and the flux ψ are all periodic functions.

Applying harmonic domain linearisation to the non-linear function (9.32) leads to a harmonic Norton equivalent given by

$$\mathbf{I} = \mathbf{B}_{abc_h}\mathbf{V} + \mathbf{I}_\mathrm{N} \tag{9.34}$$

where

$$\mathbf{I}_\mathrm{N} = \mathbf{I}_\mathrm{b} - \mathbf{B}_{abc_h}\mathbf{V}_\mathrm{b} \tag{9.35}$$

Figure 9.2 gives the equivalent circuit representation of a three-phase synchronous machine. The frequency conversion effect is included in the admittance matrix $\mathbf{Y}_\mathrm{g} = \mathbf{Y}_{abc_h}$ and the saturation effect in the Norton-type equivalent.

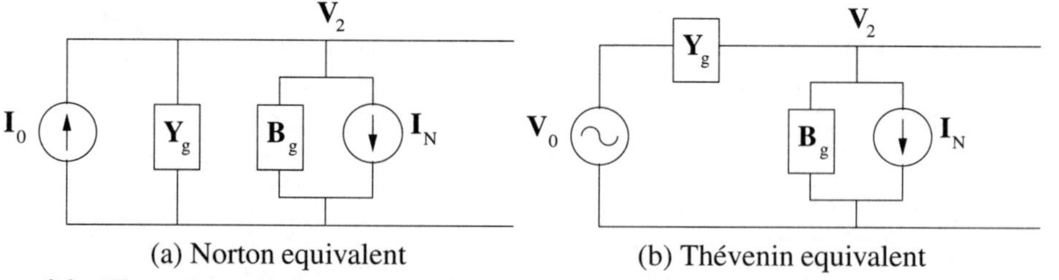

Figure 9.2 Three-phase synchronous machine model for harmonic studies including saturation effect

Example 9-1: This example illustrates the frequency conversion present in synchronous generators. In the example, a star-connected synchronous machine feeds an unbalanced load connected in star. The synchronous machine data used in this example are given in the function `dat_maq1.m`. The internal source $\mathbf{I}_0$ is a balanced three-phase sinusoidal current source with a magnitude of 1.5 p.u. The star-connected load has the following parameters: $R = 0.8$ p.u., $L = 0.8$ p.u. and $C = 0.8$ p.u., and connected to phases a, b and c, respectively.

Frequency conversion effect: Figure 9.3 shows the three-phase voltage and current waveforms at the generator's terminals. It should be mentioned that the results were generated using harmonic domain techniques and that they are presented in time domain form only as a matter of convenience.

The load connected to phase a is resistive and the voltage and current waveforms at that phase are largely in phase. Similarly, the load connected to phase b is inductive and the current lags the voltage by a considerable margin. In phase c, the load is capacitive and the opposite effect takes place: the current leads the voltage. As expected, and owing to the unbalanced nature of the load, the frequency conversion mechanism is activated, giving rise to very significant harmonic voltage and current distortion. Table 9.1 gives numeric information for the voltages in phases a, b, c, where values of around 16%, 24% and 12% are shown for the third harmonic frequency. These values are given as a percentage of the fundamental frequency value. It should be noted that harmonic values above the seventh harmonic are very small and that no saturation effect is included in this example.

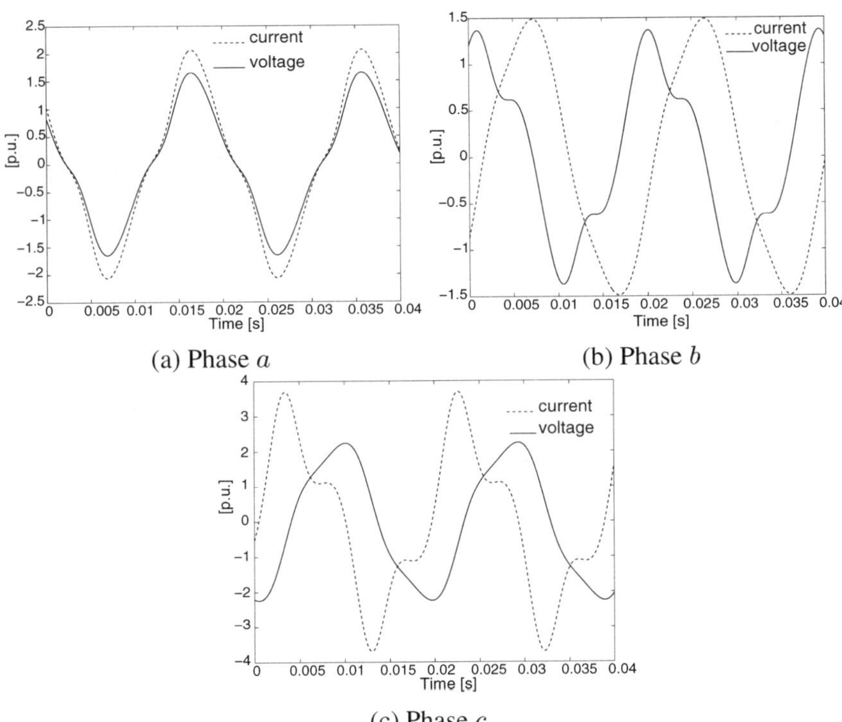

(a) Phase a (b) Phase b

(c) Phase c

Figure 9.3 Generator's waveform currents and voltages

Table 9.1 Percentage of harmonics contents

Harmonic	V_a	V_b	V_c	I_a	I_b	I_c
3	15.6856	23.8677	12.1080	15.6856	7.9559	36.3241
5	2.7745	3.5549	0.6612	2.7745	0.7110	3.3060
7	0.1675	0.2323	0.0229	0.1675	0.0332	0.1606
9	0.0056	0.0089	0.0005	0.0056	0.0010	0.0049
11	0.0001	0.0002	0.0000	0.0001	0.0000	0.0001

Table 9.1 also gives the harmonic content of the currents, where values of around 16%, 8% and 36% are shown. The large value of the third harmonic current in phase c is due to the influence of the capacitive load on the generator's winding in that phase. This result is of very considerable interest and further study shows the existence of a resonant condition in phase c, between the capacitor load and the winding inductance, when the capacitor takes a value of around 2 p.u. In this study, the load in phases a and b remain with no change but the capacitor

is varied incrementally from 0 to 3 p.u. Results are presented for the currents in Figure 9.4, where it is shown that the resonant condition in phase c affects quite significantly the currents in the other two phases.

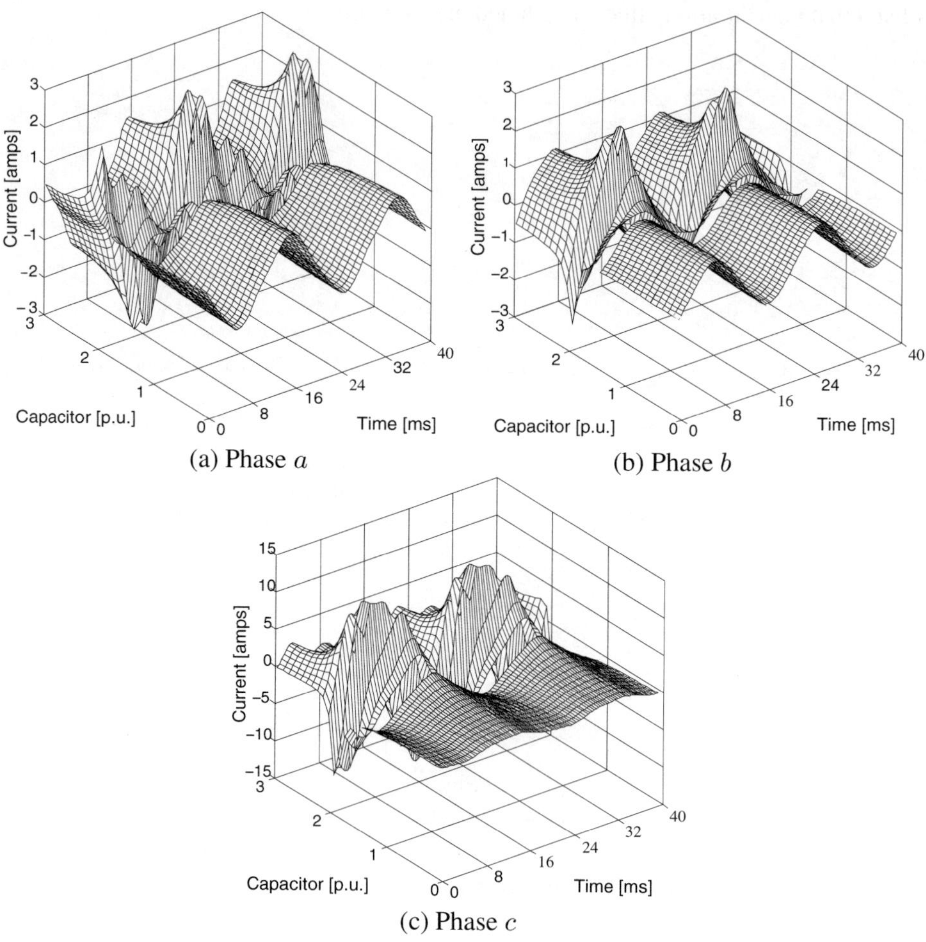

Figure 9.4 Generator's current waveforms, using different values of capacitance

Saturation effect: Figure 9.5 and Table 9.2 give similar information as above, but for the case when synchronous machine saturation is included in addition to the frequency conversion effect. The non-linear relation used to represent saturation in this example is

$$i = 0.0348\psi + 0.001526\psi^9 \text{ p.u.}$$

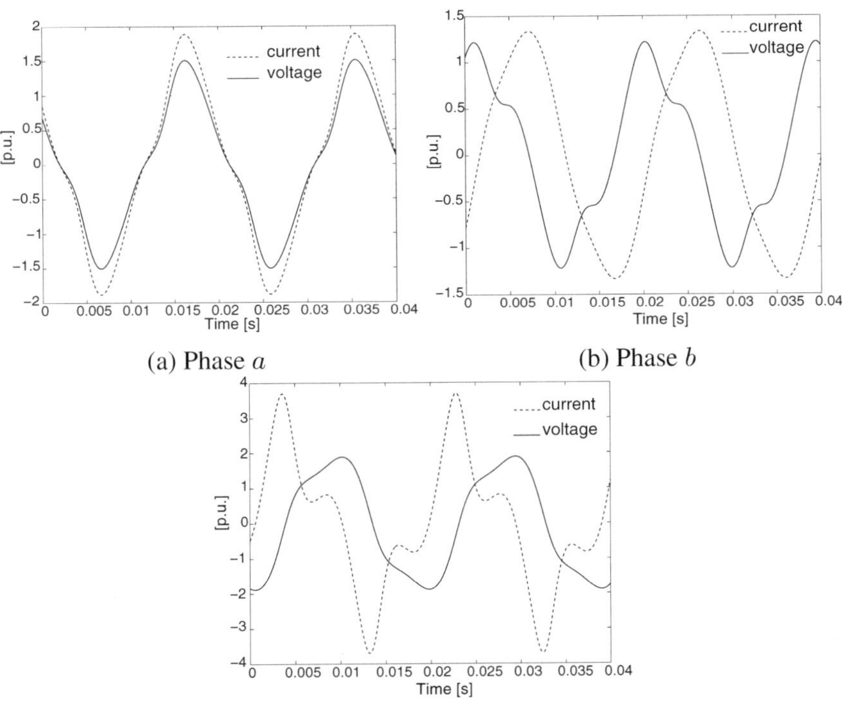

(a) Phase a (b) Phase b

(c) Phase c

Figure 9.5 Generator's waveshapes, including saturation

Table 9.2 Percentage of harmonics contents

Harmonic	V_a	V_b	V_c	I_a	I_b	I_c
3	15.7221	21.4273	15.7154	15.7221	7.1424	47.1462
5	3.4557	3.5527	1.8022	3.4557	0.7105	9.0109
7	0.3300	0.3243	0.3559	0.3300	0.0463	2.4915
9	0.0374	0.0609	0.0778	0.0374	0.0068	0.7003
11	0.0051	0.0128	0.0155	0.0051	0.0012	0.1707
13	0.0006	0.0020	0.0028	0.0006	0.0002	0.0360
15	0.0000	0.0003	0.0004	0.0000	0.0000	0.0062

9.4 Experimental Results

Tests were conducted in the laboratory using a small synchronous generator feeding an unbalanced load. The generator was a Lab-Volt EMS 8507-00, three phase, 220 V, 1.5 kVA, 120 Vdc, 4.2 A, operating at 60 Hz and unit power factor. The load comprises a resistance, an inductor and a capacitor of 60 Ω, 80 mH and 44 μF, connected in phases a, b and c , solidly grounded star. Figure 9.6 shows the currents in the three phases of the synchronous generator and their harmonic contents. Figure 9.7 shows the voltages at the terminals of the generator. For completeness, the current in the neutral wire is shown in Figure 9.8. These results clearly show the ability of the synchronous generator to produce harmonic distortion.

Owing to the inductive nature of the synchronous generator, the current in phase c presents a high level of harmonic distortion due to a possible resonant condition between the generator and the capacitor connected in phase c of the load.

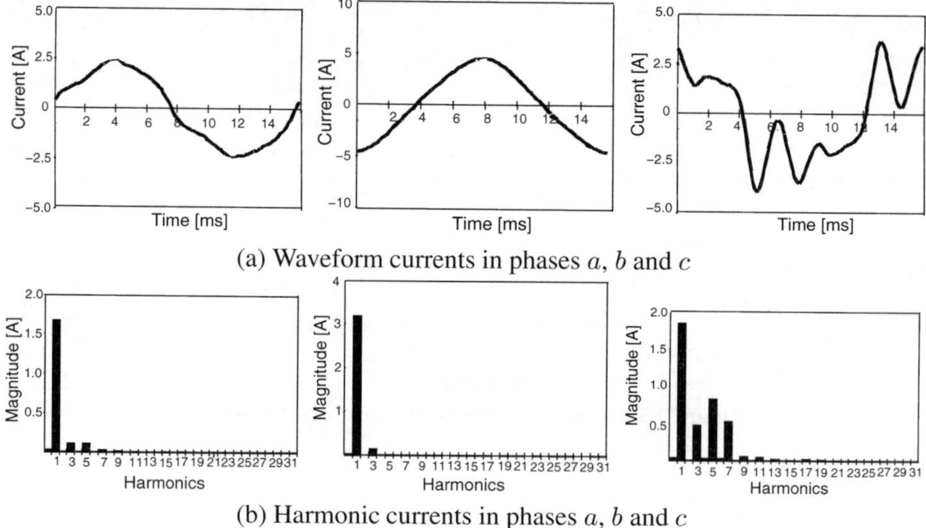

Figure 9.6 Generator currents

9.5 Summary

The periodic steady-state performance of the three-phase synchronous generator has been addressed in this chapter. A comprehensive model of the synchronous generator was developed by resorting to phase domain and harmonic domain relations expressed in terms of Hartley series. A key characteristic of the model is that the frequency conversion effect that intrinsically exists between the rotor and stator circuits, and triggered only when the generator feeds an unbalanced circuit, is elegantly and efficiently incorporated in the model. To make the model as comprehensive and realistic as possible, saturation effects in the stator's magnetic circuit were included using the stator's magnetising characteristic. The saturation model comes in the form of a harmonic Norton equivalent, and goes well with the harmonic impedance that represents the frequency conversion effects.

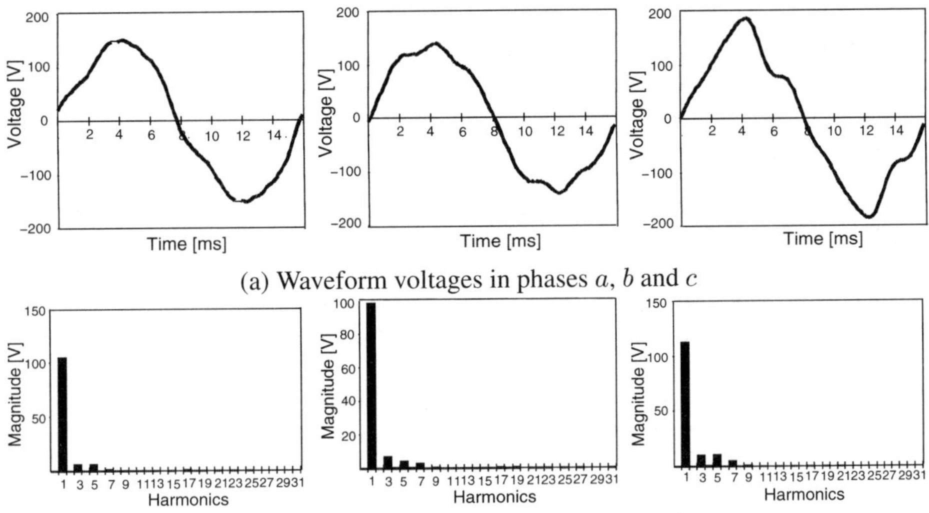

(a) Waveform voltages in phases a, b and c

(b) Harmonic voltages in phases a, b and c

Figure 9.7 Generator voltages

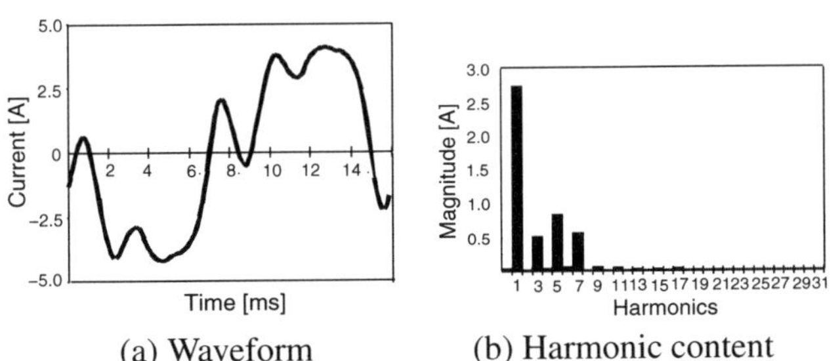

(a) Waveform (b) Harmonic content

Figure 9.8 Current in the neutral wire, THD=41.46%, RMS=2.97 A

Studies were conducted for the case when the generator feeds a star-connected unbalanced load, comprising a resistor, an inductor and a capacitor. The voltage and current waveforms behave largely as expected, with the capacitive load in on of the phases aggravating the harmonic distortion produced by the generator. Further studies showed the existence of a resonant condition in that phase for a given value of capacitive load, i.e. 2 p.u. A comparable experiment conducted in the laboratory using a small generator showed that the voltage and current in the phase, with capacitive load, had a larger amount of distortion than the other two phases.

9.6 Bibliography

1. A. Semlyen, J.F. Eggleston, J. Arrillaga, "Admittance Matrix Model of a Synchronous Machine for Harmonics Analysis", *IEEE Transactions on Power Systems*, Vol. 2, No. 4, November 1987, pp. 833–840.
2. W. Xu, H.W. Dommel, J.R. Marti, "A Synchronous Machine Model for Three-Phase Harmonic Analysis and EMTP Initialization", *IEEE Transactions on Power Systems*, Vol. 6, No. 4, November 1991, pp. 1530–1538.
3. M. Madrigal, S. Acha, "A Hartley Three Phase Synchronous Machine Model for Harmonic Analysis", *Proceedings of ICHQP-96*, Las Vegas, USA, October 1996, pp. 412–418.
4. J. Arrillaga, B.C. Smith, N.R. Watson, A.R. Wood, *Power System Harmonic Analysis*, John Wiley & Sons, Chichester, 1997.
5. D. O'Kelly, S. Simmons, *Introduction to Generalized Machine Theory*, McGraw-Hill, London, 1968.
6. P.C. Krause, O. Wasynczuk, S.D. Sudhoff, *Analysis of Electric Machinery*, IEEE Press, New York, 1994.
7. G.T. Heydt, "System Analysis Using Hartley Impedance", *IEEE Transactions on Power Delivery*, Vol.8, No. 2, April 1993, pp. 518–523.

10

Transmission Lines

10.1 Introduction

The calculation of transmission line parameters at harmonic frequencies is more involved than the calculation of transmission line parameters at power frequencies. Several effects which are normally ignored at power frequencies have to be included at harmonic frequencies [1], for instance:

- Frequency dependence
- Long-line effects
- Line imbalances
- Line transpositions
- VAR compensation plant

More specialised studies, such as inductive interference in communication circuits by power lines, may require a discontinuous representation of the earth resistivity along the path of the line. Information at several points along the line is required for the assessment of harmonic interference.

Several methods are available for the evaluation of transmission line frequency-dependent effects. Early methods required the solution of either infinite integrals or Bessel functions (infinite series) [2]. The long CPU times associated with the calculation of transmission line parameters by means of infinite integrals or infinite series, for a wide range of frequencies, triggered research into closed-form solution methods [3]. These methods are based on the use of the concept of complex penetration. They only take a small fraction of the time required by the infinite series to converge [4].

The calculation of transmission line parameters suitable for harmonic studies is covered in this chapter. The theory may be divided into two main parts:

1. Evaluation of lumped parameters: The lumped parameters are obtained from the geometric configuration of the transmission line taking into account the effect of the earth return and skin effects.
2. Evaluation of distributed parameters: Long-line effects are added to the lumped parameters, to generate an exact model of the transmission line at harmonic frequencies.

The effects of transpositions and VAR compensation on transmission lines are also covered in this chapter.

10.2 Evaluation of Lumped Parameters

The lumped series impedance and shunt admittance of a multi-conductor transmission line may be defined as follows:

$$\mathbf{Z} = \mathbf{Z}_{\text{G-E}} + \mathbf{Z}_{\text{Skin}} \;\; \Omega/\text{km} \tag{10.1}$$

$$\mathbf{Y} = \text{j}1000\omega 2\pi\epsilon_0 \mathbf{P}^{-1} \;\; 1/\Omega\text{km} \tag{10.2}$$

where $\mathbf{Z}_{\text{G-E}} = \mathbf{Z}_{\text{Geom}} + \mathbf{Z}_{\text{Earth}}$ is the impedance matrix due to the magnetic fluxes outside the conductors, including the impedance contribution due to a lossy earth return path; $\mathbf{Z}_{\text{Skin}}$ is the impedance matrix due to the magnetic fluxes inside the conductors; $\mathbf{P}$ is the matrix of potential coefficients, i.e. Maxwell coefficients, and

$\epsilon_0 = 8.854 \times 10^{-9}$ F/km is the permitivity of free space,

$\omega = \omega_0 h$, where $\omega_0 = 2\pi f_0$,

f_0 is the base operating frequency in Hz,

h is the harmonic order.

10.2.1 Matrix of potential coefficients $\mathbf{P}$

The method of images is used to calculate the potential coefficients, which are obtained solely from the geometric arrangement of the conductors in the tower.

The method of images allows the conducting plane to be replaced by a fictitious conductor located at the mirror image of the actual conductors. Figure 10.1 shows the case when the conductors l and m above ground have been replaced by two equivalent conductors and their images located at their mirror image.

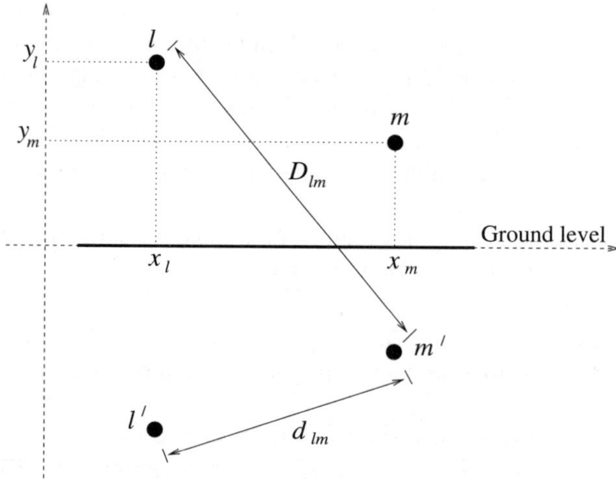

Figure 10.1 Line geometry and its image

The elements of the matrix of potential coefficients $\mathbf{P}$ are given by the self-potential coefficient P_{ll} for the lth conductor and the mutual potential coefficient P_{lm} between the lth and the mth conductors, and are defined as follows:

$$P_{ll} = \ln \frac{2y_l}{r_{\text{ext},l}} \tag{10.3}$$

$$P_{lm} = \ln \frac{D_{lm}}{d_{lm}} \tag{10.4}$$

where:

$$D_{lm} = \sqrt{(x_m - x_l)^2 + (y_m + y_l)^2} \text{ m}$$

$$d_{lm} = \sqrt{(x_m - x_l)^2 + (y_m - y_l)^2} \text{ m}$$

$r_{\text{ext},l}$ is the external radius in m of the lth conductor, and the other variables in Figure 10.1 are also given in m.

The MATLAB™ function given below is used to obtain the matrix of potential coefficients:

```
function P=calc_P(nc,rad_ext,x,y)
P=zeros(nc,nc);
for l=1:nc
 for m=1:nc
  rad_el  = rad_ext(l);
  xl      = x(l); yl  = y(l);
  xm      = x(m); ym  = y(m);
  Dlm     = sqrt((xm-xl)^2+(yl+ym)^2);
  dlm     = sqrt((xm-xl)^2+(yl-ym)^2);
  if l==m
   P(l,l)=log(2*yl/rad_el);
  else
   P(l,m)=log(Dlm/dlm);
  end
 end
end
```

10.2.2 *Impedance matrix* $\mathbf{Z}_{\text{G-E}}$

If the conductors and the earth are assumed to be equipotential surfaces then the geometric impedances can be formulated in terms of potential coefficients, and the effects of the earth return may be included by adding a correction term. Owing to the fact that the impedance of the earth return path varies with frequency in a non-linear fashion and because of the non-uniform nature of the land, as well as its lack of homogeneous conductivity, this is a problem for which an accurate solution may not exist. Nevertheless, the solution to the problem of impedances involving a perfectly flat earth, with homogeneous conductivity, is still of practical and also of theoretical significance. A satisfactory solution to the related problem of a current-carrying wire above a lossy earth was first published by Carson three-quarters of century ago [2]. The solution was given in the form of an infinite integral, which does not have a closed form or analytical answer, but it is conveniently expressed as a set of infinite series.

Ever since, and perhaps because of the existence of these infinite series, this solution has been adopted world-wide as the foundation for almost every study in the areas of electromagnetic fields, propagation characteristics and magnetic induction effects caused by power lines [6]. Carson's integral was developed with the help of Maxwell's equations and the basic concepts of circuit theory, although the use of the latter restricted the validity of the solution over the full span of frequencies.

More accurate solutions, based solely on the use of electromagnetic theory concepts (Maxwell's equations), were pursued afterwards and two equivalent answers [7,8] in the form of infinite integrals emerged. Furthermore, it has been demonstrated that Carson's integral corresponds to a particular case of the more accurate integrals of Wedepohl or Wait; but perhaps more importantly, it has been found that for most of the conditions prevailing in power systems applications, Carson's solutions are as good as the more accurate approaches. Heavy computational burdens are associated with the numerical solution of any of the three infinite integrals that prevent their use in everyday applications, and it is only in a reduced number of situations, such as conductors in close proximity to the ground and radio frequency applications, that the more accurate approaches should be used [6].

As the need arises to calculate impedances for a wide range of frequencies, and also because of the uncertainty in the available data, the tendency is to use simpler formulations aiming at a reduction in computing time, while keeping accuracy at a reasonable level. In particular, closed-form formulations which yield very elegant and accurate solutions were put forward in the late 1970s and early 1980s. The various closed-form approaches vary slightly from each other but, in power systems applications, the equations due to C. Dubanton [9] based on complex penetration have had more appeal. The concept of complex penetration, illustrated in Figure 10.2, was developed intuitively and the heuristic equations applied to actual computations. Mathematical proof followed next, showing the equations to be good approximations to the more formal approaches of Wait's or Carson's integrals [10]. The elements of $\mathbf{Z}_{\mathrm{G-E}}$ using Dubanton's equations are given by

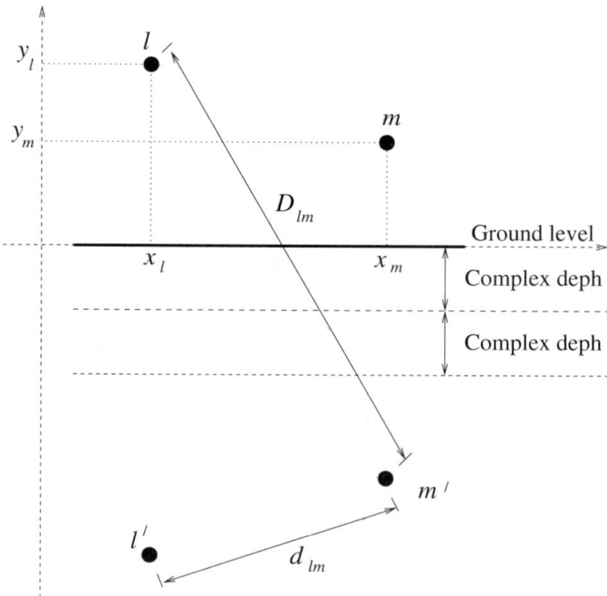

Figure 10.2 Line geometry and its image using the complex penetration approach

$$Z_{\text{G-E}ll} = \frac{\text{j}1000\omega\mu_0}{2\pi}\ln\frac{2\left(y_l + p\right)}{r_{\text{ext},l}} \quad \Omega/\text{km}$$

$$Z_{\text{G-E}lm} = \frac{\text{j}1000\omega\mu_0}{2\pi}\ln\frac{\sqrt{\left(x_l - x_m\right)^2 + \left(y_l + y_m + 2p\right)^2}}{d_{lm}} \quad \Omega/\text{km} \tag{10.5}$$

where:

$\mu_0 = 4\pi \times 10^{-7}$ H/m is the permeability of free space,

$p = \frac{1}{\sqrt{\text{j}\omega\mu_0\sigma}}$ is the complex depth,

σ is the earth conductivity in S/m.

The MATLAB™ function used to calculate $\mathbf{Z}_{\text{G-E}}$ is given below:

```
function Z_GE=calc_Z_GE(nc,rad_ext,x,y,cond,w0,h)
w =w0*h;
m0=4*pi*1e-7;
Z_GE=zeros(nc,nc);
for l=1:nc
 for m=1:nc
  rad_el= rad_ext(l);
  xl    = x(l); yl = y(l);
  xm    = x(m); ym = y(m);
  dlm   = sqrt((xl-xm)^2+(yl-ym)^2);
  p     = 1/sqrt(i*w*m0*cond);
  if l==m
   Z_GE(l,l)=i*1000*w*m0/(2*pi)*log(2*(yl+p)/rad_el);
  else
   Z_GE(l,m)=i*1000*w*m0/(2*pi)*log(sqrt((xl-xm)^2
             +(yl+ym+2*p)^2)/dlm);
  end
 end
end
```

10.2.3 Skin effect impedance matrix $\mathbf{Z}_{\text{Skin}}$

This term accounts for the internal impedance of the conductors [11]. Both resistance and inductance vary with frequency in a non-linear manner and need to be computed at each frequency of interest.

The reasons behind the non-linear variation of these parameters with frequency was recognised many years ago [12] and are attributable mainly to the fact that the current flowing in the conductor does not distribute itself uniformly over the full area available but, rather, it tends to flow on the surface. The overall effect is an increase in the resistance and a decrease in internal inductance. This trend increases with frequency and it is termed the *skin* effect [13].

In the past, extensive research, both theoretical and practical, has been carried out for conductors with regular and irregular shapes. In power systems applications the most successful approach has been the one that models the power conductors as an annular conductor (tube) [11]. The reason is that modern power transmission circuits utilise almost exclusively ACSR (aluminium conductor steel reinforced) conductors and, under the assumption that a negligible amount of current flows through the path of high resistance (the steel), the conductor can be approximated to a tube. The diameter of the conductor is the outer diameter of the tube and the diameter of the steel reinforcement is the inner diameter of the tube, as illustrated below:

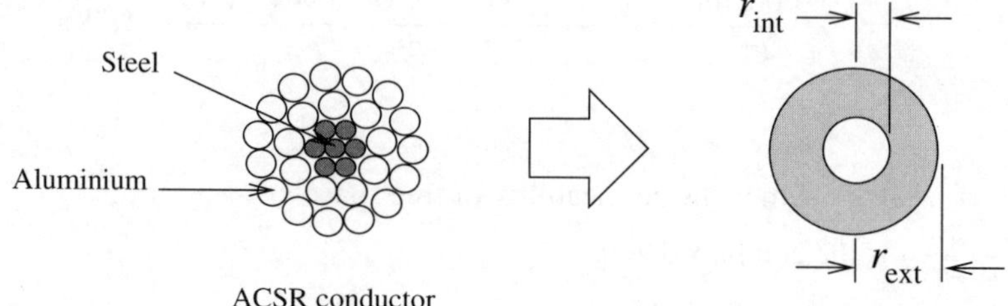

The mathematical model of an annular conductor is well described by means of Bessel functions of the first and second kinds of zero order, and their derivatives [13]. Alternatively, the complex penetration concept can also be applied to find the internal impedance of annular conductors [14]:

$$Z_{\mathrm{Skin}_{ll}} = \sqrt{R_0^2 + Z_\infty} \quad \Omega/\mathrm{km} \tag{10.6}$$

where

$$R_0 = \frac{1}{\pi\left(r_{\mathrm{ext},l}^2 - r_{\mathrm{int},l}^2\right)\sigma_{\mathrm{c}}} \tag{10.7}$$

$$Z_\infty = \frac{1}{2\pi r_{\mathrm{ext},l}\sigma_{\mathrm{c}}p_{\mathrm{c}}} \tag{10.8}$$

$$p_{\mathrm{c}} = \frac{1}{\sqrt{\mathrm{j}\omega\mu_0\sigma_{\mathrm{c}}}} \tag{10.9}$$

where:

$r_{\mathrm{int},l}$ is the internal radius of the conductor l in m,

σ_{c} is the conductivity of the conductor in S/m.

This is a closed-form formulation which overcomes difficulties experienced with the robustness of the infinite series of the Bessel functions. Although errors up to 6.5% exist at low frequencies, its accuracy increases with frequency as indicated by equations (10.7) and (10.8). The method is suitable for the two extreme cases of the impedance, the zero-frequency solution and the infinite frequency solution.

The MATLAB™ function used to calculate the impedance matrix of the skin effect is given below:

```
function Z_Skin=calc_Z_Skin(nc,rad_ext,rad_int,cond_c,w0,h)
w =w0*h;
m0=4*pi*1e-12;
Z_Skin=zeros(nc,nc);
for l=1:nc
```

```
 rad_el  = rad_ext(1);
 rad_il  = rad_int(1);
 sc       = cond_c(1);
 R0    = 1/(pi*(rad_el^2-rad_il^2)*sc);
 pc    = 1/sqrt(i*w*m0*sc/(2*pi));
 Zinf  = 1/(2*pi*rad_el*sc*pc);
 Z_Skin(1,1)=1000*sqrt(R0^2+Zinf^2);
end
```

10.2.4 *Reduced equivalent matrices* $\mathbf{Z}_{abc}$ *and* $\mathbf{Y}_{abc}$

Although AC extra high voltage (EHV) transmission lines contain a large number of conductors, i.e. earth wires and several bundle conductors per phase, as shown in Figure 10.3 below the interest of harmonic studies is not the individual conductors but rather the individual phases. Hence, steps must be taken to find reduced equivalent matrices, which should correctly account for the original configuration while retaining essential information only, i.e. one equivalent conductor per phase.

This can be achieved in at least two different ways. One of them is to use the geometric mean radius (GMR) concept, which, although it is frequency independent, provides an efficient computational solution.

The second method uses matrix reduction techniques. It includes frequency dependency for the series impedance but requires considerably more computation, with the following three steps carried out at each frequency:

1. Matrices $\mathbf{Z}$ and $\mathbf{P}$ are set up with an order equal to the total number of conductors plus earth wires.
2. By assuming that the voltage from line to ground is exactly the same for all the conductors in the bundle, the transformation matrices $\mathbf{B}^{\mathrm{T}}$ and $\mathbf{B}$ are built up, so that modified matrices $\mathbf{Z}_{\mathrm{Re}}$ and $\mathbf{P}_{\mathrm{Re}}$ are obtained,
$$\mathbf{Z}_{\mathrm{Re}} = \mathbf{B}^{\mathrm{T}}\mathbf{Z}\mathbf{B}$$
$$\mathbf{P}_{\mathrm{Re}} = \mathbf{B}^{\mathrm{T}}\mathbf{P}\mathbf{B}$$
3. Next, a partial inversion is applied to matrices $\mathbf{Z}_{\mathrm{Re}}$ and $\mathbf{P}_{\mathrm{Re}}$. All but a number of elements equal to the number of phases are inverted and each of the non-inverted locations corresponds to one conductor per phase. Thus, final reduced equivalent matrices $\mathbf{Z}_{abc}$ and $\mathbf{P}_{abc}$ are obtained, whose order equals the number of phases but implicitly accounts for the original configuration.

The reduced equivalent matrix of potential coefficients $\mathbf{P}_{abc}$ is inverted and multiplied by a constant factor to obtain the shunt admittance matrix, $\mathbf{Y}_{abc} = \mathrm{j}1000\omega 2\pi\epsilon_0 \mathbf{P}_{abc}^{-1}$. The shunt admittance parameters vary linearly with frequency.

The MATLAB™ function used to obtain the reduced equivalent impedance matrix is:

```
function Zabc=reduce_to_ThreePhase(Z,nc,nc_pp)
 B=zeros(nc,nc);
 B3=zeros(nc_pp,nc_pp);
 Zabc=zeros(3,3);
 for k=1:(nc_pp-1)
  B3(k,k)=1;
  B3(1,k+1)=-1;
```

```
end
B3(nc_pp,nc_pp)=1;
B(1:nc_pp,1:nc_pp)=B3;
B(nc_pp+1:2*nc_pp,nc_pp+1:2*nc_pp)=B3;
B(2*nc_pp+1:nc,2*nc_pp+1:nc)=B3;
Zx=B'*Z*B;
for k=2:nc
 if (k~=(nc_pp+1))&(k~=(2*nc_pp+1))
  Zx=part_inv(Zx,nc,k);
 end
end
Zabc(1,1)=Zx(1,1);                   Zabc(1,2)=Zx(1,nc_pp+1);
Zabc(1,3)=Zx(1,nc-nc_pp+1);          Zabc(2,1)=Zx(nc_pp+1,1);
Zabc(2,2)=Zx(nc_pp+1,nc_pp+1);       Zabc(2,3)=Zx(nc_pp+1,2*nc_pp+1);
Zabc(3,1)=Zx(2*nc_pp+1,1);           Zabc(3,2)=Zx(2*nc_pp+1,nc_pp+1);
Zabc(3,3)=Zx(2*nc_pp+1,2*nc_pp+1);
```

The MATLAB™ function used to obtain the partial inversion required by the above function is:

```
function A=part_inv(A,n,x)
 for m=1:n
  if m~=x
   for k=1:x-1,
    A(m,k)=A(m,k)-A(m,x)*A(x,k)/A(x,x);
   end
   for k=x+1:n
    A(m,k)=A(m,k)-A(m,x)*A(x,k)/A(x,x);
   end
  end
 end
 A(x,x)=-1/A(x,x);
 for m=1:n
  if m~=x
   A(x,m)=A(x,m)*A(x,x);
   A(m,x)=A(m,x)*A(x,x);
  end
 end
```

The MATLAB™ function used to obtain the total lumped parameters in phase quantities is:

```
function [Zabc,Yabc]=calc_Zabc_Yabc(nc,nc_pp,cond,x,y,rad_ext,
                                     rad_int,cond_c,f0,h)
w0       = 2*pi*f0;
e0       = 8.854*1e-12;
P        = calc_P(nc,rad_ext,x,y);
Z_GeEr   = calc_Z_GeEr(nc,rad_ext,x,y,cond,w0,h);
Z_Skin   = calc_Z_Skin(nc,rad_ext,rad_int,cond_c,w0,h);
Z        = Z_GeEr+Z_Skin;
Zabc     = reduce_to_ThreePhase(Z,nc,nc_pp);
Pabc     = reduce_to_ThreePhase(P,nc,nc_pp);
Yabc     = i*1000*w0*h*2*pi*e0*inv(Pabc);
```

10.2.5 500 kV power transmission line configuration

The 500 kV, three-phase transmission line of flat configuration shown in Figure 10.3 will be used in this chapter as the base configuration for all the numeric examples. The main parameters of this transmission line are given below:

Base voltage: 500 kV

Base power: 100 MVA

Conductor type: Panther (30/3.00+7/3.00 ACSR)

Conductivity of the conductor: 25785150 m/Ω

External radius: 1.049 cm

Internal radius: 0.453 cm

Conductivity of the earth: 0.01 S/m

Nominal frequency: 50 Hz

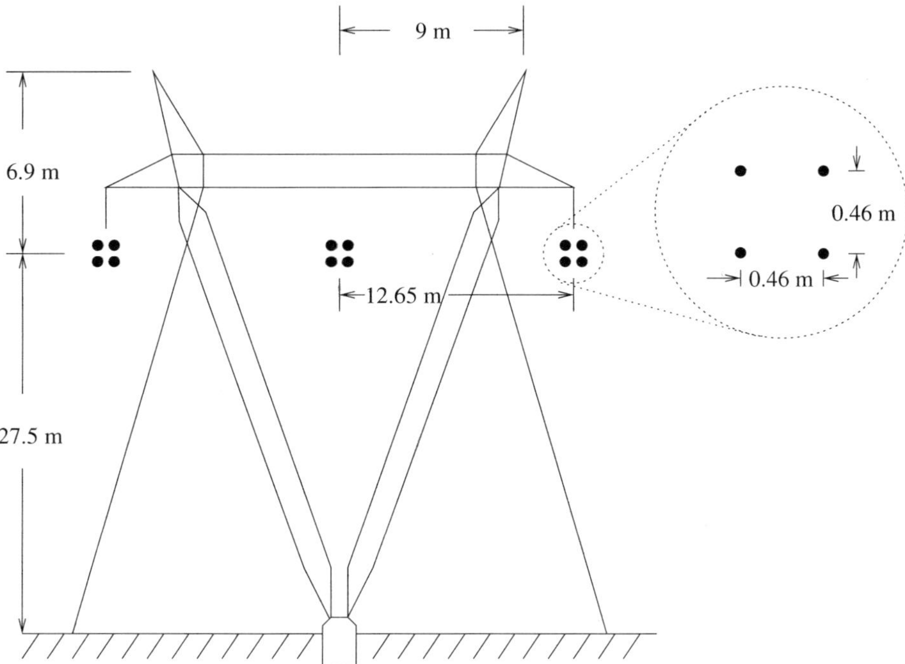

Figure 10.3 Power transmission line

The MATLAB™ file below contains the basic parameters of the transmission line shown in Figure 10.3:

```
% trans_line.m
nc      = 12;   % total number of conductors
nc_pp   = 4;    % number of conductors per phase
cond    = 0.01; % earth conductivity in S/m
% conductors coordinates in m:
```

```
% (x,y)=[ a1 a2 a3 a4 b1 b2 b3 b4 c1 c2 c3 c4]
x=[12.42 12.88 12.88 12.42 -0.23 0.23 0.23 -0.23 -12.88 -12.42
   -12.42 -12.88];
y=[27.50 27.50 27.96 27.96 27.50 27.50 27.96 27.96 27.50 27.50
   27.96 27.96];
rad_ext  = [1 1 1 1 1 1 1 1 1 1 1 1]*1.0490/100; % ext radius in m
rad_int  = [1 1 1 1 1 1 1 1 1 1 1 1]*0.4530/100; % int radius in m
cond_c   = [1 1 1 1 1 1 1 1 1 1 1 1]*25785150;   % cond in S/m
```

Example 10-1: Using the transmission line configuration in Figure 10.3, the lumped parameter $\mathbf{Z}_{abc}$ in Ω/km and $\mathbf{Y}_{abc}$ in S/km at 50 Hz are obtained using the next steps in MATLAB™:

```
>>trans_line
>>[Zabc,Yabc]=calc_Zabc_Yabc(nc,nc_pp,cond,x,y,rad_ext,rad_int,
                             cond_c,50,1)
Zabc =
   0.0815 + 0.5396i   0.0470 + 0.2774i   0.0470 + 0.2339i
   0.0470 + 0.2774i   0.0815 + 0.5396i   0.0470 + 0.2774i
   0.0470 + 0.2339i   0.0470 + 0.2774i   0.0815 + 0.5396i
Yabc =
   1.0e-05 *
        0 + 0.3357i        0 - 0.0811i        0 - 0.0307i
        0 - 0.0811i        0 + 0.3525i        0 - 0.0811i
        0 - 0.0307i        0 - 0.0811i        0 + 0.3357i
>>
```

10.3 Evaluation of Distributed Parameters

The incorporation of long-line effects into multi-conductor transmission lines is not as simple as for single-phase lines, because it involves matrices rather than scalar operations [15,16].

In a multi-conductor transmission line the voltages and the currents existing in an incremental section Δx may be expressed as

$$\frac{\partial^2 \mathbf{V}}{\partial x^2} = \mathbf{Z}_{abc}\mathbf{Y}_{abc}\mathbf{V} \tag{10.10}$$

$$\frac{\partial^2 \mathbf{I}}{\partial x^2} = \mathbf{Y}_{abc}\mathbf{Z}_{abc}\mathbf{I} \tag{10.11}$$

The solution of these equations for $\mathbf{V}$ and $\mathbf{I}$ may be expressed in a variety of ways, which are all useful for conducting harmonic studies involving multi-conductor transmission lines [17]. For instance, for multi-phase transmission lines one possible solution is given by

$$\begin{bmatrix} \mathbf{V}_{ABC} \\ \mathbf{V}_{abc} \end{bmatrix} = \begin{bmatrix} \mathbf{Z}_{\mathrm{SS}} & \mathbf{Z}_{\mathrm{SR}} \\ \mathbf{Z}_{\mathrm{RS}} & \mathbf{Z}_{\mathrm{RR}} \end{bmatrix} \begin{bmatrix} \mathbf{I}_{ABC} \\ \mathbf{I}_{abc} \end{bmatrix} \tag{10.12}$$

where the subindices ABC and abc are the sending and receiving ends of the transmission line.

The operations involved in the calculation of these impedance parameters may be difficult to achieve directly in the phases. Because operations such as square roots, logarithms, circular and hyperbolic functions, etc., are not directly defined in matrix theory, the solution adopted to obtain (10.12) has been to diagonalise the matrices [15] using modal analysis, and then normal scalar operations are carried out for the decoupled matrices. Once this stage has been completed, inverse transformations are used to obtain phase domain parameters, which are given by

$$\mathbf{Z}_{\mathrm{SS}} = \mathbf{Z}_{\mathrm{RR}} = \mathbf{T}_v \mathbf{Z}_{c,m} \tanh^{-1} \left(\mathrm{diag}\left(\Gamma_m \right) l \right) \mathbf{T}_i^{-1} \tag{10.13}$$

$$\mathbf{Z}_{SR} = \mathbf{Z}_{RS} = \mathbf{T}_v \mathbf{Z}_{c,m} \sinh^{-1} \left(\mathrm{diag}\left(\Gamma_m \right) l \right) \mathbf{T}_i^{-1} \tag{10.14}$$

where m includes the two aerial modes and the ground mode, α, β and 0, respectively.

Moreover, the following key information is also defined:

$$\begin{aligned}
\mathbf{T}_v &= \mathrm{eig}\left\{ \mathbf{Z}_{abc} \mathbf{Y}_{abc} \right\} & \mathbf{T}_i &= \mathrm{eig}\left\{ \mathbf{Y}_{abc} \mathbf{Z}_{abc} \right\} \\
\Gamma_m &= \sqrt{\mathrm{diag}\left(\mathbf{Z}_m \mathbf{Y}_m \right)} & \mathbf{Z}_{c,m} &= \sqrt{\mathrm{diag}\left(\mathbf{Z}_m \mathbf{Y}_m^{-1} \right)} \\
\mathbf{Z}_m &= \mathbf{T}_v^{-1} \mathbf{Z}_{abc} \mathbf{T}_i & \mathbf{Y}_m &= \mathbf{T}_i^{-1} \mathbf{Y}_{abc} \mathbf{T}_v \\
\mathbf{T}_i^T &= \mathbf{T}_v^{-1}
\end{aligned}$$

$\mathbf{T}_v$ and $\mathbf{T}_i$ contain the eigenvectors of $\mathbf{Z}_{abc}\mathbf{Y}_{abc}$ and $\mathbf{Y}_{abc}\mathbf{Z}_{abc}$, respectively.

Γ_m is a diagonal matrix which contains the propagation constants $\gamma_m = \sqrt{z_m y_m}$ per km.

$\mathbf{Z}_{c,m}$ is a diagonal matrix which contains the characteristic impedances $z_{c,m} = \sqrt{z_m / y_m}$ in Ω where $z_m \in \mathbf{Z}_m$ and $y_m \in \mathbf{Y}_m$.

$\mathbf{Z}_m$ and $\mathbf{Y}_m$ are diagonal matrices of modal impedances and modal admittances, respectively.

l is the length of the transmission line in km.

Also,

$$\mathbf{V}_{abc} = \mathbf{T}_v \mathbf{V}_m$$

$$\mathbf{I}_{abc} = \mathbf{T}_i \mathbf{I}_m$$

where $\mathbf{V}_m$ and $\mathbf{I}_m$ are vectors of modal quantities.

Another representation of the transmission line is given in the form of a transfer admittance matrix,

$$\begin{aligned}
\begin{bmatrix} \mathbf{I}_{ABC} \\ \mathbf{I}_{abc} \end{bmatrix} &= \begin{bmatrix} \mathbf{Y}_{\mathrm{SS}} & \mathbf{Y}_{\mathrm{SR}} \\ \mathbf{Y}_{\mathrm{RS}} & \mathbf{Y}_{\mathrm{RR}} \end{bmatrix} \begin{bmatrix} \mathbf{V}_{ABC} \\ \mathbf{V}_{abc} \end{bmatrix} \\
&= \begin{bmatrix} \mathbf{Z}_{\mathrm{SS}} & \mathbf{Z}_{\mathrm{SR}} \\ \mathbf{Z}_{\mathrm{RS}} & \mathbf{Z}_{\mathrm{RR}} \end{bmatrix}^{-1} \begin{bmatrix} \mathbf{V}_{ABC} \\ \mathbf{V}_{abc} \end{bmatrix}
\end{aligned} \tag{10.15}$$

Equation (10.15) can be used to obtain an equivalent π-circuit of the transmission line, as shown in Figure 10.4, with the following parameters:

$$\mathbf{Y}_{\text{SS}} = \mathbf{Y}_{\text{RR}} = \mathbf{Z}_{\text{series}}^{-1} + \frac{1}{2}\mathbf{Y}_{\text{shunt}}$$

$$\mathbf{Y}_{\text{SR}} = \mathbf{Y}_{\text{RS}}^{\text{T}} = -\mathbf{Z}_{\text{series}}^{-1}$$

where

$$\mathbf{Z}_{\text{series}} = \mathbf{Z}_{abc}\mathbf{T}_i\Gamma_m^{-1}\sinh\left(\text{diag}\left(\Gamma_m\right)l\right)\mathbf{T}_i^{-1} \tag{10.16}$$

$$\mathbf{Y}_{\text{shunt}} = 2\mathbf{T}_i\Gamma_m^{-1}\tanh\left(\frac{1}{2}\text{diag}\left(\Gamma_m\right)l\right)\mathbf{T}_i^{-1}\mathbf{Y}_{abc} \tag{10.17}$$

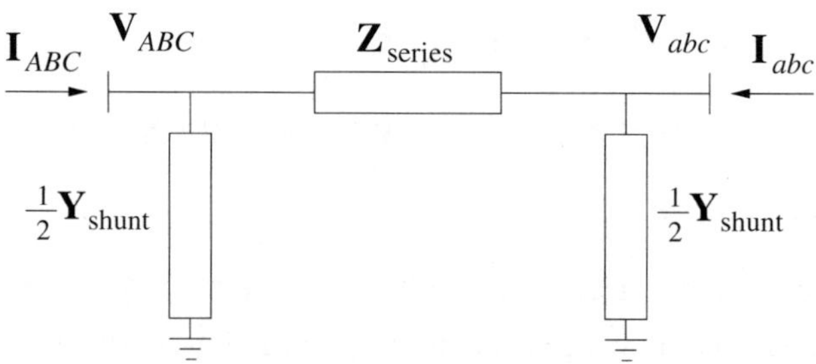

Figure 10.4 Transmission equivalent line π-circuit

The representation of the multi-phase transmission line in the form of **ABCD** parameters also provides a very useful analysis tool:

$$\begin{bmatrix} \mathbf{V}_{ABC} \\ \mathbf{I}_{ABC} \end{bmatrix} = \begin{bmatrix} \mathbf{A} & \mathbf{B} \\ \mathbf{C} & \mathbf{D} \end{bmatrix} \begin{bmatrix} \mathbf{V}_{abc} \\ -\mathbf{I}_{abc} \end{bmatrix} \tag{10.18}$$

where

$$\begin{aligned}
\mathbf{A} &= \mathbf{T}_v\cosh\left(\text{diag}\left(\Gamma_m\right)l\right)\mathbf{T}_v^{-1} \\
\mathbf{B} &= \mathbf{T}_v\mathbf{Z}_{c,m}\sinh\left(\text{diag}\left(\Gamma_m\right)l\right)\mathbf{T}_i^{-1} \\
\mathbf{C} &= \mathbf{T}_i\mathbf{Z}_{c,m}^{-1}\sinh\left(\text{diag}\left(\Gamma_m\right)l\right)\mathbf{T}_v^{-1} \\
\mathbf{D} &= \mathbf{T}_i\cosh\left(\text{diag}\left(\Gamma_m\right)l\right)\mathbf{T}_i^{-1} = \mathbf{A}^{\text{T}}
\end{aligned}$$

The MATLAB™ function used to obtain the distributed parameters of a transmission line is given below:

```
function [Zss,Zsr,Zrs,Zrr,Zseries,Yshunt,A,B,C,D]=
                   calc_DistParam(Zabc,Yabc,long)
[Tv,Av]  = eig(Zabc*Yabc);
Ti       = transpose(inv(Tv));
Zm       = inv(Tv)*Zabc*Ti;
```

```
Ym       = inv(Ti)*Yabc*Tv;
p_c      = sqrt(Zm*Ym);
Zcm      = sqrt(Zm*inv(Ym));
Zss      = Tv*Zcm*inv(tanh(p_c*long))*inv(Ti);
Zsr      = Tv*Zcm*inv(sinh(p_c*long))*inv(Ti);
Zrs      = Zsr;
Zrr      = Zss;
Zseries  = Zabc*Ti*inv(p_c)*sinh(p_c*long)*inv(Ti);
Yshunt   = 2*Ti*inv(p_c)*tanh(p_c*long/2)*inv(Ti)*Yabc;
A        = Tv*inv(tanh(p_c*long))*sinh(p_c*long)*inv(Tv);
B        = Tv*Zcm*sinh(p_c*long)*inv(Ti);
C        = Ti*inv(Zcm)*sinh(p_c*long)*inv(Tv);
D        = transpose(A);
```

Example 10-2: Calculate the fundamental frequency distributed parameters of the transmission line of Section 10.2.5 assuming a line length of 230 km. The following results are obtained using MATLAB™:

```
>>trans_line
>>[Zabc,Yabc]=calc_Zabc_Yabc(nc,nc_pp,cond,x,y,rad_ext,rad_int,cond_c,
                                   50,1);
>>[Zss,Zsr,Zrs,Zrr,Zseries,Yshunt,A,B,C,D]=calcDistParam(Zabc,Yabc,
                                                              230);

>>Zseries
Zseries =
   1.0e+02 *
   0.1810 + 1.2213i   0.1033 + 0.6240i   0.1032 + 0.5250i
   0.1033 + 0.6240i   0.1811 + 1.2212i   0.1033 + 0.6240i
   0.1032 + 0.5250i   0.1033 + 0.6240i   0.1810 + 1.2213i
>>Yshunt
Yshunt =
   1.0e-03 *
   0.0007 + 0.7768i   0.0000 - 0.1867i   0.0002 - 0.0700i
   0.0000 - 0.1867i   0.0007 + 0.8154i   0.0000 - 0.1867i
   0.0002 - 0.0700i   0.0000 - 0.1867i   0.0007 + 0.7768i
>>A
A =
   0.9602 + 0.0058i  -0.0091 + 0.0016i  -0.0103 + 0.0024i
  -0.0106 + 0.0020i   0.9619 + 0.0055i  -0.0106 + 0.0020i
  -0.0103 + 0.0024i  -0.0091 + 0.0016i   0.9602 + 0.0058i
>>B
B =
   1.0e+02 *
   0.1810 + 1.2213i   0.1033 + 0.6240i   0.1032 + 0.5250i
   0.1033 + 0.6240i   0.1811 + 1.2212i   0.1033 + 0.6240i
   0.1032 + 0.5250i   0.1033 + 0.6240i   0.1810 + 1.2213i
>>C
C =
   1.0e-03 *
```

```
   -0.0013 + 0.7627i  -0.0000 - 0.1864i  -0.0004 - 0.0716i
   -0.0000 - 0.1864i  -0.0013 + 0.8016i  -0.0000 - 0.1864i
   -0.0004 - 0.0716i  -0.0000 - 0.1864i  -0.0013 + 0.7627i
>>
```

Fundamental frequency equivalent impedances

The transmission line model can be used for fundamental and harmonic frequency analyses. In studies such as power flows and fault analysis, the use of equivalent parameters in the sequence domain is normally preferred. Transmission line sequence parameters may be derived from phase domain parameters by means of the following relations:

$$\mathbf{Z}_{012} = \mathbf{T}^{-1}\mathbf{Z}_{abc}\mathbf{T}$$
$$\mathbf{Y}_{012} = \mathbf{T}^{-1}\mathbf{Y}_{abc}\mathbf{T}$$

where

$$\mathbf{T} = \begin{bmatrix} 1 & 1 & 1 \\ 1 & a^2 & a \\ 1 & a & a^2 \end{bmatrix}$$

and $a = -1/2 + \mathrm{j}\sqrt{3}/2$ and $a^2 = -1/2 - \mathrm{j}\sqrt{3}/2$.

The resulting sequence parameters are multiplied by the length of the line to obtain the total parameters in Ω and S, respectively.

For the transmission line of Example 10-1, and assuming a line length of 300 km, the following equivalent impedances in the sequences domain are obtained using MATLAB ™:

```
>> Z012 =
    52.63 + 319.63i   03.77 -   2.18i   -3.77 -   2.17i
    -3.77 -    2.17i  10.35 +  83.00i   -7.54 +   4.36i
     3.77 -    2.18i   7.54 +   4.35i   10.35 +  83.00i
>> Y012 =
         0 + 0.0006i  -0.0000 + 0.0000i   0.0000 + 0.0000i
    0.0000 + 0.0000i         0 + 0.0012i   0.0001 - 0.0001i
   -0.0000 + 0.0000i  -0.0001 - 0.0001i   0.0000 + 0.0012i
>>
```

The results show that these matrices are not diagonal, but only diagonally dominant. This is an expected result since the transmission line is untransposed.

10.4 Harmonic Voltage Wave Propagation in Transmission Lines

The characteristics of propagation and attenuation of the voltage waves travelling along a transmission line above ground depend heavily on the length of the line and the number of phases involved. The velocity of propagation of the voltage wave associated with a single line conductor will differ from those of a three-phase transmission line of the same length, and so does its attenuation.

The analysis of the propagation of the voltage wave can be carried out in either the phase domain or the modal domain. Mutual couplings exist between the phases in the phase domain analysis and changes of voltage in one phase are a function of the changes of voltage in the other phases. In general, a complex interaction takes place which becomes more complicated as the number of phases increases. In the modal domain, a number of modal voltage waves equal to the number of active conductors is assumed to exist. Each mode is independent of the others, with its own characteristics of propagation and attenuation.

In transmission circuits with more than one conductor, one of the modes will travel with a smaller velocity and higher attenuation than the rest. This is mainly due to ground penetration effects, and it is called the ground mode. In cases of a continuously transposed, three-phase transmission line, the ground mode is also known as the zero sequence component.

For the case of harmonic voltage excitation at the sending end of the line, $\mathbf{V}_{ABC}$, and open-ended line $\mathbf{I}_{abc} = 0$, the receiving end voltage is given by

$$\mathbf{V}_{abc} = \mathbf{A}^{-1}\mathbf{V}_{ABC}$$

while for the case of harmonic current excitation at the receiving end, $\mathbf{I}_{abc}$, and short circuit ended line at the sending end, $\mathbf{V}_{ABC} = 0$, the receiving end voltage is given by,

$$\mathbf{V}_{abc} = \mathbf{A}^{-1}\mathbf{B}\mathbf{I}_{abc}$$

The latter case is likely to be of more practical interest because most of the non-linear power plant components act as harmonic current sources, i.e. static converters and power transformers. However, balanced voltage excitations offer a better opportunity for assessing voltage imbalances at the far end of the line.

The transmission line is expected to show a resonant peak at a harmonic order which is a function of both the length of the line and the fundamental frequency,

$$\text{harmonic order} = \frac{\lambda}{kl} \tag{10.19}$$

where:

λ is the wavelength at fundamental frequency,

l is the length of the transmission line and

k is equal to two when the excitation is produced by a current source and equal to four when the excitation is produced by a voltage source.

An alternating series of resonant and antiresonant peaks will also be present, with a regular span equal to the harmonic order given above. In the case of a 500 km long line and a fundamental frequency of 50 Hz, resonant peaks will be observed for the 3rd, 9th, 15th, 21st,... harmonics. However, the answer provided by equation (10.19) is an idealised, theoretical harmonic order which, strictly speaking, is valid only for voltage waves travelling with zero delay. In the case of a single phase above earth, the voltage wave travelling along the conductor will be considerably retarded and attenuated, mainly because of ground penetration effects.

Example 10-3: The transmission line of Section 10.2.5 with a length of 300 km, and open ended, was fed with a 1 p.u. balanced harmonic voltage. Figure 10.5(a) shows the receiving end voltage where two resonant peaks appear, and Figures 10.5(b) and (c) show the two resonance peaks using a finer resolution. The following MATLAB™ program can be used to calculate and plot the results shown in Figure 10.5:

```
% Exa_10_3.m
clear all
trans_line;
long = 300;
f0   = 50;
VA   =  1;
VB   = -1/2-i*sqrt(3)/2;
VC   = -1/2+i*sqrt(3)/2;
VABC =   [VA;VB;VC];
pos=1;hi=1;dh=1;hf=20;
for h=hi:dh:hf
[Z,Y]=calc_Zabc_Yabc(nc,nc_pp,cond,x,y,rad_ext,rad_int,cond_c,f0,h);
[Zss,Zsr,Zrs,Zrr,Zseries,Yshunt,A,B,C,D]=calc_DistParam(Z,Y,long);
 Vabc=inv(A)*VABC;
 Va(pos)=Vabc(1);Vb(pos)=Vabc(2);Vc(pos)=Vabc(3);
 pos=pos+1;
end
freq=(hi:dh:hf)*f0;
plot(freq,abs(Va),'-',freq,abs(Vb),'--',freq,abs(Vc),'-.')
```

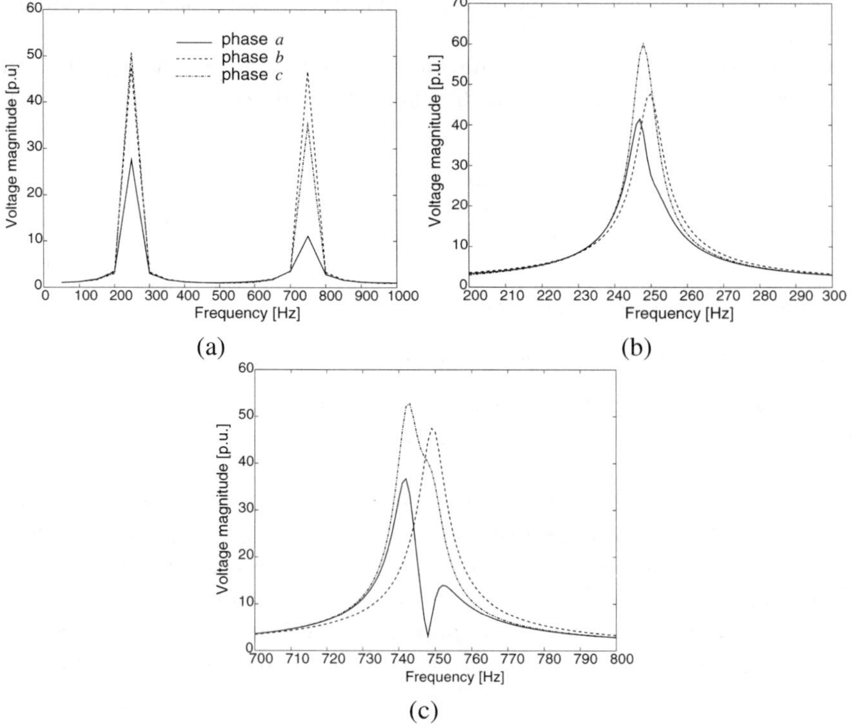

(a) (b)

(c)

Figure 10.5 Receiving end harmonic voltages

Example 10-4: The sending end of the transmission line of Section 10.2.5 is fed with fundamental frequency, 1 p.u. voltage excitation and the receiving end is in open circuit. Different transmission line lengths are considered, in the range 50–1650 km long lines. The exercise is repeated but for the case of third harmonic voltage excitation.

Figure 10.6(a) shows the receiving end voltage at the fundamental frequency and Figure 10.6(b) shows similar information but for the third harmonic voltage. The former result has no practical meaning since 1500 km long transmission lines do not exist without compensation, but its study is important because shorter lines will exhibit a similar behaviour at harmonic frequencies. For instance, the 1500 km long line under consideration will exhibit voltage magnification behaviour at the far end of the line due to the Ferranti effect. This is the result shown in Figure 10.6(a). Likewise, a 500 km long line but excited with third harmonic voltage, as opposed to fundamental frequency voltage, will exhibit third harmonic voltage magnification at the far end of the line. This is the result shown in Figure 10.6(b). In this result we also see that a 1500 km long line excited with third harmonic voltage will exhibit third harmonic voltage magnification at the far end of the line. By the same token, a 300 km long line excited with fifth harmonic voltage will exhibit fifth harmonic voltage magnification at the far end of the line. The basic relationship (10.19) may be used to predict such behaviour.

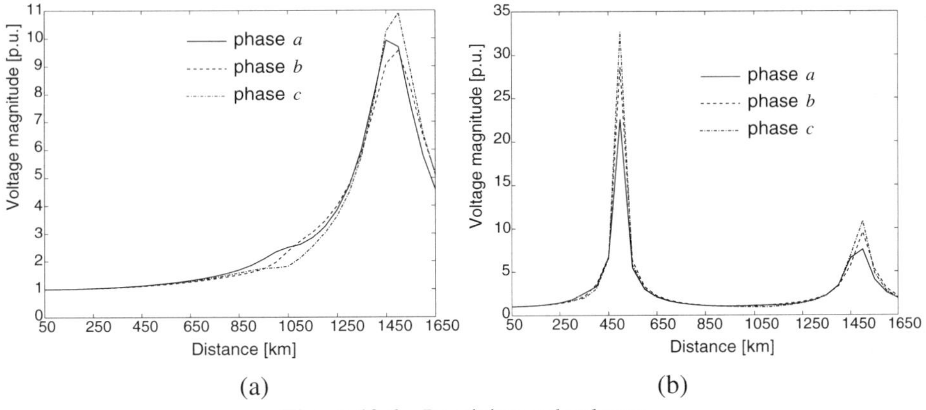

Figure 10.6 Receiving end voltages

10.5 Transmission Line Transpositions

The conductor geometries of high-voltage transmission lines produce considerable impedance asymmetry, which in turn causes a corresponding voltage imbalance at the far end of the line, and transpositions are often used in long-distance transmission as a means of balancing the impedances of the line.

With the rapidly increasing numbers and ratings of static converters and thyristor-controlled compensators, the harmonic voltage and current levels present in high-voltage transmission systems are growing considerably. It is thus important to assess the effect that conventional voltage balancing techniques have on the propagation of harmonics, particularly in long-distance transmission lines.

In fundamental frequency studies the effect of transposition is generally accounted for by averaging the lumped parameters of the transposed sections and using them in a single nominal π-circuit. Such a method, however, assumes that the line geometry is perfectly symmetrical at all points, whereas transpositions occur at discrete distances. In reality each transposed section is an independent subsystem which should be modelled separately and then cascaded with the **ABCD** matrices representing the other sections to obtain an overall equivalent matrix.

A transmission line with a full set of transpositions included consists of three subsystems, as shown in Figure 10.7, where each section can be viewed as an equivalent π-circuit, where each section is calculated from the lumped parameters $\mathbf{Z}_{abc}$ and $\mathbf{Y}_{abc}$ as follows:

$$
\begin{aligned}
\mathbf{Z}_{abc_1} &= \mathbf{Z}_{abc} \\
\mathbf{Y}_{abc_1} &= \mathbf{Y}_{abc} \\
\mathbf{Z}_{cab_2} &= \mathbf{R}_\phi^{-1}\mathbf{Z}_{abc}\mathbf{R}_\phi \\
\mathbf{Y}_{cab_2} &= \mathbf{R}_\phi^{-1}\mathbf{Y}_{abc}\mathbf{R}_\phi \\
\mathbf{Z}_{bca_3} &= \mathbf{R}_\phi\mathbf{Z}_{abc}\mathbf{R}_\phi^{-1} \\
\mathbf{Y}_{bca_3} &= \mathbf{R}_\phi\mathbf{Y}_{abc}\mathbf{R}_\phi^{-1}
\end{aligned}
$$

where $\mathbf{R}_\phi$ is the rotation matrix given by

$$
\mathbf{R}_\phi = \begin{bmatrix} 0 & 1 & 0 \\ 0 & 0 & 1 \\ 1 & 0 & 0 \end{bmatrix}
$$

The transposed line of Figure 10.7 is easily obtained by using the impedances and admittances derived above to obtain the **ABCD** parameters of the individual elements. These are then cascaded to find the overall **ABCD** transfer parameters as follows:

$$
\begin{bmatrix} \mathbf{V}_{ABC} \\ \mathbf{I}_{ABC} \end{bmatrix} = \begin{bmatrix} \mathbf{A}_1 & \mathbf{B}_1 \\ \mathbf{C}_1 & \mathbf{D}_1 \end{bmatrix} \begin{bmatrix} \mathbf{A}_2 & \mathbf{B}_2 \\ \mathbf{C}_2 & \mathbf{D}_2 \end{bmatrix} \begin{bmatrix} \mathbf{A}_3 & \mathbf{B}_3 \\ \mathbf{C}_3 & \mathbf{D}_3 \end{bmatrix} \begin{bmatrix} \mathbf{V}_{abc} \\ -\mathbf{I}_{abc} \end{bmatrix} \tag{10.20}
$$

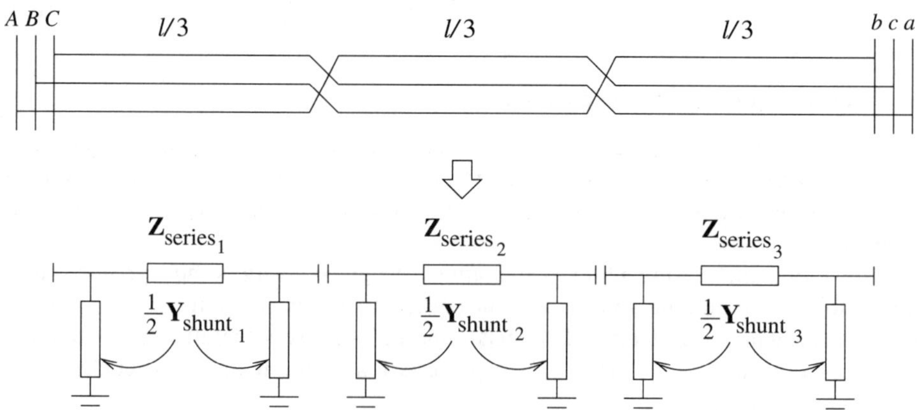

Figure 10.7 Transposed line layout and equivalent π sections

Fundamental frequency equivalent

For fundamental frequency studies the transposed line equivalent is given by

$$\begin{aligned}\mathbf{Z}_{\text{trans}} &= \frac{1}{3}\left(\mathbf{Z}_{abc_1} + \mathbf{Z}_{cab_2} + \mathbf{Z}_{bca_3}\right) \\ \mathbf{Y}_{\text{trans}} &= \frac{1}{3}\left(\mathbf{Y}_{abc_1} + \mathbf{Y}_{cab_2} + \mathbf{Y}_{bca_3}\right)\end{aligned}$$

and in the sequences,

$$\begin{aligned}\mathbf{Z}_{012} &= \mathbf{T}^{-1}\mathbf{Z}_{\text{trans}}\mathbf{T} \\ \mathbf{Y}_{012} &= \mathbf{T}^{-1}\mathbf{Y}_{\text{trans}}\mathbf{T}\end{aligned}$$

These parameters are multiplied by the transmission line length to obtain the total parameters in Ω and S, respectively. This equivalent corresponds to a nominal π-circuit. For Example 10-1, at fundamental frequency, and using a fully transposed line, the following sequence parameters are obtained using MATLAB™:

```
>> Z012 =
   52.63 + 319.63i     0.00 +   0.00i      0.00 +   0.00i
    0.00 +    0.00i   10.35 + 83.00i       0.00 +   0.00i
    0.00 +    0.00i     0.00 +  0.00i     10.35 + 83.00i
>> Y012 =
            0 + 0.0006i            0                   0
            0                      0 + 0.0012i         0
            0                      0                   0 + 0.0012i
>>
```

It should be noted that due to the fact that the transmission line is fully transposed, the three sequence quantities are fully decoupled. These quantities may be used to form three decoupled nominal π-circuits as follows:

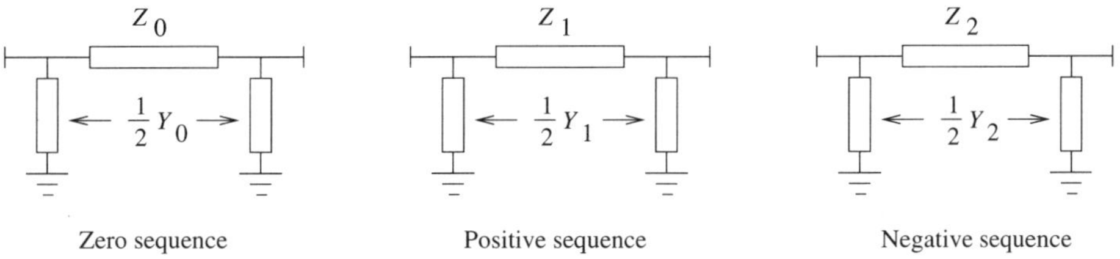

Example 10-5: The transmission line of Section 10.2.5 with a length of 300 km, a full set of transpositions, and open ended, was fed with a balanced 1 p.u. harmonic voltage source. Figure 10.8(a) shows the receiving end voltage, where two resonant peaks appear, and Figures 10.8(b) and (c) show the two resonant peaks in more detail.

The following MATLAB™ function may be used to calculate and to plot these results:

```
% Exa_10_5.m
clear all
trans_line;
long = 300;
f0   = 50;
VA   =  1;
VB   = -1/2-i*sqrt(3)/2;
VC   = -1/2+i*sqrt(3)/2;
VABC =  [VA;VB;VC];
pos=1;hi=14;dh=0.02;hf=16;
Rf=[0 1 0;0 0 1;1 0 0];
for h=hi:dh:hf
  [Zabc,Yabc]=calc_Zabc_Yabc(nc,nc_pp,cond,x,y,rad_ext,rad_int,cond_c,
                            f0,h);
  Z1=Zabc; Z2=inv(Rf)*Zabc*Rf; Z3=Rf*Zabc*inv(Rf);
  Y1=Yabc; Y2=inv(Rf)*Yabc*Rf; Y3=Rf*Yabc*inv(Rf);
  l =long/3;
  [Zs1,Zsr1,Zrs1,Zr1,Zse1,Ysh1,A1,B1,C1,D1]=calc_DistParam(Z1,Y1,l);
  [Zs2,Zsr2,Zrs2,Zr2,Zse2,Ysh2,A2,B2,C2,D2]=calc_DistParam(Z2,Y2,l);
  [Zs3,Zsr3,Zrs3,Zr3,Zse3,Ysh3,A3,B3,C3,D3]=calc_DistParam(Z3,Y3,l);
  ABCD1=[A1 B1;C1 D1]; ABCD2=[A2 B2;C2 D2]; ABCD3=[A3 B3;C3 D3];
  ABCD =ABCD1*ABCD2*ABCD3;
  A    =ABCD(1:3,1:3);
  Vbca =inv(A)*VABC;
  Vb(pos)=Vbca(1);
  Vc(pos)=Vbca(2);
  Va(pos)=Vbca(3);
  pos=pos+1;
end
freq=(hi:dh:hf)*f0;
plot(freq,abs(Va),'-',freq,abs(Vb),'--',freq,abs(Vc),'-.')
```

Example 10-6: The transmission line of Section 10.2.5, with one set of transpositions, was fed with a balanced 1 p.u. fundamental frequency voltage source and with a balanced 1 p.u. third harmonic voltage source. The length of the transmission line was increased gradually from 50 to 1650 km, using 50 km intervals. Figure 10.9(a) shows the response for the 1 p.u. fundamental voltage source and Figure 10.9(b) shows the response for the 1 p.u. third harmonic voltage source.

These figures indicate that in the absence of voltage compensation, the effectiveness of transpositions is limited to line distances under one-eighth of the wavelength, i.e. 750 km at 50 Hz. For distances approaching the quarter wavelength the transposed line produces considerably greater imbalances than the untransposed line.

Although such transmission distances are impractical without compensation, the results provide an indication of the behaviour to be expected with shorter lines at harmonic frequencies. By comparing the results of Figures 10.6 and 10.9, it can be seen that the region of resonant distances increases with the use of transpositions.

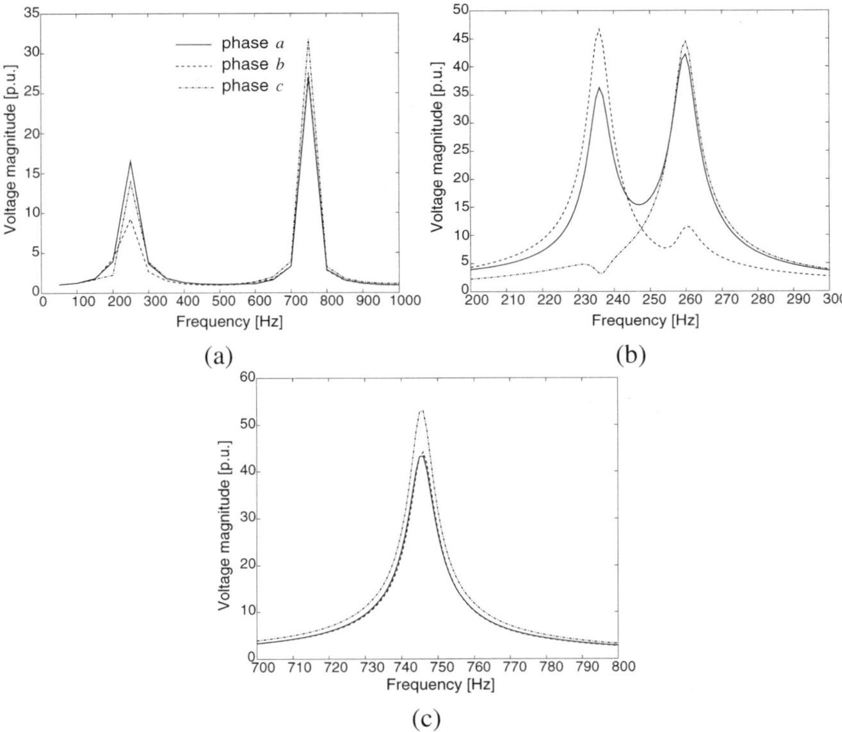

Figure 10.8 Receiving end harmonic voltages for a transposed line

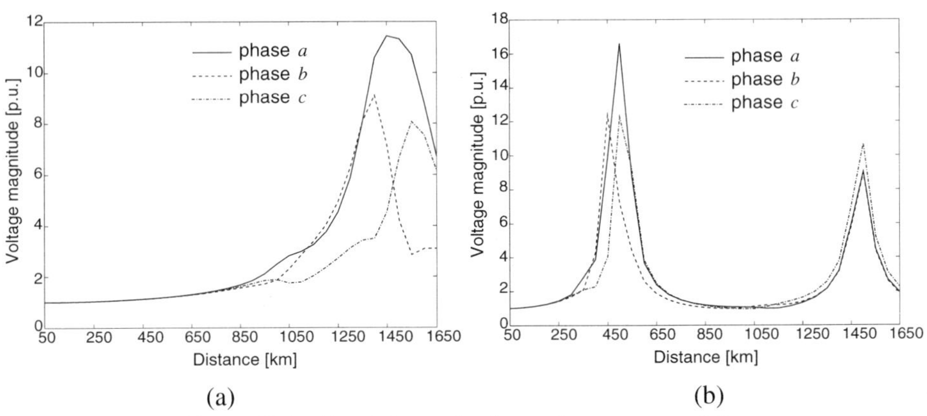

Figure 10.9 Receiving end voltages for a transposed line

10.6 VAR Compensation

Transmission systems will often include series and shunt compensation as well as transpositions. In such cases, each individual section of the line needs to be represented as an independent unit by means of its **ABCD** parameters, where the compensation unit shown below

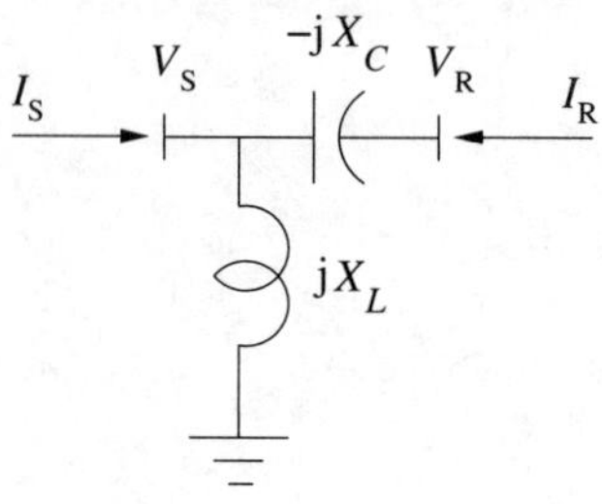

has the following **ABCD** representation:

$$\begin{bmatrix} \mathbf{V}_S \\ \mathbf{I}_S \end{bmatrix} = \begin{bmatrix} \mathbf{A}_c & \mathbf{B}_c \\ \mathbf{C}_c & \mathbf{D}_c \end{bmatrix} \begin{bmatrix} \mathbf{V}_R \\ -\mathbf{I}_R \end{bmatrix}$$

where

$$\mathbf{A}_c = \mathbf{U} \qquad \mathbf{B}_c = -j\frac{1}{h}\mathbf{X}_C$$

$$\mathbf{C}_c = -j\frac{1}{h}\mathbf{X}_L^{-1} \qquad \mathbf{D}_c = -\frac{1}{h^2}\mathbf{X}_L^{-1}\mathbf{X}_C + \mathbf{U}$$

and $\mathbf{U}$ is the identity matrix.

It must be noticed that transmission lines will contain distributed parameters whilst VAR compensating plant will contain lumped parameters. All the units are then cascaded to obtain an equivalent **ABCD** matrix, which relates the input to the output taking into account all the elements in the transmission system,

$$\begin{bmatrix} \mathbf{V}_S \\ \mathbf{V}_R \end{bmatrix} = \begin{bmatrix} \mathbf{A}_1 & \mathbf{B}_1 \\ \mathbf{C}_1 & \mathbf{D}_1 \end{bmatrix} \begin{bmatrix} \mathbf{A}_{c,1} & \mathbf{B}_{c,1} \\ \mathbf{C}_{c,1} & \mathbf{D}_{c,1} \end{bmatrix} \begin{bmatrix} \mathbf{A}_2 & \mathbf{B}_2 \\ \mathbf{C}_2 & \mathbf{D}_2 \end{bmatrix} \times \begin{bmatrix} \mathbf{A}_{c,2} & \mathbf{B}_{c,2} \\ \mathbf{C}_{c,2} & \mathbf{D}_{c,2} \end{bmatrix} \begin{bmatrix} \mathbf{A}_3 & \mathbf{B}_3 \\ \mathbf{C}_3 & \mathbf{D}_3 \end{bmatrix} \begin{bmatrix} \mathbf{I}_S \\ -\mathbf{I}_R \end{bmatrix} \tag{10.21}$$

10.6.1 Network nodal analysis

If only input–output voltage information is required, the cascading approach described in the previous section is sufficient. However, if extra information along the line is required then appropriate fictitious nodes are created at specified points or at regular intervals, and nodal techniques are applied. This approach offers a systematic way of solving any number of three-phase sections (including discontinuities). The following matrix equation is formed, inverted (factorised) and solved. The resultant vectors provide the harmonic voltage profile along the line. The analysis applies to both homogeneous and non-homogeneous transmission lines.

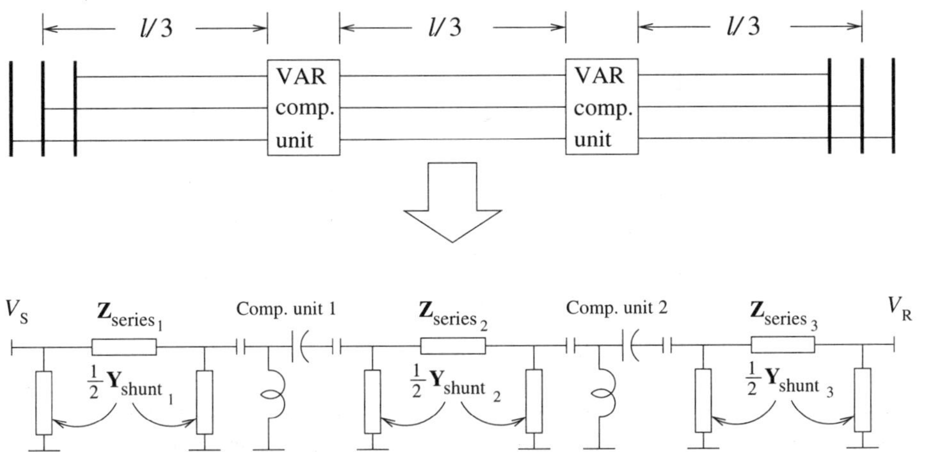

Figure 10.10 Layout and equivalent circuit of a non-homogeneous long-distance transmission line

$$\begin{bmatrix} \mathbf{I}_{\mathrm{S}} \\ \mathbf{I}_1 \\ \vdots \\ \mathbf{I}_n \\ \mathbf{I}_{\mathrm{R}} \end{bmatrix} = \begin{bmatrix} \mathbf{Y}_{\mathrm{SS}_1} & \mathbf{Y}_{\mathrm{SR}_1} & & & \\ \mathbf{Y}_{\mathrm{RS}_1} & \mathbf{Y}_{\mathrm{RR}_1} + \mathbf{Y}_{\mathrm{SS}_2} & & & \\ & & \ddots & & \\ & & & \mathbf{Y}_{\mathrm{RR}_{n-1}} + \mathbf{Y}_{\mathrm{SS}_n} & \mathbf{Y}_{\mathrm{SR}_n} \\ & & & \mathbf{Y}_{\mathrm{RS}_n} & \mathbf{Y}_{\mathrm{RR}_n} \end{bmatrix} \begin{bmatrix} \mathbf{V}_{\mathrm{S}} \\ \mathbf{V}_1 \\ \vdots \\ \mathbf{V}_n \\ \mathbf{V}_{\mathrm{R}} \end{bmatrix} \tag{10.22}$$

Example 10-7: This example illustrates the case of a fully loaded long-distance non-homogeneous transmission system. It relates to El Chacon [20], a 1000 km double line operating at 500 kV, 50 Hz. The layout and equivalent circuit (per line), shown in Figure 10.10, include two identical compensation units symmetrically placed along the line. The configuration of the transmission line is taken to be that in Figure 10.3.

When the line carries maximum power (i.e. 1650 MW) it requires two series capacitors each of −j0.0188 p.u. reactance (to provide a total of 40% series compensation) combined with two sets of shunt inductive compensation (to provide 30% shunt compensation).

The voltage profile of the transmission system for phase a is shown in Figure 10.11 for the fundamental, second, third and sixth harmonics.

Example 10-8: The realistic assessment of line transposition effects in long-distance transmission requires the incorporation of VAR compensation plant, because transmission distances of one-eighth of the wavelength are not practical without compensation. In this example a VAR compensating plant and line transpositions are combined together and applied to the solution of the lightly loaded case of the transmission system of Figure 10.10. Under these operating conditions a single line is used to transmit 300 MW and two sets of shunt inductive compensation are included to provide 40% shunt compensation.

Figure 10.12(a) presents the three-phase voltages along the line for the fundamental and second harmonic of the untransposed transmission system, while Figure 10.12(b) presents the case when the transmission system has been transposed. This result shows the balancing

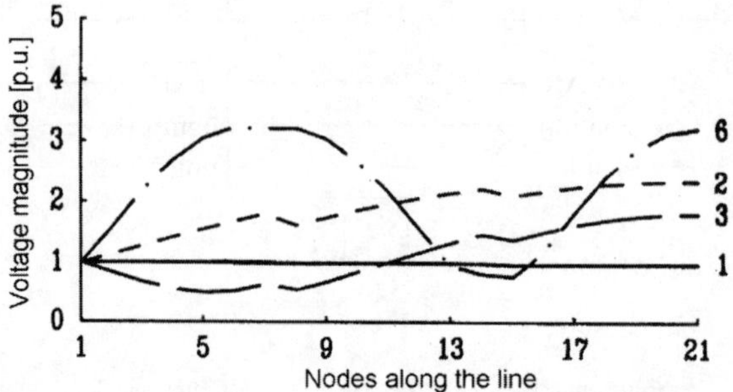

Figure 10.11 Standing waves along the compensated line

effects taking place at the fundamental voltage; in contrast, the voltage profile of the second harmonic becomes more unbalanced than the case containing no transpositions. It is clear that 1 p.u. second harmonic voltage injection is unrealistic and the results would need to be scaled down accordingly.

10.7 Single-phase Transmission Line Equivalents

Even though the electrical power network is three phase in nature, it is very illustrative to look at key aspects of an equivalent single-phase transmission line circuit. The **ABCD** representation is given by

$$\begin{bmatrix} V_S \\ I_S \end{bmatrix} = \begin{bmatrix} \cosh \gamma l & Z_c \sinh \gamma l \\ \frac{1}{Z_c} \sinh \gamma l & \cosh \gamma l \end{bmatrix} \begin{bmatrix} V_R \\ -I_R \end{bmatrix} \tag{10.23}$$

where the propagation constant and the characteristic impedance are $\gamma = \sqrt{zy}$ and $Z_c = \sqrt{z/y}$, respectively.

For a lossless transmission line $\gamma = jh\omega_0\sqrt{LC}$ per km and $Z_c = \sqrt{L/C}\ \Omega$, where L is the series inductance in henries and C the shunt capacitance in farads per km, respectively.

From equation (10.23) an equivalent π-circuit may be obtained, given in the form of a series impedance Z_{series},

$$Z_{\text{series}} = Z_c \sinh \gamma l \tag{10.24}$$

and the shunt admittance, Y_{shunt}, split into two equal parts and placed at both ends of Z_{series},

$$Y_{\text{shunt}} = \frac{2}{Z_c} \tanh \frac{\gamma l}{2} \tag{10.25}$$

The equivalent π-circuit is given in Figure 10.13.

Example 10-9: The test circuit used in this example is very simple. It consists of an unloaded reactor fed from a sinusoidal 1 p.u., 50 Hz voltage source via a lossless 400 km long transmission line with $L = 1.2$ mH/km and $C = 8.4093$ nF/km. Figure 10.14 shows the test system.

The propagation constant at fundamental frequency for the lossless line is

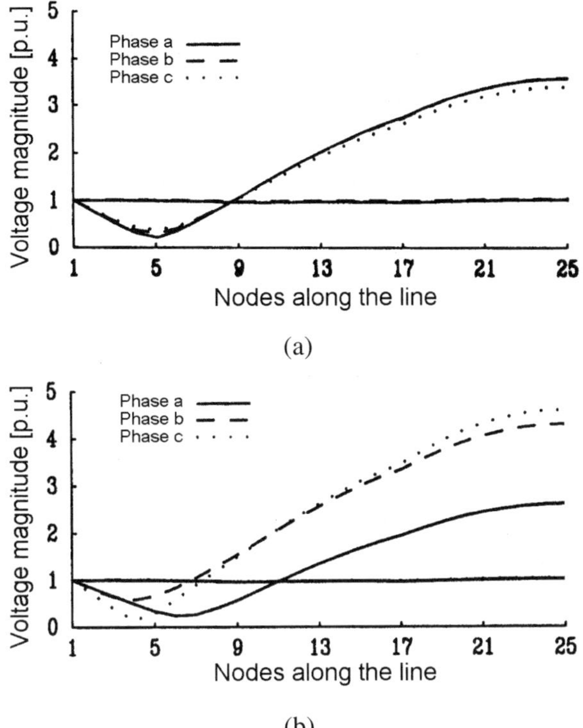

Figure 10.12 Three-phase voltages along the compensated line with (a) no transposition and (b) one set of transpositions

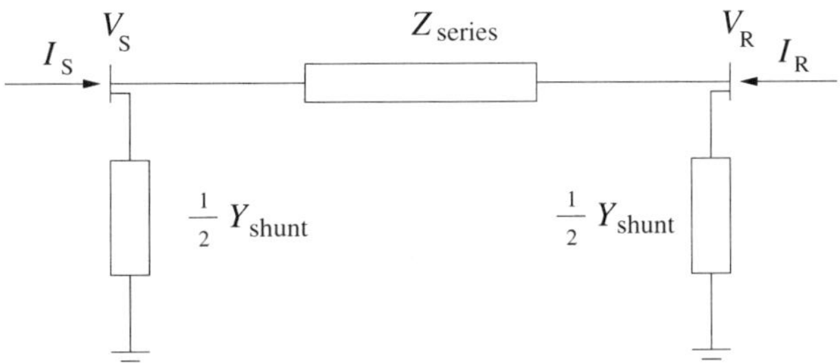

Figure 10.13 π long-line circuit equivalent

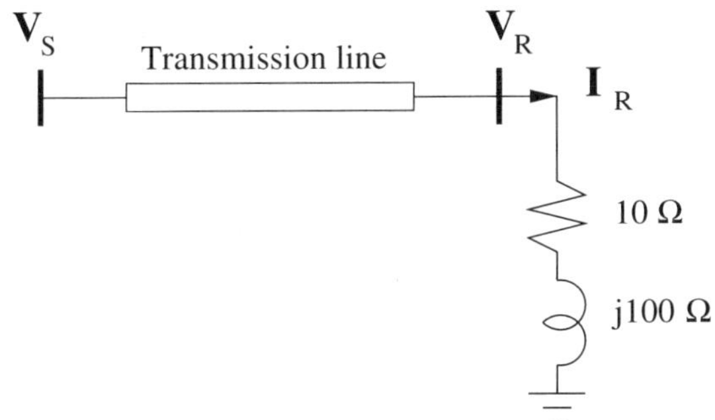

Figure 10.14 Unloaded reactor supplied via a 400 km transmission line

$$\gamma = \mathrm{j}h\omega_0\sqrt{LC} = \mathrm{j}h0.001 \text{ per km}$$

and the characteristic impedance of the line is

$$Z_\mathrm{c} = \sqrt{\frac{L}{C}} = 378.52 \ \ \Omega$$

With reference to (10.23) and Figure 10.14, the sending end voltage for a given harmonic, h, is

$$V_{\mathrm{S},h} = \cosh(\gamma l)V_{\mathrm{R},h} + Z_\mathrm{c}\sinh(\gamma l)I_{\mathrm{R},h} \tag{10.26}$$

where, at the fundamental frequency,

$$V_{\mathrm{S},1} = \frac{\mathrm{e}^{\mathrm{j}\omega_0 t} + \mathrm{e}^{-\mathrm{j}\omega_0 t}}{2}$$

The numeric evaluation of equation (10.26) in the harmonic domain, up to the third harmonic, for the 400 km long transmission line with $f = 50$ Hz is

$$\begin{bmatrix} 0 \\ 0 \\ 0.5 \\ 0 \\ 0.5 \\ 0 \\ 0 \end{bmatrix} = \begin{bmatrix} 0.3624 & & & & & & \\ & 0.6967 & & & & & \\ & & 0.9211 & & & & \\ & & & 1 & & & \\ & & & & 0.9211 & & \\ & & & & & 0.6967 & \\ & & & & & & 0.3624 \end{bmatrix} \begin{bmatrix} V_{-3} \\ V_{-2} \\ V_{-1} \\ V_0 \\ V_1 \\ V_2 \\ V_3 \end{bmatrix}$$
$$+ \begin{bmatrix} -\mathrm{j}352.80 & & & & & & \\ & -\mathrm{j}271.53 & & & & & \\ & & -\mathrm{j}147.40 & & & & \\ & & & 0 & & & \\ & & & & \mathrm{j}147.40 & & \\ & & & & & \mathrm{j}271.53 & \\ & & & & & & \mathrm{j}352.80 \end{bmatrix} \begin{bmatrix} I_{-3} \\ I_{-2} \\ I_{-1} \\ I_0 \\ I_1 \\ I_2 \\ I_3 \end{bmatrix}$$

In short-form notation,

$$\mathbf{V}_\mathrm{S} = \mathbf{A}\mathbf{V}_\mathrm{R} + \mathbf{B}\mathbf{I}_\mathrm{R} \tag{10.27}$$

The impedance equation of the load in the harmonic domain is

$$\mathbf{Z}_L = \mathbf{D}(R + \mathrm{j}h\omega_0 L)$$

where $R = 10\,\Omega$ and $\omega_0 L = 100\,\Omega$. The voltage in the load is given by

$$\mathbf{V}_\mathrm{R} = \mathbf{Z}_L\mathbf{I}_\mathrm{R} \tag{10.28}$$

which when numerically evaluated becomes

$$\begin{bmatrix} V_{-3} \\ V_{-2} \\ V_{-1} \\ V_0 \\ V_1 \\ V_2 \\ V_3 \end{bmatrix} =$$
$$\begin{bmatrix} 10-\text{j}300 & & & & & & \\ & 10-\text{j}200 & & & & & \\ & & 10-\text{j}100 & & & & \\ & & & 0 & & & \\ & & & & 10+\text{j}100 & & \\ & & & & & 10+\text{j}200 & \\ & & & & & & 10+\text{j}300 \end{bmatrix} \begin{bmatrix} I_{-3} \\ I_{-2} \\ I_{-1} \\ I_0 \\ I_1 \\ I_2 \\ I_3 \end{bmatrix}$$

Using (10.27) in (10.28),

$$\mathbf{V}_{\mathrm{S}} = \mathbf{A}\mathbf{V}_{\mathrm{R}} + \mathbf{B}\mathbf{Z}_L^{-1}\mathbf{V}_{\mathrm{R}}$$

then the voltage at the load point is

$$\mathbf{V}_{\mathrm{R}} = \left(\mathbf{A} + \mathbf{B}\mathbf{Z}_L^{-1}\right)^{-1}\mathbf{V}_{\mathrm{S}}$$

giving

$$\mathbf{V}_{\mathrm{R}} = \begin{bmatrix} 0 \\ 0 \\ 0.2093+\text{j}0.0128 \\ 0 \\ 0.2093-\text{j}0.0128 \\ 0 \\ 0 \end{bmatrix}$$

Example 10-10: A test circuit consists of an ideal non-linear inductor fed from an infinite busbar via a lossless transmission line. The non-linear inductor has the following magnetising characteristic:

$$f(\psi) = 0.001\psi + 0.0743\psi^3 \text{ p.u.} \tag{10.29}$$

The winding resistance and inductance are negligibly small, i.e. zero. Furthermore, the idealised non-linear reactor is fed from a sinusoidal 1 p.u. voltage source via a lossless, 400 km long transmission line. Figure 10.15 shows the test system. Owing to unloaded conditions in the transmission circuit, the magnetising branch of the transformer is pushed into saturation and becomes a source of harmonic distortion.

The propagation constant and characteristic impedance of the line at fundamental frequency are

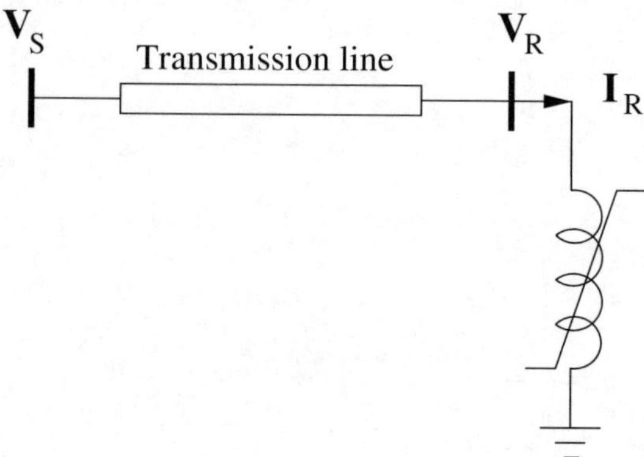

Figure 10.15 Unloaded reactor supplied via a 400 km transmission line

$$\gamma = jh\omega_0\sqrt{LC} = jh0.001 \text{ per km}$$

$$Z_c = \sqrt{\frac{L}{C}} = 378.52 \ \ \Omega$$

The equation of the line, for a given harmonic h, is

$$V_{S,h} = \cosh(\gamma l)V_{R,h} + jZ_c \sinh(\gamma l)I_{R,h} \tag{10.30}$$

where

$$V_{S,1} = \frac{e^{j\omega_0 t} + e^{-j\omega_0 t}}{2}$$

The numeric evaluation of equation (10.30), up to the third harmonic, for the 400 km long transmission line is (see Example 10-9)

$$\begin{bmatrix} 0 \\ 0 \\ 0.5 \\ 0 \\ 0.5 \\ 0 \\ 0 \end{bmatrix} = \begin{bmatrix} 0.3624 & & & & & & \\ & 0.6967 & & & & & \\ & & 0.9211 & & & & \\ & & & 1 & & & \\ & & & & 0.9211 & & \\ & & & & & 0.6967 & \\ & & & & & & 0.3624 \end{bmatrix} \begin{bmatrix} V_{-3} \\ V_{-2} \\ V_{-1} \\ V_0 \\ V_1 \\ V_2 \\ V_3 \end{bmatrix}$$

$$+ \begin{bmatrix} -j352.80 & & & & & & \\ & -j271.53 & & & & & \\ & & -j147.40 & & & & \\ & & & 0 & & & \\ & & & & j147.40 & & \\ & & & & & j271.53 & \\ & & & & & & j352.80 \end{bmatrix} \begin{bmatrix} I_{-3} \\ I_{-2} \\ I_{-1} \\ I_0 \\ I_1 \\ I_2 \\ I_3 \end{bmatrix}$$

or, in compact form,

$$\mathbf{V}_{\mathrm{S}} = \mathbf{A}\mathbf{V}_{\mathrm{R}} + \mathbf{B}\mathbf{I}_{\mathrm{R}} \tag{10.31}$$

The Norton equivalent which represent the non-linear inductor is given by,

$$\mathbf{I}_{\mathrm{R}} = \mathbf{Y}\mathbf{V}_{\mathrm{R}} + \mathbf{I}_{\mathrm{N}} \tag{10.32}$$

Using (10.31) and (10.32) the voltage at the end of the transmission line is given by

$$\mathbf{V}_{\mathrm{R}} = (\mathbf{A} + \mathbf{B}\mathbf{Y})^{-1}(\mathbf{V}_{\mathrm{S}} - \mathbf{B}\mathbf{I}_{\mathrm{N}}) \tag{10.33}$$

where $\mathbf{V}_{\mathrm{S}}$, $\mathbf{A}$ and $\mathbf{B}$ remain constant. The elements of the Norton equivalent $\mathbf{Y}$ and $\mathbf{I}_{\mathrm{N}}$, are evaluated using the following initial condition, $\mathbf{V}_{\mathrm{R}} = \mathbf{V}_{\mathrm{S}}$:

$$\mathbf{Y} = \begin{bmatrix} \mathrm{j}0.0375 & 0 & -\mathrm{j}0.0557 & & & & \\ 0 & \mathrm{j}0.0563 & 0 & \infty & & & \\ -\mathrm{j}0.0186 & 0 & \mathrm{j}0.1124 & \infty & \mathrm{j}0.0557 & & \\ & -\mathrm{j}0.0278 & 0 & \infty & 0 & \mathrm{j}0.0278 & \\ & & -\mathrm{j}0.0557 & \infty & -\mathrm{j}0.1124 & 0 & \mathrm{j}0.0186 \\ & & & \infty & 0 & -\mathrm{j}0.0562 & 0 \\ & & & & \mathrm{j}0.0557 & 0 & -\mathrm{j}0.0375 \end{bmatrix}$$

and

$$\mathbf{I}_{\mathrm{N}} = \begin{bmatrix} \mathrm{j}0.0186 \\ 0 \\ -\mathrm{j}0.0557 \\ 0 \\ \mathrm{j}0.0557 \\ 0 \\ -\mathrm{j}0.0186 \end{bmatrix}$$

where the symbol ∞ indicates very a large value.

Equation (10.33) is solved and new, improved values of $\mathbf{Y}$ and $\mathbf{I}_{\mathrm{N}}$ are obtained at the end of the iteration. A harmonic Newton–Raphson algorithm involves the joint solution of the linear and linearised matrix equations at each iterative step, i.e. a unified approach, and the overall procedure is carried out as shown in Figure 10.16.

The solution gives the following harmonic voltage vector result:

$$\mathbf{V}_{\mathrm{R}} = \begin{bmatrix} 0.1207 \\ 0 \\ 0.2130 \\ 0 \\ 0.2130 \\ 0 \\ 0.1207 \end{bmatrix}$$

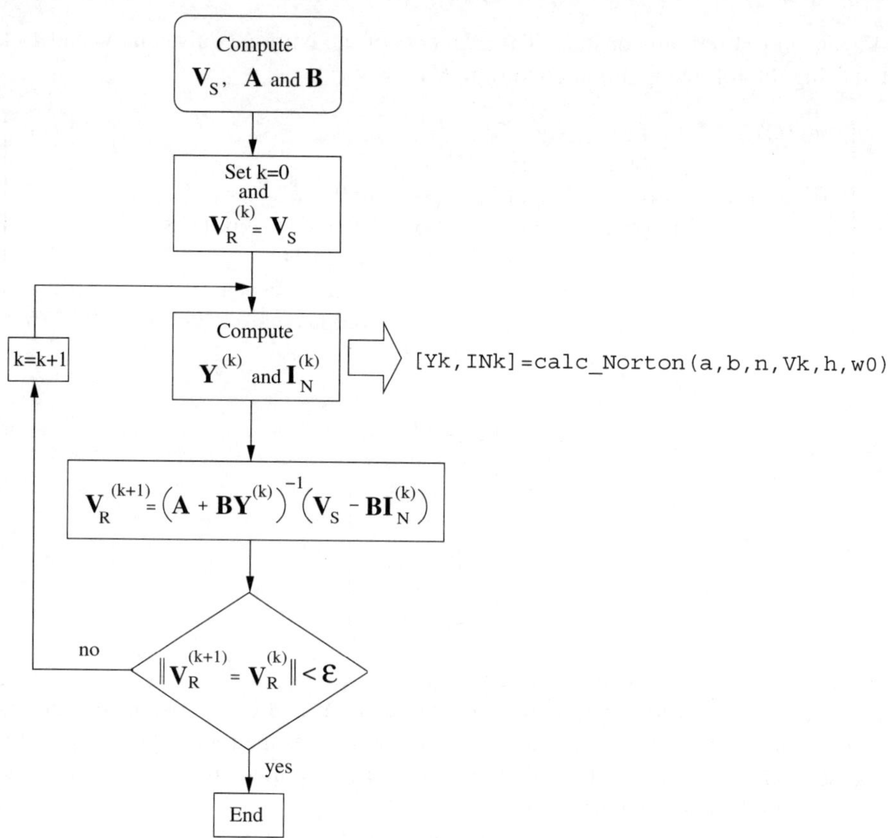

Figure 10.16 Flow diagram for equation (10.33)

The following MATLAB™ program was used in this example:

```
clear all;
h=3;
w0=1; % in p.u.
Vinf=[0;0;0.5;0;0.5;0;0];
a=0.001; b=0.0743; n=3;
A=[0.3624 0 0 0 0 0 0
   0 0.6967 0 0 0 0 0
   0 0 0.9211 0 0 0 0
   0 0 0 1.0000 0 0 0
   0 0 0 0 0.9211 0 0
   0 0 0 0 0 0.6967 0
   0 0 0 0 0 0 0.3624];
B=i*[-352.80 0 0 0 0 0 0
     0 -271.53 0 0 0 0 0
     0 0 -147.40 0 0 0 0
     0 0        0 0 0 0 0
     0 0 0 0  147.40 0 0
     0 0 0 0 0  271.53 0
     0 0 0 0 0 0  352.80];
Ve=Vinf;
error=1;
k=0;
while error>1e-3
 [Y,IN] = calc_Norton(a,b,n,Ve,h,w0);
 V = inv(A+B*Y)*(Vinf-B*IN);
 error = norm(V-Ve);
 Ve = V;
 k = k+1
end
V
```

10.8 Summary

The theory of phase domain modelling of multi-conductor transmission lines at harmonic frequencies has been presented. Frequency dependence due to earth return and skin effects requires accurate representation. The Carson and Dubanton equations are two well-known methods for calculating earth return impedances. Arguably, the latter is amenable to more elegant and simpler calculations and has been the only method covered in this chapter. Similarly, the method used for calculating the internal impedance of ACSR conductors also uses the complex depth concept.

High-voltage transmission circuits contain considerable impedance imbalances at harmonic frequencies and the phase domain is the most attractive frame of reference for conducting harmonic propagation studies. Furthermore, the effects of line transpositions and VAR compensation are efficiently included by using matrix cascading techniques. Long-line effects are more difficult to handle because hyperbolic functions are not directly defined in matrix theory. Accordingly, transformation matrices based on eigenvalues and eigenvectors need to be calculated at each frequency point of interest.

The most relevant MATLAB™ functions required for the calculation of transmission line parameters have been given in this chapter. They have been applied to the solution of a wide range of transmission line harmonic problems.

10.9 Bibliography

1. J. Arrillaga, B.C. Smith, N.R. Watson, A.R. Wood, *Power System Harmonic Analysis*, John Wiley & Sons, Chichester, 1997.
2. J.R. Carson, "Wave Propagation in Overhead Wires with Ground Return", *Bell System Technical Journal*, Vol. 5, 1926, pp. 539–554.
3. A. Deri, G. Tean, A. Semlyen, A. Castanheira, "The Complex Ground Return Plane: A Simplified Model for Homogeneous and Multi-Layer Earth Return", *IEEE Transactions on Power Apparatus and Systems*, Vol. PAS-100, 1981, pp. 3686–3693.
4. E. Acha, "Modelling of Power System Transformers in the Complex Conjugate Harmonic Space", *PhD Thesis*, University of Canterbury, Christchurch, New Zealand, 1988.
5. J.R. Wait, K.P. Spies, "On the Image Representation of the Quasi-Static Fields of a Line Current Source Above the Ground", *Canadian Journal of Physics*, Vol. 47, 1969, pp. 2730–2731.
6. R.G. Olsen, T.A. Pankaskie, "On the Exact, Carson and Image Theories for Wire at or above the Earth's Interface", *IEEE Transactions on Power Apparatus and Systems*, Vol. PAS-102, 1983, pp. 769–777.
7. J.R. Wait, "Theory of Wave Propagation Along a Thin Wire Parallel to an Interface", *Radio Science*, Vol. 7, June 1972, pp. 675–679.
8. L.M. Wedepohl, A.E. Efthymiadis, "Wave Propagation in Transmission Lines Over Lossy Ground: A New, Complete Field Solution", *Proceedings of the IEE*, Vol. 125, No. 6, June 1978, pp. 505–510.
9. C. Gary, "Approache Complète de la Propagation Multifilaire en Haute Fréquency par Utilisation des Matrices Complexes", *EDF Bulletin de la Direction des Éutedus et Recherches Series B*, Vol. 4, No. 3, 1976, pp. 5–20.
10. G. Tevan, A. Deri, "Some Remarks About the Accurate Evaluation of the Carson Integral for Mutual Impedance of Lines with Earth Return", *Arkiv fűr Elektrotechnik*, Vol. 67, 1984, pp. 83–90.
11. V.A. Lewis, P.D. Tuttle, "The Resistance and Reactance of Aluminium Conductors Steel Reinforced", *AIEE Transactions*, Vol. PAS-77, 1958, pp. 1189–1215.
12. A.E. Kennelly, F.A. Laws, P.H. Pierce, "Experimental Researches on Skin Effect in Condition", *AIEE Transactions*, Vol. 34, 1915, pp. 1935–2018.
13. T.J. Densem, "Three Phase Power System Harmonic Penetration", *PhD Thesis*, University of Canterbury, Christchurch, New Zealand, 1983.
14. A. Semlyen, A. Deri, "Time Domain Modeling of Frequency Dependent Three Phase Transmission Line Impedance", *IEEE Transactions on Power Apparatus and Systems*, Vol. PAS-104, 1985, pp. 1549–1555.
15. L.M. Wedepohl, "Application of Matrix Methods to the Solution of Travelling Wave Phenomena in Poly-Phase Systems", *Proceeding of the IEE*, Vol. 110, No. 12, 1963, pp. 2200–2212.
16. J.A. Parle, "Phase Domain Transmission Line Modelling for EMTP-Type Studies with Application to Real-Time Digital Simulation", *PhD Thesis*, University of Glasgow, 2000.
17. W.I. Bowman, J.M. McNamee, "Development of Equivalent Pi and T Matrix Circuits for Long Untransposed Transmission Lines", *IEEE Transactions on Power Apparatus and Systems*, Vol. PAS-84, 1964, pp. 625–632.
18. J. Arrillaga, E. Acha, T.J. Densem, P.S. Bodger, "Ineffectiveness of Transmission Line Transpositions at Harmonic Frequencies", *Proceedings of the IEE*, Part C, Vol. 133, No. 2,

March 1986, pp. 99–104.

19. H.W. Dommel, "Digital Computer Solution of Electromagnetic Transients in Single and Multiphase Networks", *IEEE Transactions on Power Apparatus and Systems*, Vol. PAS-88, 1969, pp. 388–399.
20. K.M. Jones, "AC System Network Control Methods Compensation", *Symposium on Power System Dynamics, Control and Compensation*, IEE North Eastern Centre, Newcastle upon Tyne, 1974.
21. A. Semlyen, M.H. Abdel-Rahman, "Transmission Line Modelling by Rational Transfer Functions", *IEEE Transactions on Power Apparatus and Systems*, Vol. PAS-101, 1982, pp. 3576–3584.

11

Magnetic Non-linearities

11.1 Introduction

An important class of non-linear characteristics in electrical power circuits is the magnetising impedance of saturated power transformers and shunt reactors. Saturated magnetic cores may generate harmonic currents during steady-state operation, as well as transient harmonic currents and temporary over-voltages following a major switching operation in the transformer's vicinity, with a critical case being the energisation of the transformer itself. The steady-state magnetising currents of power transformers are only 1–2% of the rated current but they may reach 10–20 times their rated value when transformers are switched on to source, exhibiting a current spectrum which is rich in harmonics. This transient phenomenon is termed inrush currents.

Owing to the slow attenuation of this transient, its effect may persist for several seconds before the steady state is reached, causing many adverse operating conditions. Transients are known to have caused unnecessary trippings of differential protection relays. Inrush currents may also give rise to long-term over-voltages in system configurations with pronounced parallel resonant points and a low degree of damping. In industrial systems such situations may occur when large power factor correction capacitors exist at the secondary side of the transformer which, when combined with the predominantly inductive impedance of the system at low harmonic frequencies, lead to a parallel circuit of high impedance. Moreover, if the ensuing parallel circuit is tuned to a harmonic frequency component of the inrush current then a voltage magnification will take place. In the long term, if over-voltages occur frequently then the life expectancy of the capacitor bank will be very much reduced.

Modern high-voltage power transformers with grain-oriented steel cores are designed to operate in the region of 1.6–1.7 T, and have a sharply defined knee. Hence, over-excitation of 20%, and even 10% in some cases, above the rated value will push the transformer deep into saturation. The magnetising current of single-phase iron cores will be symmetric in most cases, with the harmonic current spectrum containing no even harmonics and no DC term, just the fundamental frequency component and odd harmonics. For most practical purposes, harmonic terms in the magnetising current above the 15th are negligibly small and are not cause for no concern. Moreover, in most three-phase transformer applications at least one of the three-phase windings is delta connected to confine the zero-sequence-like harmonic currents within this winding, and only the 5th, 7th, 11th and 13th harmonic currents would normally deserve attention. However, some caution should be exercised because the delta connection will only be 100% effective if the power circuit is perfectly balanced. It is unavoidable that practical power networks will always contain asymmetries of one type or another.

A practical problem of greater concern is the case when a DC component finds its way into the magnetic core, causing the transformer to saturate asymmetrically. The end result is that the harmonic spectrum of the magnetising current becomes richer, containing in addition to the fundamental frequency component and odd harmonics, the even harmonics and a DC term. In practice, the DC term causing the asymmetrical core behaviour may come from a power electronics circuit fed directly from the transformer, such as a half-wave rectifier or a three-phase converter with unequal firing pulses. In the case of furnace transformers, it may come from asymmetric operation of the arc. It has been observed using simulations that the magnitude of the low-order harmonic currents increases linearly with the amount of direct current at the secondary of the transformer. It has also been observed that the harmonic currents generated as a result of the spurious DC excitation are largely independent of AC excitation [1].

At a more detailed level, transformer designers have been concerned for many years with the evaluation of iron core losses, mainly under sinusoidal excitation. Since nowadays power transformers, intended for general use, are more likely than ever before to operate in environments which are polluted with distorted waveform voltages, power systems researchers have taken a keen interest in the problem. Analysis tools and methods that deal with such a problem have been reported in the open literature over the past decade [2]. One approach taken has been to analyse the space and time distribution of the field variables B, H and E inside saturating laminations, using the discretised Maxwell's equations for the plane geometry of the laminations. These equations go well with harmonic domain techniques and are addressed in this chapter [3].

11.2 Inrush Currents

To gain some insight into magnetising inrush currents and how they are calculated, the unloaded, single-phase transformer equivalent circuit shown in Figure 11.1 is used.

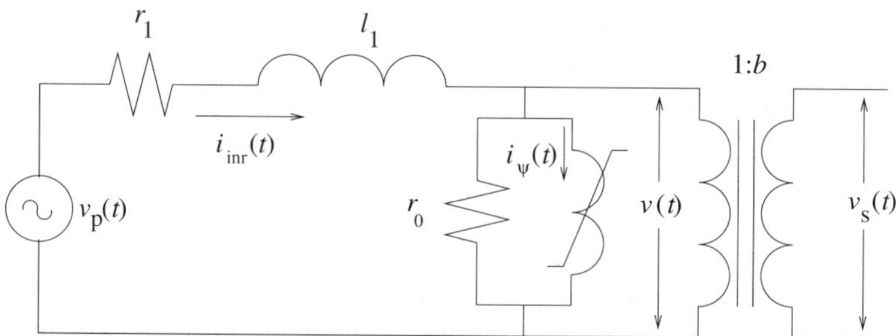

Figure 11.1 Unloaded single-phase transformer

In this particular case the non-linear characteristic is represented by the following polynomial equation:

$$i_\psi(t) = 0.7576\psi(t) + 1.03 \times 10^7 \psi^{19}(t)$$

The other transformer parameters are $r_1 = 0.192\,\Omega$, $r_0 = 612.86\,\Omega$ and $l_1 = 0.9$ mH.

DC pre-magnetisation is strictly speaking the effect of a remanent flux. However, the energisation of a winding at a point different from the voltage peak will also result in a flux which is equivalent to the flux due to remanence. Inrush currents can be calculated following this basic premise. It is important to mention at this point that the attenuation of inrush currents is not significantly affected by eddy current and hysteresis losses in the iron core [2]. The state equation below describes the circuit:

$$\begin{bmatrix} \dot{i}_{\mathrm{inr}}(t) \\ \dot{\psi}(t) \end{bmatrix} = \begin{bmatrix} -\left(r_1 + r_0\right)/l_1 & 0 \\ r_0 & 0 \end{bmatrix} \begin{bmatrix} i_{\mathrm{inr}}(t) \\ \psi(t) \end{bmatrix} + \begin{bmatrix} 1/l_1 & r_0/l_1 \\ 0 & -r_0 \end{bmatrix} \begin{bmatrix} v_{\mathrm{p}}(t) \\ i_{\psi}(t) \end{bmatrix} \tag{11.1}$$

For this study $v_{\mathrm{p}}(t) = 110\cos(\omega_0 t - 42.97°)$ V, $f_0 = 60$ Hz was selected, and 200 cycles of simulation were selected. Figure 11.2 shows the inrush current evolution during the full simulation time (3.33 s). Further detail is provided at three different stages of the inrush current evolution. This result shows that the initial magnetising current is much larger than the steady-state current, and takes a considerably long time to reach steady state. This behaviour of the inrush current has been exploited to conduct the simulation using harmonic domain techniques [5], arguing that the inrush current may be considered like a non-sinusoidal current in a quasi-steady-state condition if one only looks at a reduced number of cycles. This is particularly the case after the first few cycles have elapsed.

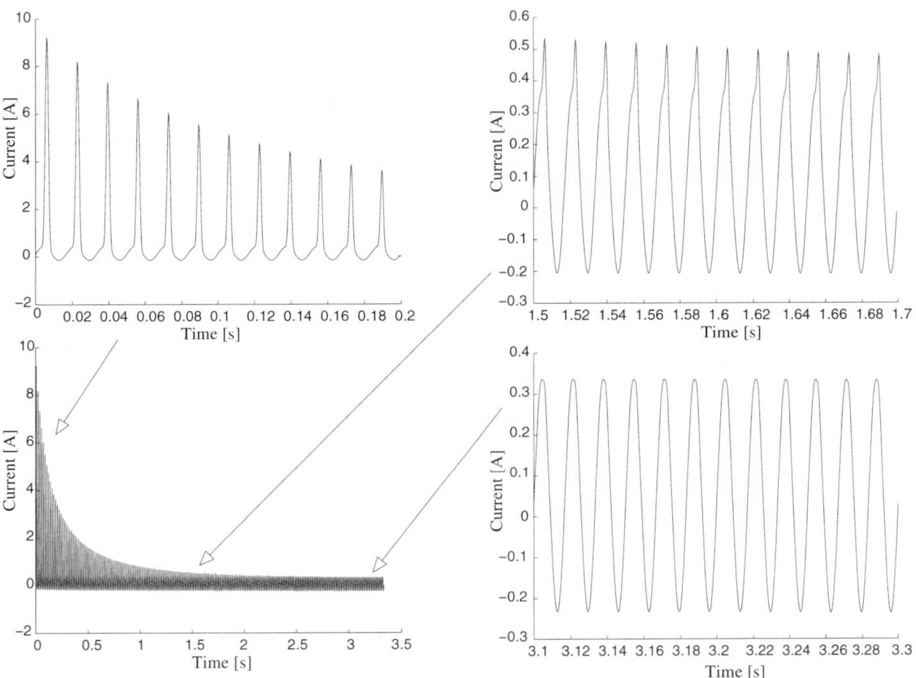

Figure 11.2 Inrush current

Figure 11.3 shows the harmonic content of the inrush current at each cycle. This result shows the decay of the inrush current and its DC component, together with the harmonic terms. The slow decay of the harmonics suggests that the inrush current may be represented by a constant harmonic source for a limited number of consecutive cycles, i.e. window. This characteristic of slow attenuation has been used to establish a harmonic domain model [5] for the inrush current, where a pre-determined sequence of steady-state "images" is used to capture the complete picture of the inrush phenomenon. A more general concept based on operational matrices and harmonic domain techniques [4] may also be used to obtain the complete response of the inrush current using a single steady-state "image".

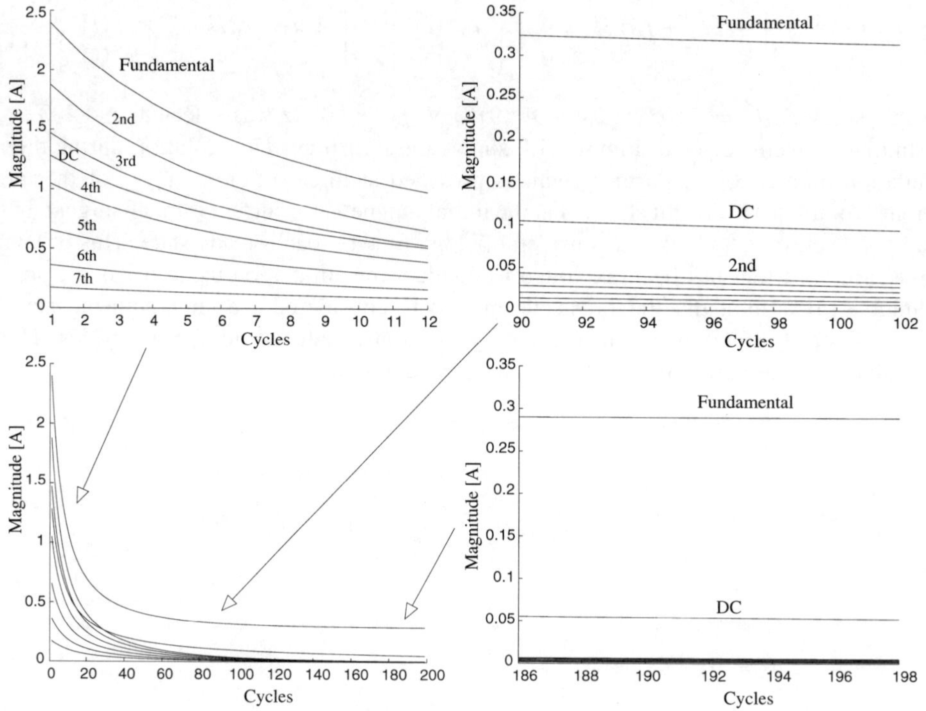

Figure 11.3 Harmonic content of the inrush current

The following MATLAB™ program may be used to solve this example:

```
clear all
global A B w
r1 = 0.192; r0 = 612.86; l1 = 0.9e-3;
f  = 60;     w  = 2*pi*f;
A  = [-(r1+r0)/l1 0
              r0 0];
B  = [1/l1 r0/l1
        0   -r0];
X0 = [0;0];
m     = 10000;
tf    = 200/f;
ts    = tf/m;
t1    = 0:ts:tf-ts;
[t,X] = ode23t('inrush_current',t1,X0);
Iinrt = X(:,1);
Fluxt = X(:,2);
[Im,Ia,Ic]=fft_windowing(Iinrt,50,10000);
plot(t,Iinrt)
```

```
plot(Im)
```

This MATLAB™ function contains the state equation (11.1):

```
function dXdt=inrush_current(t,X);
global A B w
a     = 0.7576; b = 1.03e7; n = 19;
Vp    = 110*cos(w*t-42.97*pi/180);
Flux  = X(2);
Iflux = a*Flux+b*Flux^n;
U     = [Vp;Iflux];
dXdt  = A*X+B*U;
```

This MATLAB™ function applies windowing FFT to a discrete signal:

```
function [Vn,an,Vc]=fft_windowing(v,T,NP)
% Apply the windowing FFT to the vector v,
% i.e. v(t)=Vdc+sum{Vn*cos(nwt+an)}
% T  : is the number of points per cycle
% NP : is the total number of points
% Vn : contain the harmonic magnitudes
% an : contain the angles in degrees
% Vc : contain the complex form of the harmonics
c   = round((T+1)/2);
res = zeros(c-T+1,1);
H=0:c-2;H=H';
N=1:T;
wP = 1-abs((N-1/2*(T-1))/(1/2*(T+1))); % Parzen window
wH = 1/2*(1-cos(2*pi*N./(T-1)));       % Hanning window
wW = 1-((N-1/2*(T-1))/(1/2*(T+1))).^2; % Welch window
wS = ones(1,T);                        % Square window
w  = wS';
for k=1:T:NP-T-1
 vv   = v(k:T+k-1).*w;
 res1 = fft(vv)/T;
 res1 = fftshift(res1);
 res1 = res1(c:T);
 shift_f  = exp(-i.*H*2*pi*(k-1)/(T));
 res1     = res1.*shift_f;
 res1(1)  = res1(1)/2;
 res      = [res res1];
end
Vc = res;
Vn = abs(res)'*2;
an = angle(res)'*180/pi;
```

11.3 Single-phase Transformer

Single-phase transformers operating at steady state under no saturating conditions are well represented by the classic transformer model shown in Figure 11.4. The reactance $\omega_0 l_1$ is obtained from standard short-circuit tests and the resistance r_1 from measurements of the windings' resistances. The shunt parameters r_0 and $\omega_0 l_{mag}$, are obtained from standard open-circuit tests, to form the impedance of the iron core.

On the other hand, when the transformer is operated above the knee point of the magnetising characteristic the magnetic core is pushed into saturation and the simple transformer model

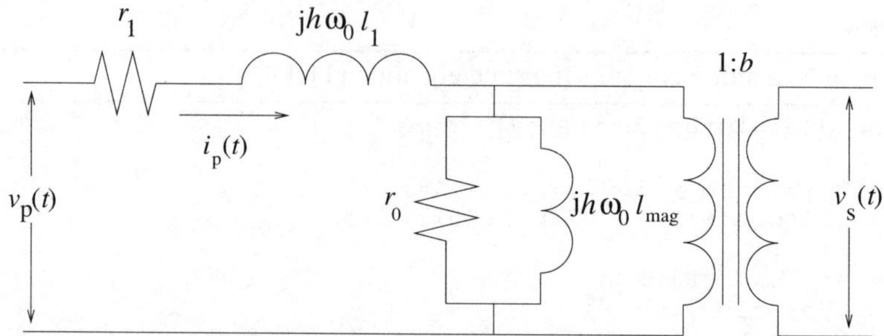

Figure 11.4 Harmonic equivalent circuit of a single-phase transformer under no saturating conditions

shown in Figure 11.4 would need to be replaced by a fuller model, to reflect the more complex electromagnetic phenomenon that takes place in the saturated transformer, as shown in Figure 11.1. If the aim is to study the periodic steady-state behaviour of the transformer then one option open to us is to represent the magnetic core as a harmonic current source [6]. Alternatively, a more powerful representation of the single-phase magnetic core is to use a harmonic Norton equivalent to represent the magnetising branch as opposed to a harmonic current source [7].

11.3.1 Lattice equivalent circuit representation

A more flexible transformer representation than the one given in Figure 11.1 may be achieved by taking the following steps: (1) To combine the impedances of the primary and secondary windings into a single impedance, and to express the resulting impedance as a leakage admittance, i.e. y_t. (2) To split the magnetising impedance into two equal parts and to connect one-half at each terminal of the transformer. (3) To introduce off-nominal tap settings at both the primary and the secondary sides of the transformer.

The resultant equivalent circuit is shown in Figure 11.5, from which the following fundamental equations are derived:

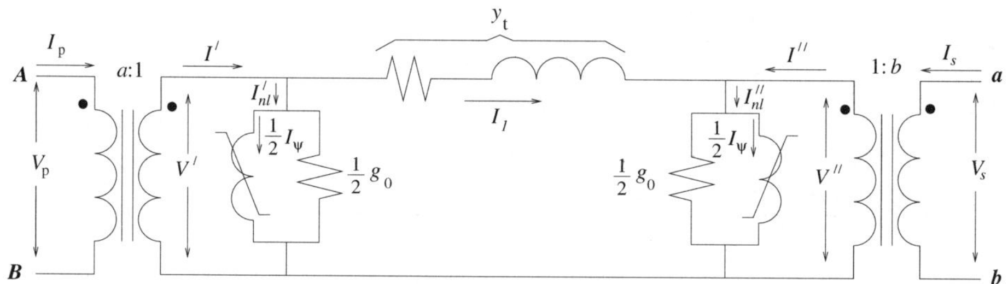

Figure 11.5 Single-phase transformer with two tapping mechanisms

$$\frac{V_p}{V'} = a \qquad \frac{I_p}{I'} = \frac{1}{a}$$
$$\frac{V_s}{V''} = b \qquad \frac{I_s}{I''} = \frac{1}{b} \tag{11.2}$$

Also,

$$
\begin{aligned}
I_1 &= y_t (V' - V'') \\
I' &= I_1 + I'_{nl} \\
I'' &= -I_1 + I''_{nl}
\end{aligned} \tag{11.3}
$$

The primary current is given by

$$
\begin{aligned}
I_p &= \frac{1}{a} I' \\
&= \frac{y_t}{a} (V' - V'') + \frac{1}{a} I'_{nl} \\
&= \frac{y_t}{a} \left(\frac{1}{a} V_p - \frac{1}{b} V_s \right) + \frac{1}{a} \left(\frac{1}{2} g_0 V' + \frac{1}{2} I_\psi \right) \\
&= \frac{y_t}{a^2} V_p - \frac{y_t}{ab} V_s + \frac{g_0}{2a^2} V_p + \frac{1}{2a} I_\psi
\end{aligned} \tag{11.4}
$$

Following a similar approach, the secondary current I_s is,

$$
I_s = -\frac{y_t}{ab} V_p + \frac{y_t}{b^2} V_s + \frac{g_0}{2b^2} V_s + \frac{1}{2b} I_\psi \tag{11.5}
$$

The circuit in Figure 11.5 may be expanded to incorporate four current injection points, namely A, B, a and b, with the following attributes:

$$
\begin{aligned}
I_A &= I_p & \qquad I_B &= -I_p \\
I_a &= I_s & \qquad I_b &= -I_s \\
V_p &= V_A - V_B & \qquad V_s &= V_a - V_b
\end{aligned}
$$

The injection current I_A is given by:

$$
I_A = \frac{y_t}{a^2} V_A - \frac{y_t}{a^2} V_B - \frac{y_t}{ab} V_a + \frac{y_t}{ab} V_b + \frac{g_0}{2a^2} V_A - \frac{g_0}{2a^2} V_B + \frac{1}{2a} I_\psi \tag{11.6}
$$

Using the same line of reasoning, the four current injections can be obtained and represented in matrix form as follows:

$$
\begin{bmatrix} I_A \\ I_B \\ I_a \\ I_b \end{bmatrix} = y_t \begin{bmatrix} 1/a^2 & -1/a^2 & -1/ab & 1/ab \\ -1/a^2 & 1/a^2 & 1/ab & -1/ab \\ -1/ab & 1/ab & 1/b^2 & -1/b^2 \\ 1/ab & -1/ab & -1/b^2 & 1/b^2 \end{bmatrix} \begin{bmatrix} v_A \\ v_B \\ v_a \\ v_b \end{bmatrix}
+ \frac{1}{2} g_0 \begin{bmatrix} 1/a^2 & -1/a^2 & 0 & 0 \\ -1/a^2 & 1/a^2 & 0 & 0 \\ 0 & 0 & 1/b^2 & -1/b^2 \\ 0 & 0 & -1/b^2 & 1/b^2 \end{bmatrix} \begin{bmatrix} v_A \\ v_B \\ v_a \\ v_b \end{bmatrix}
$$

$$+\frac{1}{2}I_{\psi}\begin{bmatrix} 1/a \\ -1/a \\ 1/b \\ -1/b \end{bmatrix} \tag{11.7}$$

This equation may be interpreted in electric circuit form, resulting in the lattice circuit shown in Figure 11.6.

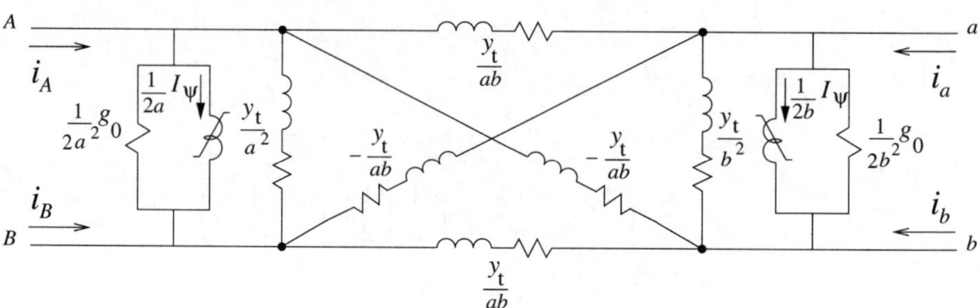

Figure 11.6 Single-phase transformer lattice circuit

By induction, the harmonic lattice counterpart of the circuit in Figure 11.6 is as shown in Figure 11.7. Where the harmonic Norton equivalent of the magnetising branch of the transformer is linearised around a base operating point,

$$\begin{aligned} \mathbf{I}_{\psi} &= \mathbf{B}\mathbf{V}'' + \mathbf{I}_{\mathrm{N}} \\ &= \frac{1}{b}\mathbf{B}\mathbf{V}_{s} + \mathbf{I}_{\mathrm{N}} \end{aligned} \tag{11.8}$$

where

$$\mathbf{B} = \mathbf{F}'\mathbf{D}(\mathrm{j}h\omega_0)^{-1} \tag{11.9}$$

$$\mathbf{I}_{\mathrm{N}} = \mathbf{I}_{\psi_e} - \frac{1}{b}\mathbf{B}\mathbf{V}_{s_e} \tag{11.10}$$

$\mathbf{F}'$ is a Toeplitz matrix assembled with the harmonic inductance of the magnetising characteristic, whilst $\mathbf{I}_{\psi_e}$ is a vector assembled with the harmonic current coefficients of the magnetising characteristic.

An authoritative source has recommended that unless knowledge of construction details is available, the per unit magnetising admittance should be divided equally to all terminals [8]. Information on transformer construction details is seldom available when conducting system-level studies and transformer models that use such a recommendation have a very practical appeal. Accordingly, in Figure 11.7 one-half of the harmonic Norton equivalent is placed on the primary side and other half on the secondary side of the equivalent circuit.

The following nodal equation is derived by inspection of the lattice circuit shown in Figure 11.7:

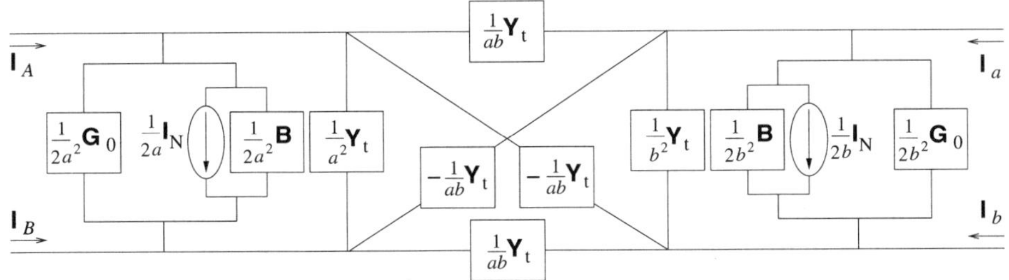

Figure 11.7 Lattice circuit in the harmonic domain

$$
\begin{aligned}
\begin{bmatrix} \mathbf{I}_A \\ \mathbf{I}_B \\ \mathbf{I}_a \\ \mathbf{I}_b \end{bmatrix} &= \mathbf{Y}_t \begin{bmatrix} 1/a^2 & -1/a^2 & -1/ab & 1/ab \\ -1/a^2 & 1/a^2 & 1/ab & -1/ab \\ -1/ab & 1/ab & 1/b^2 & -1/b^2 \\ 1/ab & -1/ab & -1/b^2 & 1/b^2 \end{bmatrix} \begin{bmatrix} \mathbf{V}_A \\ \mathbf{V}_B \\ \mathbf{V}_a \\ \mathbf{V}_b \end{bmatrix} \\
&\quad + \frac{1}{2}(\mathbf{G}_0 + \mathbf{B}) \begin{bmatrix} 1/a^2 & -1/a^2 & 0 & 0 \\ -1/a^2 & 1/a^2 & 0 & 0 \\ 0 & 0 & 1/b^2 & -1/b^2 \\ 0 & 0 & -1/b^2 & 1/b^2 \end{bmatrix} \begin{bmatrix} \mathbf{V}_A \\ \mathbf{V}_B \\ \mathbf{V}_a \\ \mathbf{V}_b \end{bmatrix} \\
&\quad + \frac{1}{2}\mathbf{I}_N \begin{bmatrix} 1/a \\ -1/a \\ 1/b \\ -1/b \end{bmatrix}
\end{aligned} \tag{11.11}
$$

where $\mathbf{B}$, $\mathbf{G}_0$ and $\mathbf{Y}_t$ are matrices of order $(2h + 1)(2h + 1)$. $\mathbf{B}$ is obtained from the linearisation exercise. $\mathbf{G}_0$ is a diagonal matrix with entries g_0. $\mathbf{Y}_t$ is a diagonal matrix with entries $y_t = 1/\left(R\sqrt{h} + \mathrm{j}X_t h\right)$, [9,14]. R is derived from the transformer power losses and X_t is the transformer's short-circuit reactance.

11.3.2 Transformer parameters in per unit values

One of the main advantages of using per unit values in a transformer circuit is that the per unit magnitudes of voltage and current are the same on both sides of the transformer. For instance, in the single-phase transformer used in this section $V_p/V_s = a/b = 1$ p.u.

In three-phase transformer applications where the unit consists of three single-phase units, the circuit rating is determined by the rating of each individual single-phase transformer. The per unit impedance of the three-phase unit is the same as that of each individual transformer. Since the impedance of the single-phase transformer is in per units referred to its bases, we have that for a star–star-connected unit the following relation applies: $a/b = 1$. For a star–delta-connected unit we have $a/b = 1/\sqrt{3}$. For a delta–star-connected unit the relation is $a/b = \sqrt{3}$. For a delta–delta-connected unit the relation is $a/b = 1$. These relations are very important to ensure that the per unit magnitudes of voltage and current are the same on both sides of the three-phase unit.

11.4 Three-phase Transformer Bank

Suitable connection of three single-phase transformer units, say T1, T2 and T3, produces a three-phase bank of transformers. For example, Figure 11.8(a) shows the connection of a three-phase bank with the primary winding star connected and the secondary winding delta connected. The star point is grounded through a grounding admittance of value y_{g}. Figure 11.8(b) gives three lattice equivalent circuits together with the grounding admittance and six harmonic Norton equivalents. It should be noted that the nodes' labels in the various components incorporate the effect of the star–delta connection.

By inspection of the three-phase circuit in Figure 11.8, the following nodal equation is assembled:

$$\begin{bmatrix} \mathbf{I}_{ABC} \\ \mathbf{I}_{abc} \\ \mathbf{I}_{N} \end{bmatrix} = \begin{bmatrix} \mathbf{Y}_{\mathrm{S}} & \mathbf{Y}_{\mathrm{SD}} & \mathbf{Y}_{\mathrm{SN}} \\ \mathbf{Y}_{\mathrm{SD}}^{\mathrm{T}} & \frac{1}{3}\mathbf{Y}_{\mathrm{D}} & 0 \\ \mathbf{Y}_{\mathrm{SN}}^{\mathrm{T}} & 0 & \mathbf{Y}_{\mathrm{N}} \end{bmatrix} \begin{bmatrix} \mathbf{V}_{ABC} \\ \mathbf{V}_{abc} \\ \mathbf{V}_{N} \end{bmatrix} + \begin{bmatrix} \mathbf{B}_{\mathrm{S}} & 0 & \mathbf{B}_{\mathrm{SN}} \\ 0 & \frac{1}{3}\mathbf{B}_{\mathrm{D}} & 0 \\ \mathbf{B}_{\mathrm{SN}}^{\mathrm{T}} & 0 & \mathbf{B}_{\mathrm{N}} \end{bmatrix} \begin{bmatrix} \mathbf{V}_{ABC} \\ \mathbf{V}_{abc} \\ \mathbf{V}_{N} \end{bmatrix} + \begin{bmatrix} \mathbf{I}_{\mathrm{N}_{ABC}} \\ \mathbf{I}_{\mathrm{N}_{abc}} \\ \mathbf{I}_{\mathrm{N}_{N}} \end{bmatrix} \tag{11.12}$$

where the individual matrices are given below in a more explicit form. It should be noted that owing to the star–delta connection, the tap ratios $a = 1$ and $b = \sqrt{3}$ are used as part of the solution.

$$\mathbf{Y}_{\mathrm{D}} = \begin{bmatrix} 2\mathbf{Y}_{\mathrm{t}} & -\mathbf{Y}_{\mathrm{t}} & -\mathbf{Y}_{\mathrm{t}} \\ -\mathbf{Y}_{\mathrm{t}} & 2\mathbf{Y}_{\mathrm{t}} & -\mathbf{Y}_{\mathrm{t}} \\ -\mathbf{Y}_{\mathrm{t}} & -\mathbf{Y}_{\mathrm{t}} & 2\mathbf{Y}_{\mathrm{t}} \end{bmatrix}; \quad \mathbf{Y}_{\mathrm{SD}} = \begin{bmatrix} -\frac{1}{\sqrt{3}}\mathbf{Y}_{\mathrm{t}} & \frac{1}{\sqrt{3}}\mathbf{Y}_{\mathrm{t}} & 0 \\ 0 & -\frac{1}{\sqrt{3}}\mathbf{Y}_{\mathrm{t}} & \frac{1}{\sqrt{3}}\mathbf{Y}_{\mathrm{t}} \\ \frac{1}{\sqrt{3}}\mathbf{Y}_{\mathrm{t}} & 0 & -\frac{1}{\sqrt{3}}\mathbf{Y}_{\mathrm{t}} \end{bmatrix}$$

$$\mathbf{Y}_{\mathrm{S}} = \begin{bmatrix} \mathbf{Y}_{\mathrm{t}} & 0 & 0 \\ 0 & \mathbf{Y}_{\mathrm{t}} & 0 \\ 0 & 0 & \mathbf{Y}_{\mathrm{t}} \end{bmatrix}; \quad \mathbf{Y}_{\mathrm{SN}} = \begin{bmatrix} -\mathbf{Y}_{\mathrm{t}} \\ -\mathbf{Y}_{\mathrm{t}} \\ -\mathbf{Y}_{\mathrm{t}} \end{bmatrix}; \quad \mathbf{Y}_{\mathrm{N}} = 3\mathbf{Y}_{\mathrm{t}} + \mathbf{Y}_{\mathrm{g}}$$

For the non-linear part,

$$\mathbf{B}_{\mathrm{D}} = \frac{1}{2}\begin{bmatrix} \mathbf{B}_1 + \mathbf{B}_3 & -\mathbf{B}_1 & -\mathbf{B}_3 \\ -\mathbf{B}_1 & \mathbf{B}_2 + \mathbf{B}_1 & -\mathbf{B}_2 \\ -\mathbf{B}_3 & -\mathbf{B}_2 & \mathbf{B}_3 + \mathbf{B}_2 \end{bmatrix}; \quad \mathbf{B}_{\mathrm{S}} = \frac{1}{2}\begin{bmatrix} \mathbf{B}_1 & 0 & 0 \\ 0 & \mathbf{B}_2 & 0 \\ 0 & 0 & \mathbf{B}_3 \end{bmatrix}$$

$$\mathbf{B}_{\mathrm{SN}} = \frac{1}{2}\begin{bmatrix} -\mathbf{B}_1 \\ -\mathbf{B}_2 \\ -\mathbf{B}_3 \end{bmatrix}; \quad \mathbf{B}_{\mathrm{N}} = \frac{1}{2}(\mathbf{B}_1 + \mathbf{B}_2 + \mathbf{B}_3)$$

$$\mathbf{I}_{\mathrm{N}_{ABC}} = \frac{1}{2}\begin{bmatrix} -\mathbf{I}_{1_{\mathrm{N}}} \\ -\mathbf{I}_{2_{\mathrm{N}}} \\ -\mathbf{I}_{3_{\mathrm{N}}} \end{bmatrix}; \quad \mathbf{I}_{\mathrm{N}_{abc}} = \frac{1}{2\sqrt{3}}\begin{bmatrix} \mathbf{I}_{3_{\mathrm{N}}} - \mathbf{I}_{1_{\mathrm{N}}} \\ \mathbf{I}_{1_{\mathrm{N}}} - \mathbf{I}_{2_{\mathrm{N}}} \\ \mathbf{I}_{2_{\mathrm{N}}} - \mathbf{I}_{3_{\mathrm{N}}} \end{bmatrix}; \quad \mathbf{I}_{\mathrm{N}_{N}} = \frac{1}{2}(\mathbf{I}_{1_{\mathrm{N}}} + \mathbf{I}_{2_{\mathrm{N}}} + \mathbf{I}_{3_{\mathrm{N}}})$$

It should be noticed that in the above derivations, the terms $\mathbf{G}_0$ and $\mathbf{B}$ are both combined, e.g. $\mathbf{B}_1 = \mathbf{B}_1 + \mathbf{G}_0$.

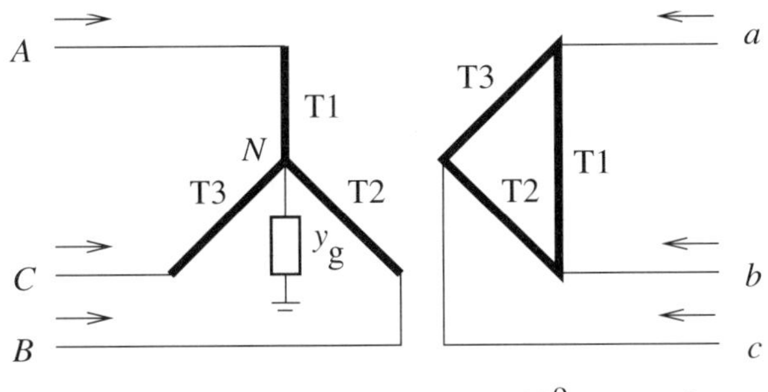

(a) British standard for star delta -30° connection

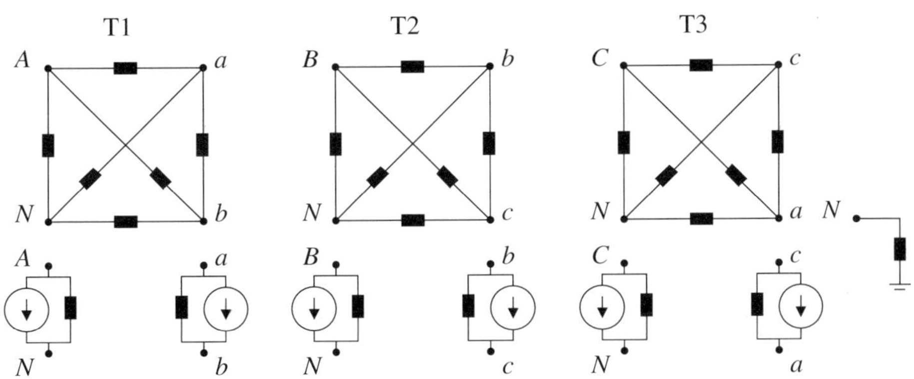

(b) Lattice circuit connection

Figure 11.8 Grounded star–delta transformer bank connection

The following practical aspects should be taken into account in the mathematical model given above: the injection current $\mathbf{I}_N$ is always zero since node N is not directly connected to the external system, i.e. a current injection source $\mathbf{I}_N$ does not exist.

Voltage $\mathbf{V}_N$ would have a value of zero if the star point were solidly connected to ground. If the star point is floating and circuit and operating conditions are balanced then $\mathbf{V}_N$ would also have a value of zero, and otherwise a value different from zero. Incorporating such facts in equation (11.12) leads to equivalent systems of reduced dimensions: (1) If $\mathbf{V}_N = 0$, the row and column associated with node N are deleted. (2) If $\mathbf{V}_N \neq 0$, the row and column associated with node N are eliminated mathematically via Kron's reduction.

The most common transformer connections are shown in Table 11.1 together with the transfer admittance matrices associated with each transformer connection. It should be noted that the non-linear part of the transformer is not included in these matrix representations and that they are, strictly speaking, only valid for banks of transformers. Equivalent, slightly more elaborate matrix representations exist for single-core, multi-legged transformers [10].

Table 11.1 Transformers bank admittance equivalent

CASE	CONNECTION	ADMITTANCE
1	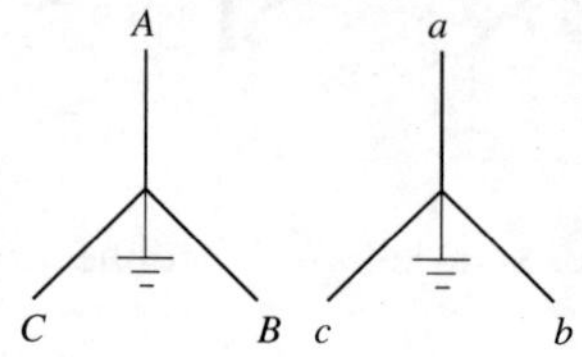 A, C, B a, c, b	$\begin{bmatrix} \mathbf{Y}_S & -\mathbf{Y}_S \\ -\mathbf{Y}_S & \mathbf{Y}_S \end{bmatrix}$
2	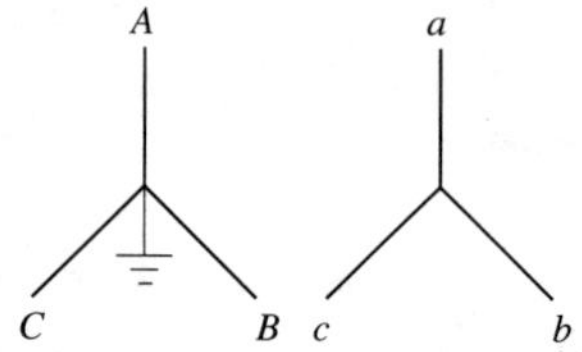 A, C, B a, c, b	$\begin{bmatrix} \frac{1}{3}\mathbf{Y}_D & -\frac{1}{3}\mathbf{Y}_D \\ -\frac{1}{3}\mathbf{Y}_D & \frac{1}{3}\mathbf{Y}_D \end{bmatrix}$
3	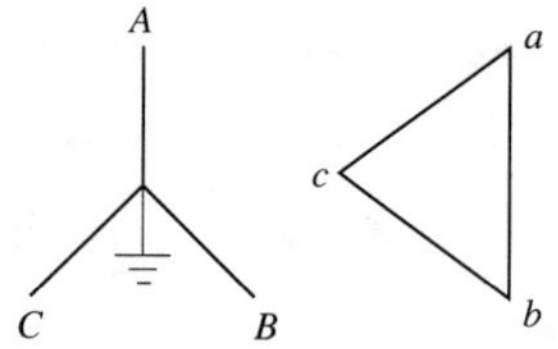 A, C, B a, c, b	$\begin{bmatrix} \mathbf{Y}_S & \mathbf{Y}_{SD} \\ \mathbf{Y}_{SD}^T & \frac{1}{3}\mathbf{Y}_D \end{bmatrix}$
4	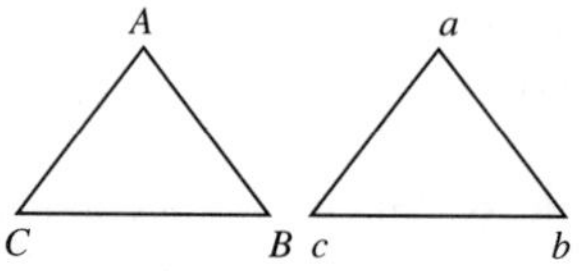 A, C, B a, c, b	$\begin{bmatrix} \mathbf{Y}_D & -\mathbf{Y}_D \\ -\mathbf{Y}_D & \mathbf{Y}_D \end{bmatrix}$
5	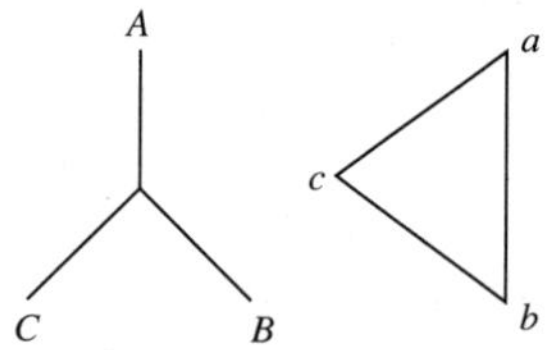 A, C, B a, c, b	$\begin{bmatrix} \frac{1}{3}\mathbf{Y}_D & \mathbf{Y}_{SD} \\ \mathbf{Y}_{SD}^T & \frac{1}{3}\mathbf{Y}_D \end{bmatrix}$

Example 11-1: Consider the simple, three-phase circuit shown in Figure 11.9 comprising a balanced, three-phase voltage source, three single-phase transformers and three single-phase non-linear loads. Two transformer connections are considered in this example: (i) the star–delta and (ii) the delta–star connections. In both cases, the star point is solidly grounded.

The voltage source has a value of 1 p.u. magnitude in each phase and the phase voltage angles are 0°, −120° and 120°, respectively. The three single-phase transformers have equal impedance values: $R = 0.001$ p.u. and $X_{\mathrm{t}} = 0.01$ p.u. Each non-linear load is given as the parallel combination of a 10 p.u. resistance and a non-linear inductor.

The inverse of the magnetic inductances of the non-linear inductor is given below for the three phases using MATLAB™:

```
Fa=[         0    Fb=[                  0     Fc=[                  0
       -0.0045           0.0022 + 0.0039i             0.0022 - 0.0039i
             0                          0                           0
        0.0046           0.0046 - 0.0000i             0.0046 + 0.0000i
             0                          0                           0
       -0.0022           0.0011 - 0.0019i             0.0011 + 0.0019i
             0                          0                           0
       -0.0027           0.0014 + 0.0024i             0.0014 - 0.0024i
             0                          0                           0
        0.0092           0.0092 - 0.0000i             0.0092 + 0.0000i
             0                          0                           0
       -0.0157           0.0078 - 0.0136i             0.0078 + 0.0136i
             0                          0                           0
        0.0205          -0.0102 - 0.0177i            -0.0102 + 0.0177i
             0                          0                           0
        0.0778                     0.0778                      0.0778
             0                          0                           0
        0.0205          -0.0102 + 0.0177i            -0.0102 - 0.0177i
             0                          0                           0
       -0.0157           0.0078 + 0.0136i             0.0078 - 0.0136i
             0                          0                           0
        0.0092           0.0092 + 0.0000i             0.0092 - 0.0000i
             0                          0                           0
       -0.0027           0.0014 - 0.0024i             0.0014 + 0.0024i
             0                          0                           0
       -0.0022           0.0011 + 0.0019i             0.0011 - 0.0019i
             0                          0                           0
        0.0046           0.0046 + 0.0000i             0.0046 - 0.0000i
             0                          0                           0
       -0.0045           0.0022 - 0.0039i             0.0022 + 0.0039i
             0 ]                        0 ]                         0 ]
```

These harmonic coefficients are used to assemble the equivalent admittance matrix of the three non-linear elements using (5.24), e.g. $\mathbf{Y}_a = \mathbf{F}_a\mathbf{D}(\mathrm{j}h\omega_0)^{-1}$.

The harmonic matrix equation, in nodal admittance form, that describes the system is

$$\begin{bmatrix} \mathbf{I}_{ABC} \\ 0 \end{bmatrix} = \begin{bmatrix} \mathbf{Y}_{ABC} & \mathbf{Y}_{ABCabc} \\ \mathbf{Y}_{abcABC} & \mathbf{Y}_{abc} + \mathbf{Y}_{\mathrm{load}} \end{bmatrix} \begin{bmatrix} \mathbf{V}_{ABC} \\ \mathbf{V}_{abc} \end{bmatrix}$$

The solution is obtained for $\mathbf{V}_{ABC}$, $\mathbf{V}_{abc}$ and $\mathbf{I}_{abc}$,

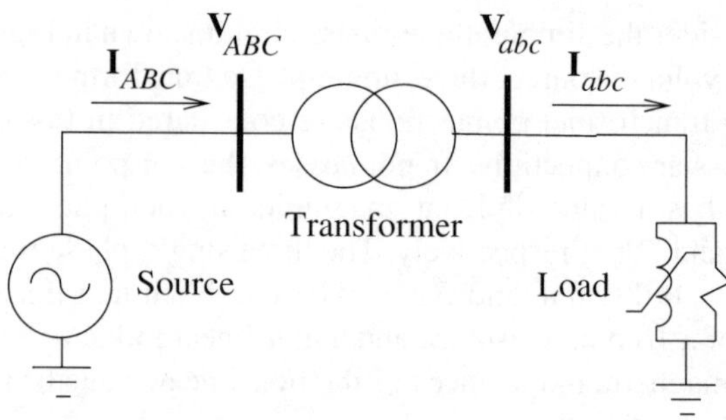

Figure 11.9 Simple test system

$$\mathbf{V}_{abc} = -\left(\mathbf{Y}_{abc} + \mathbf{Y}_{\text{load}}\right)^{-1} \mathbf{Y}_{abcABC} \mathbf{V}_{ABC}$$

$$\mathbf{I}_{ABC} = \mathbf{Y}_{ABC} \mathbf{V}_{ABC} + \mathbf{Y}_{ABCabc} \mathbf{V}_{abc}$$

$$\mathbf{I}_{abc} = \mathbf{Y}_{\text{load}} \mathbf{V}_{abc}$$

where $\mathbf{Y}_{ABC}$, $\mathbf{Y}_{ABCabc}$, $\mathbf{Y}_{abcABC}$ and $\mathbf{Y}_{abc}$ are selected to suit the relevant transformer bank connection. Also, $\mathbf{Y}_{\text{load}}$ is given by,

$$\mathbf{Y}_{\text{load}} = \begin{bmatrix} \mathbf{F}_a \mathbf{D}(jh\omega_0)^{-1} + \mathbf{Y}_\text{R} & 0 & 0 \\ 0 & \mathbf{F}_b \mathbf{D}(jh\omega_0)^{-1} + \mathbf{Y}_\text{R} & 0 \\ 0 & 0 & \mathbf{F}_c \mathbf{D}(jh\omega_0)^{-1} + \mathbf{Y}_\text{R} \end{bmatrix}$$

(i) Star–delta connection. As expected, only the fundamental frequency component and odd harmonics appear in the spectrum generated by the MATLAB ™ function used to solve this example. Also, the three-phase voltage and current magnitudes are balanced throughout the system. The harmonic voltages and currents are given in Tables 11.2 and 11.3, respectively.

Table 11.2 Voltages on the primary and secondary sides of a star–delta connection in p.u.

h	V_{ABC}	$\angle\theta_A^\circ$	$\angle\theta_B^\circ$	$\angle\theta_C^\circ$	V_{abc}	$\angle\theta_a^\circ$	$\angle\theta_b^\circ$	$\angle\theta_c^\circ$
1	1.0	0.0	-120.0	120.0	0.9993	-30.04	-150.04	89.95
3	0.0	0.0	0.0	0.0	0.3046	101.78	101.78	101.78
5	0.0	0.0	0.0	0.0	0.0011	-15.77	104.22	-135.77
7	0.0	0.0	0.0	0.0	0.0008	151.87	31.87	-88.12
9	0.0	0.0	0.0	0.0	0.0383	-166.32	-166.32	-166.32
11	0.0	0.0	0.0	0.0	0.0007	15.38	135.38	-104.61
13	0.0	0.0	0.0	0.0	0.0010	175.39	55.39	-64.60
15	0.0	0.0	0.0	0.0	0.0482	-120.86	-120.86	-120.86

The three-phase non-linear loads are responsible for generating the harmonic currents, giving rise to harmonic voltages and currents on the secondary side of the transformer. It should be noted that no harmonic voltages appear on the primary side of the transformer, which behaves very much like an infinite busbar. A phase shift of $-30°$ appears between the voltages on the primary and secondary sides of the transformer because of the star–delta connection. As discussed in Chapter 2, the 7th and 13th harmonics are termed positive sequence harmonics and the 5th and 11th are termed negative sequence harmonics. By extension, the 3rd, 9th and 15th harmonics are zero sequence harmonics.

No zero sequence harmonic currents appear on the secondary side because a path to ground does not exist owing to the delta connection. Hence, no zero sequence harmonic currents will appear on the primary side either. The positive and negative sequence harmonic currents present a phase shift of $-30°$ and $30°$, respectively.

Table 11.3 Currents on the primary and secondary sides of a star–delta connection in p.u.

h	I_{ABC}	$\angle\theta_A^\circ$	$\angle\theta_B^\circ$	$\angle\theta_C^\circ$	I_{abc}	$\angle\theta_a^\circ$	$\angle\theta_b^\circ$	$\angle\theta_c^\circ$
1	0.1059	-39.05	-159.05	80.94	0.1059	-69.05	170.94	50.94
3	0.0000	-173.53	-17.95	80.99	0.0000	-53.93	176.96	156.57
5	0.0225	46.78	166.78	-73.21	0.0225	76.78	-163.21	-43.21
7	0.0116	-85.96	154.03	34.03	0.0116	-115.96	124.03	4.03
9	0.0000	49.34	-132.65	-44.39	0.0000	14.92	-92.05	49.22
11	0.0062	77.10	-162.89	-42.89	0.0062	107.10	-132.89	-12.89
13	0.0079	-63.01	176.98	56.98	0.0079	-93.01	146.98	26.98
15	0.0000	24.62	165.65	-72.97	0.0000	75.36	-55.89	-27.31

(ii) Delta–star connection. This time a phase shift of $+30°$ appears between the voltages on the primary and secondary sides of the transformer. Also, the zero sequence harmonic voltages, on the secondary side, are much smaller than in case (i). The results are shown in Tables 11.4 and 11.5.

Table 11.4 Voltages on the primary and secondary sides of a delta–star connection in p.u.

h	V_{ABC}	$\angle\theta_A^\circ$	$\angle\theta_B^\circ$	$\angle\theta_C^\circ$	V_{abc}	$\angle\theta_a^\circ$	$\angle\theta_b^\circ$	$\angle\theta_c^\circ$
1	1.0	0.0	-120.0	120.0	0.9992	29.93	-90.06	149.93
3	0.0	0.0	0.0	0.0	0.0009	-179.10	-179.10	-179.10
5	0.0	0.0	0.0	0.0	0.0011	5.69	125.69	-114.30
7	0.0	0.0	0.0	0.0	0.0008	-165.20	74.79	-45.20
9	0.0	0.0	0.0	0.0	0.0002	76.58	76.58	76.58
11	0.0	0.0	0.0	0.0	0.0007	-13.98	106.01	-133.98
13	0.0	0.0	0.0	0.0	0.0010	178.08	58.08	-61.91
15	0.0	0.0	0.0	0.0	0.0007	27.56	27.56	27.56

Zero sequence harmonic currents appear on the secondary side of the transformer because in this case there is a path to ground, via the solidly grounded star point. These currents are induced in the delta but cannot flow into the system source, i.e. sink at harmonic frequencies.

The following MATLAB™ function is used to obtain a transformer bank:

Table 11.5 Currents on the primary and secondary sides of a delta–star connection in p.u.

h	I_{ABC}	$\angle\theta_A^\circ$	$\angle\theta_B^\circ$	$\angle\theta_C^\circ$	I_{abc}	$\angle\theta_a^\circ$	$\angle\theta_b^\circ$	$\angle\theta_c^\circ$
1	0.1356	-29.89	-149.89	90.10	0.1356	0.10	-119.89	120.10
3	0.0000	4.29	-22.06	165.68	0.0314	-85.80	-85.80	-85.80
5	0.0218	128.25	-111.74	8.25	0.0218	98.25	-141.74	-21.74
7	0.0108	-103.03	136.96	16.96	0.0108	-73.03	166.96	46.96
9	0.0000	-72.94	-76.33	106.21	0.0025	168.49	168.49	168.49
11	0.0060	107.74	-132.25	-12.25	0.0060	77.74	-162.25	-42.25
13	0.0078	-120.32	119.67	-0.32	0.0078	-90.32	149.67	29.67
15	0.0000	-154.91	-150.99	27.22	0.0045	119.04	119.04	119.04

```
function [YABC,YABCabc,YabcABC,Yabc]=transf_bank(R,Xt,h,connection)
% R          : Resistance of the single-phase transformer
% Xt         : Reactance of the single-phase transformer
% h          : Number of harmonics
% connection : Ss-Ss Ss-Sf D-D Ss-D Sf-D
% Sf         : star with floating connection to ground
% Ss         : star with solid connection to ground
% D          : Delta connection
Yt = zeros(2*h+1,2*h+1);
O  = zeros(2*h+1,2*h+1);
k  = 1;
for n=-h:h
 if n~=0
  yt = 1/(R*sqrt(abs(n))+j*Xt*n);
 else
  yt = 1/R;
 end
 Yt(k,k) = yt;
 k = k+1;
end
YD  = [2*Yt  -Yt  -Yt
        -Yt 2*Yt  -Yt
        -Yt  -Yt 2*Yt ];
YSD = [ -Yt   Yt    O
         O  -Yt   Yt
        Yt    O  -Yt]/sqrt(3);
YS  = [  Yt    O    O
         O   Yt    O
         O    O   Yt];
switch connection
 case 'Ss-Ss',
  YABC    =  YS;
  YABCabc = -YS;
  YabcABC = -YS;
  Yabc    =  YS;
 case 'Ss-Sf'
  YABC    =  YD/3;
  YABCabc = -YD/3;
  YabcABC = -YD/3;
  Yabc    =  YD/3;
```

```
case 'D-D',
 YABC    =  YD;
 YABCabc = -YD;
 YabcABC = -YD;
 Yabc    =  YD;
case 'Ss-D',
 YABC    = YS;
 YABCabc = YSD;
 YabcABC = transpose(YSD);
 Yabc    = YD/3;
case 'Sf-D',
 YABC    = YD/3;
 YABCabc = YSD;
 YabcABC = transpose(YSD);
 Yabc    = YD/3;
end
```

The following MATLAB™ function is used to obtain a three-phase sinusoidal voltage source:

```
function [Vah,Vbh,Vch]=source_h(Va,Vb,Vc,fa,fb,fc,h)
% Va,Vb,Vc : pick value of the sinusoidal waveform
% fa,fb,fc : phase angle in degrees
% h : number of harmonics
% V : three-phase voltage source
Vah       = zeros(2*h+1,1);
Vah(h)    = i*Va/2*exp(-i*fa*pi/180);
Vah(h+2)  = -i*Va/2*exp(i*fa*pi/180);
Vbh       = zeros(2*h+1,1);
Vbh(h)    = i*Vb/2*exp(-i*fb*pi/180);
Vbh(h+2)  = -i*Vb/2*exp(i*fb*pi/180);
Vch       = zeros(2*h+1,1);
Vch(h)    = i*Vc/2*exp(-i*fc*pi/180);
Vch(h+2)  = -i*Vc/2*exp(i*fc*pi/180);
```

11.5 Laminated Iron Cores

Transformer iron cores are normally laminated in order to minimise the significance of the skin effect at power frequencies. However, at harmonic frequencies, a pronounced skin effect may take place and the accurate calculation of iron losses will require that the harmonic analysis be carried out at the lamination level.

11.5.1 Modelling of cores

Maxwell's equations for the plane geometry of a lamination are

$$\frac{\partial H}{\partial x} = \sigma E \tag{11.13}$$

$$\frac{\partial E}{\partial x} = \dot{B} \tag{11.14}$$

which can be combined into

$$\frac{\partial^2 H}{\partial x^2}\sigma\dot{B} \tag{11.15}$$

where:

E is the electric field in V/m,
σ is the conductivity of the laminations in S/m,
H is the magnetic field in A/m,
B is the flux density in T.

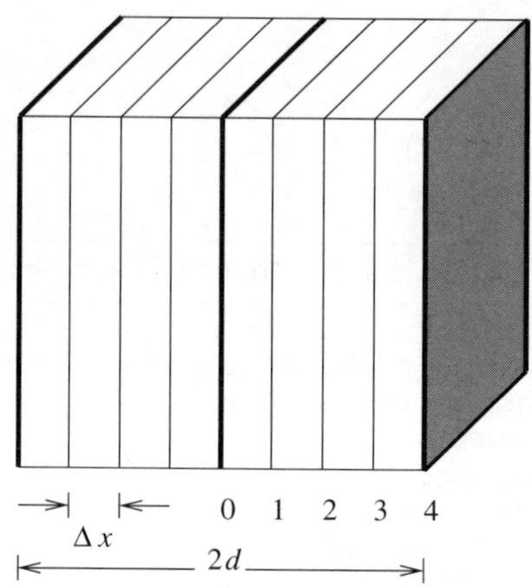

Figure 11.10 A lamination subdivides into eight sub-laminations

With reference to Figure 11.10, if a lamination is discretised in $2n$ sub-laminations, say eight, the rhs half of the lamination will give rise to the discrete points 0, 1, 2, 3, 4. For most practical purposes, symmetry may be assumed in the lamination, and the discrete point 0 will be at the centre of the lamination, playing the role of a mirroring surface. In this case, the discrete points are separated by

$$\Delta x = \frac{2d}{8} \tag{11.16}$$

In general, the magnetic field density, B_n, and the electric field density, E_n, at the surface of the lamination (point $n = 4$) are the input variables of interest. E_n is proportional to the time derivative of the total flux in the lamination,

$$E_n \propto \frac{\mathrm{d}}{\mathrm{d}t} B_{\mathrm{eff}} \tag{11.17}$$

Equations (11.13) and (11.14) are discretised for the surface layer by means of the trapezoidal rule of integration,

$$\frac{1}{\Delta x}(H_n - H_{n-1}) = \frac{\sigma}{2}(E_n + E_{n-1}) \tag{11.18}$$

$$\frac{1}{\Delta x}(E_n - E_{n-1}) = \frac{1}{2}\left(\dot{B}_n + \dot{B}_{n-1}\right) \tag{11.19}$$

Elimination of E_{n-1} from equations (11.18) and (11.19) leads to

$$-\sigma \Delta x^2 \left(\dot{B}_n + \dot{B}_{n-1} \right) - 4 \left(H_n - H_{n-1} \right) + 4 \sigma \Delta x E_n = 0 \tag{11.20}$$

Equation (11.15) is used for the internal points of the lamination, and is discretised by using the Stormer–Numerov formula,

$$\frac{1}{\Delta x^2} \left(H_{k+1} - 2H_k + H_{k-1} \right) = \frac{\sigma}{12} \left(\dot{B}_{k+1} + 10\dot{B}_k + \dot{B}_{k-1} \right) \tag{11.21}$$

or

$$-\sigma \Delta x^2 \left(\dot{B}_{k+1} + 10\dot{B}_k + \dot{B}_{k-1} \right) + 12 \left(H_{k+1} - 2H_k + H_{k-1} \right) = 0 \tag{11.22}$$

This procedure involves three discrete points and two sub-laminations at any one time, giving a more accurate result.

In this application, the magnetic field is a non-linear function of the flux density,

$$H = aB + bB^p + cB^q \tag{11.23}$$

which for the discrete point i leads to

$$H_i = aB_i + bB_i^p + cB_i^q$$

Since symmetry around the centre of the lamination is assumed, i.e. $B_{-i} = B_i$, the $n+1$ unknown variables $B_0, B_1, \ldots, B_n$ may be determined by using the following $n+1$ equations: $f_0, f_1, \ldots, f_{n-1}, f_n$. Equation (11.22) is used to derive the first $n-1$ equations whereas (11.20) is used to derive f_n:

$$\begin{aligned}
f_0 &= -\sigma \Delta x^2 \left(2\dot{B}_1 + 10\dot{B}_0 \right) + 12 \left(2H_1 - 2H_0 \right) = 0 \\
f_1 &= -\sigma \Delta x^2 \left(\dot{B}_2 + 10\dot{B}_1 + \dot{B}_0 \right) + 12 \left(H_2 - 2H_1 + H_0 \right) = 0 \\
&\vdots \\
f_k &= -\sigma \Delta x^2 \left(\dot{B}_{k+1} + 10\dot{B}_k + \dot{B}_{k-1} \right) + 12 \left(H_{k+1} - 2H_k + H_{k-1} \right) = 0 \\
&\vdots \\
f_n &= -\sigma \Delta x^2 \left(\dot{B}_n + \dot{B}_{n-1} \right) - 4 \left(H_n - H_{n-1} \right) + 4\sigma \Delta x E_n = 0
\end{aligned} \tag{11.24}$$

If the aim is to carry out steady-state periodic studies of saturated laminations then these equations may be expressed in the harmonic domain as follows:

$$\begin{aligned}
\mathbf{F}_0 &= -\sigma \Delta x^2 \mathbf{D}(jh\omega_0) \left(2\mathbf{B}_1 + 10\mathbf{B}_0 \right) + 12 \left(2\mathbf{H}_1 - 2\mathbf{H}_0 \right) = 0 \\
\mathbf{F}_1 &= -\sigma \Delta x^2 \mathbf{D}(jh\omega_0) \left(\mathbf{B}_2 + 10\mathbf{B}_1 + \mathbf{B}_0 \right) + 12 \left(\mathbf{H}_2 - 2\mathbf{H}_1 + \mathbf{H}_0 \right) = 0 \\
&\vdots \\
\mathbf{F}_k &= -\sigma \Delta x^2 \mathbf{D}(jh\omega_0) \left(\mathbf{B}_{k+1} + 10\mathbf{B}_k + \mathbf{B}_{k-1} \right) + 12 \left(\mathbf{H}_{k+1} - 2\mathbf{H}_k + \mathbf{H}_{k-1} \right) = 0
\end{aligned}$$

$$\vdots$$

$$\mathbf{F}_n = -\sigma\Delta x^2\mathbf{D}(jh\omega_0)(\mathbf{B}_n + \mathbf{B}_{n-1}) - 4(\mathbf{H}_n - \mathbf{H}_{n-1}) + 4\sigma\Delta x\mathbf{E}_n = 0 \tag{11.25}$$

with

$$\mathbf{H}_i = a\mathbf{B}_i + b\mathbf{B}_i^p + c\mathbf{B}_i^q \tag{11.26}$$

where the polynomial evaluation is carried out via self- and mutual convolution operations, guided by the exponents p and q, as explained in Section 4.3.1.

The following MATLAB™ functions may be used to evaluate the harmonic domain expression $\mathbf{F}_i$, given by (11.25), and to carry out the polynomial evaluation of (11.26):

```
function F0=calc_F0(a,b,c,p,q,B0,B1,sigma,dx,h,w)
D  =form_Zm(0,w,h);
H0 =calc_H(a,b,c,p,q,B0,h);
H1 =calc_H(a,b,c,p,q,B1,h);
F0 =-sigma*dx^2*D*(2*B1+10*B0)+12*(2*H1-2*H0);
```

```
function Fk=calc_Fk(a,b,c,p,q,Bk_1,Bk,Bk1,sigma,dx,h,w)
D    =form_Zm(0,w,h);
Hk_1=calc_H(a,b,c,p,q,Bk_1,h);
Hk   =calc_H(a,b,c,p,q,Bk,h);
Hk1  =calc_H(a,b,c,p,q,Bk1,h);
Fk   =-sigma*dx^2*D*(Bk1+10*Bk+Bk_1)+12*(Hk1-2*Hk+Hk_1);
```

```
function Fn=calc_Fn(a,b,c,p,q,Bn_1,Bn,En,sigma,dx,h,w)
D    =form_Zm(0,w,h);
Hn_1=calc_H(a,b,c,p,q,Bn_1,h);
Hn   =calc_H(a,b,c,p,q,Bn,h);
Fn   =-sigma*dx^2*D*(Bn+Bn_1)-4*(Hn-Hn_1)+4*sigma*dx*En;
```

```
function H=calc_H(a,b,c,p,q,B,h)
H=a*B+b*self_conv(B,h,p)+c*self_conv(B,h,q);
```

```
function Xsc=self_conv(X,h,n)
c=h*n+1;
Xsc=X;
for k=1:(n-1)
 Xsc=conv(X,Xsc);
end
Xsc=Xsc(c-h:c+h);
```

Equation (11.25) may be written in compact form as

$$\mathbf{F}(\mathbf{B}) = 0$$

This is a non-linear equation that may be solved very efficiently using Newton's method,

$$
\begin{aligned}
\Delta\mathbf{B}^{(\mathrm{k}+1)} &= -\mathbf{J}^{-1}\mathbf{F}^{(\mathrm{k})} \\
\mathbf{B}^{(\mathrm{k}+1)} &= \mathbf{B}^{(\mathrm{k})} + \Delta\mathbf{B}^{(\mathrm{k}+1)}
\end{aligned}
\tag{11.27}
$$

where k is the iteration counter.

The elements of the Jacobian are given by

$$
\begin{aligned}
\frac{\partial \mathbf{F}_0}{\partial \mathbf{B}_0} &= -10\sigma\Delta x^2\mathbf{D}(\mathrm{j}h\omega_0) - 24\frac{\partial \mathbf{H}_0}{\partial \mathbf{B}_0} \\
\frac{\partial \mathbf{F}_0}{\partial \mathbf{B}_1} &= -2\sigma\Delta x^2\mathbf{D}(\mathrm{j}h\omega_0) + 24\frac{\partial \mathbf{H}_1}{\partial \mathbf{B}_1} \\
\frac{\partial \mathbf{F}_1}{\partial \mathbf{B}_0} &= -\sigma\Delta x^2\mathbf{D}(\mathrm{j}h\omega_0) + 12\frac{\partial \mathbf{H}_0}{\partial \mathbf{B}_0} \\
\frac{\partial \mathbf{F}_1}{\partial \mathbf{B}_1} &= -10\sigma\Delta x^2\mathbf{D}(\mathrm{j}h\omega_0) - 24\frac{\partial \mathbf{H}_1}{\partial \mathbf{B}_1} \\
\frac{\partial \mathbf{F}_1}{\partial \mathbf{B}_2} &= -\sigma\Delta x^2\mathbf{D}(\mathrm{j}h\omega_0) + 12\frac{\partial \mathbf{H}_2}{\partial \mathbf{B}_2} \\
&\vdots \\
\frac{\partial \mathbf{F}_k}{\partial \mathbf{B}_{k-1}} &= -\sigma\Delta x^2\mathbf{D}(\mathrm{j}h\omega_0) - 12\frac{\partial \mathbf{H}_{k-1}}{\partial \mathbf{B}_{k-1}} \\
\frac{\partial \mathbf{F}_k}{\partial \mathbf{B}_k} &= -10\sigma\Delta x^2\mathbf{D}(\mathrm{j}h\omega_0) - 24\frac{\partial \mathbf{H}_k}{\partial \mathbf{B}_k} \\
\frac{\partial \mathbf{F}_k}{\partial \mathbf{B}_{k+1}} &= -\sigma\Delta x^2\mathbf{D}(\mathrm{j}h\omega_0) + 12\frac{\partial \mathbf{H}_{k+1}}{\partial \mathbf{B}_{k+1}} \\
&\vdots \\
\frac{\partial \mathbf{F}_n}{\partial \mathbf{B}_{n-1}} &= -\sigma\Delta x^2\mathbf{D}(\mathrm{j}h\omega_0) + 4\frac{\partial \mathbf{H}_{n-1}}{\partial \mathbf{B}_{n-1}} \\
\frac{\partial \mathbf{F}_n}{\partial \mathbf{B}_n} &= -\sigma\Delta x^2\mathbf{D}(\mathrm{j}h\omega_0) - 4\frac{\partial \mathbf{H}_n}{\partial \mathbf{B}_n}
\end{aligned}
\tag{11.28}
$$

where

$$
\frac{\partial \mathbf{H}_i}{\partial \mathbf{B}_i} = \mathbf{a} + bp\mathbf{B}_i^{p-1} + cq\mathbf{B}_i^{q-1} \tag{11.29}
$$

The expanded form of (11.27) gives

$$
\Delta\mathbf{B}^{(\mathrm{k}+1)} = \begin{bmatrix} \Delta\mathbf{B}_0 \\ \vdots \\ \Delta\mathbf{B}_{k-1} \\ \Delta\mathbf{B}_k \\ \Delta\mathbf{B}_{k+1} \\ \vdots \\ \Delta\mathbf{B}_n \end{bmatrix}^{(\mathrm{k}+1)} \quad ; \mathbf{F}^{(\mathrm{k})} = \begin{bmatrix} \mathbf{F}_0 \\ \vdots \\ \mathbf{F}_{k-1} \\ \mathbf{F}_k \\ \mathbf{F}_{k+1} \\ \vdots \\ \mathbf{F}_n \end{bmatrix}^{(\mathrm{k})}
$$

with the following three band-diagonal Jacobian matrix:

$$\mathbf{J} = \begin{bmatrix} \frac{\partial \mathbf{F}_0}{\partial \mathbf{B}_0} & \frac{\partial \mathbf{F}_0}{\partial \mathbf{B}_1} & 0 & 0 & 0 & 0 & 0 \\ \frac{\partial \mathbf{F}_1}{\partial \mathbf{B}_0} & \ddots & \ddots & 0 & 0 & 0 & 0 \\ 0 & \ddots & \frac{\partial \mathbf{F}_{k-1}}{\partial \mathbf{B}_{k-1}} & \frac{\partial \mathbf{F}_{k-1}}{\partial \mathbf{B}_k} & 0 & 0 & 0 \\ 0 & 0 & \frac{\partial \mathbf{F}_k}{\partial \mathbf{B}_{k-1}} & \frac{\partial \mathbf{F}_k}{\partial \mathbf{B}_k} & \frac{\partial \mathbf{F}_k}{\partial \mathbf{B}_{k+1}} & 0 & 0 \\ 0 & 0 & 0 & \frac{\partial \mathbf{F}_{k+1}}{\partial \mathbf{B}_k} & \frac{\partial \mathbf{F}_{k+1}}{\partial \mathbf{B}_{k+1}} & \ddots & 0 \\ 0 & 0 & 0 & 0 & \ddots & \ddots & \frac{\partial \mathbf{F}_{n-1}}{\partial \mathbf{B}_n} \\ 0 & 0 & 0 & 0 & 0 & \frac{\partial \mathbf{F}_n}{\partial \mathbf{B}_{n-1}} & \frac{\partial \mathbf{F}_n}{\partial \mathbf{B}_n} \end{bmatrix}$$

The following MATLAB™ functions are used to obtain the Jacobian elements:

```
function [dF0,dF01]=calc_dF0(a,b,c,p,q,B0,B1,sigma,dx,h,w)
D    =form_Zm(0,w,h);
dH0 =calc_dH(a,b,c,p,q,B0,h);
dH0 =calc_Fm(dH0,h);
dH1 =calc_dH(a,b,c,p,q,B1,h);
dH1 =calc_Fm(dH1,h);
dF0 = -10*sigma*dx^2*D-24*dH0;
dF01=  -2*sigma*dx^2*D+24*dH1;
```

```
function [dFk_1,dFk,dFk1]=calc_dFk(a,b,c,p,q,Bk_1,Bk,Bk1,sigma,dx,h,w)
D     =form_Zm(0,w,h);
dHk_1=calc_dH(a,b,c,p,q,Bk_1,h);
dHk_1=calc_Fm(dHk_1,h);
dHk   =calc_dH(a,b,c,p,q,Bk,h);
dHk   =calc_Fm(dHk,h);
dHk1  =calc_dH(a,b,c,p,q,Bk1,h);
dHk1  =calc_Fm(dHk1,h);
dFk_1=      -sigma*dx^2*D+12*dHk_1;
dFk   = -10*sigma*dx^2*D-24*dHk;
dFk1  =     -sigma*dx^2*D+12*dHk1;
```

```
function [dFn_1,dFn]=calc_dFn(a,b,c,p,q,Bn_1,Bn,sigma,dx,h,w)
D     =form_Zm(0,w,h);
dHn_1=calc_dH(a,b,c,p,q,Bn_1,h);
dHn_1=calc_Fm(dHn_1,h);
dHn   =calc_dH(a,b,c,p,q,Bn,h);
dHn   =calc_Fm(dHn,h);
dFn_1=-sigma*dx^2*D+4*dHn_1;
dFn   =-sigma*dx^2*D-4*dHn;
```

```
function dH=calc_dH(a,b,c,p,q,B,h)
U =zeros(2*h+1,1); U(h+1)=1;
dH=a*U+b*p*self_conv(B,h,p-1)+c*q*self_conv(B,h,q-1);
```

Example 11-2: Using the sub-lamination arrangement shown in Figure 11.10, a harmonic computer solution has been conducted for one lamination with the following parameters: thickness $2d = 0.33$ mm and conductivity $\sigma = 6 \times 10^6$ S/m. The operating frequency is 60 Hz and the number of harmonics considered in the solution is 50. The MATLAB ™ functions given above were used to carry out the calculations.

Table 11.6 gives the harmonic content of B at the surface and at two discrete points inside the lamination, whereas Table 11.7 gives the harmonic content of H.

The polynomial characteristic chosen for this example is

$$H = 51.8025B + 0.2181B^{17} + 0.1353B^{21} \tag{11.30}$$

A sinusoidal input is selected, $E_4 = -E_{4\max} \sin \omega_0 t$ and $B_{\text{eff}} = 1.65$ T. Also, it should be remarked that $E_{4\max}$ and B_{eff} are related at the surface of the lamination by

$$E_{4\max} = -j\omega_0 d B_{\text{eff}} \tag{11.31}$$

Table 11.6 Harmonics of B in the laminations

h	B_0 [T]	B_2 [T]	B_4 [T]
1	1.6506∠-8.40°	1.6585∠-1.78°	1.6996∠16.24°
3	0.1061∠75.26°	0.0298∠111.38°	0.2227∠-82.80°
5	0.0150∠101.32°	0.0032∠-127.56°	0.0037∠-76.24°
7	0.0291∠-119.24°	0.0120∠-37.29°	0.0476∠119.45°
9	0.0070∠53.67°	0.0043∠117.23°	0.0132∠12.84°
11	0.0118∠47.73°	0.0063∠161.09°	0.0115∠-59.67°
13	0.0052∠127.98°	0.0040∠-47.63°	0.0103∠-128.52°
15	0.0057∠-142.29°	0.0033∠-4.31°	0.0068∠105.37°

Table 11.7 Harmonics of H in the lamination

h	H_0 [kA/m]	H_2 [kA/m]	H_4 [kA/m]
1	2.1673∠-0.42°	2.1689∠-0.08°	2.1711∠0.91°
3	1.7231∠-0.00°	1.7210∠0.01°	1.7184∠0.01°
5	1.1740∠-0.02°	1.1736∠-0.02°	1.1733∠-0.03°
7	0.6559∠0.03°	0.6571∠-0.01°	0.6587∠-0.01°
9	0.2979∠0.03°	0.2981∠0.08°	0.2982∠0.09°
11	0.1083∠-0.18°	0.1076∠0.03°	0.1068∠0.05°
13	0.0308∠-0.09°	0.0306∠-0.57°	0.0305∠-0.61°
15	0.0066∠2.27°	0.0070∠-0.41°	0.0075∠-0.61°

Figure 11.11 shows the penetration delay of $B(t)$ at the zero crossing, which disappears

as soon as saturation levels are reached. Consequently, B peaks in all sub-laminations at the same time. As expected, $H(t)$ shows a high degree of saturation, which remains almost with no change at different depths of the lamination. For completeness, the time variation of E is shown at the surface of the lamination, which is sinusoidal.

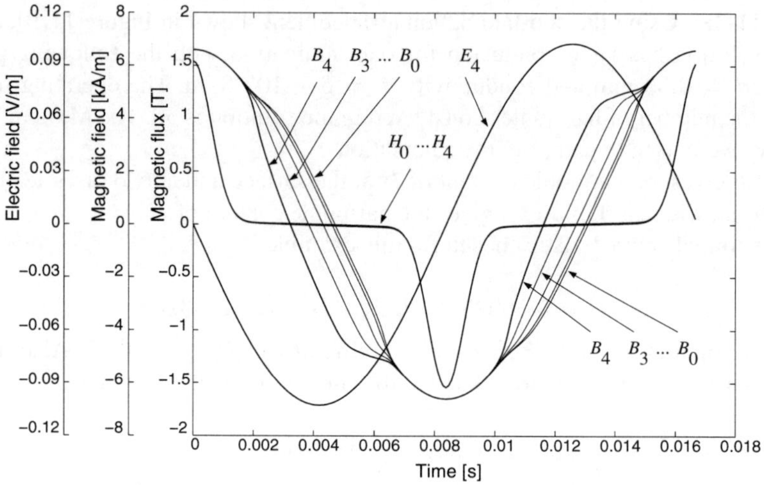

Figure 11.11 Waveforms of E_4, B and H as a function of time

A better appreciation of the electromagnetic phenomenon taking place in the lamination is shown in Figure 11.12, using a three-dimensional representation of B and H as a function of time and depth inside the lamination.

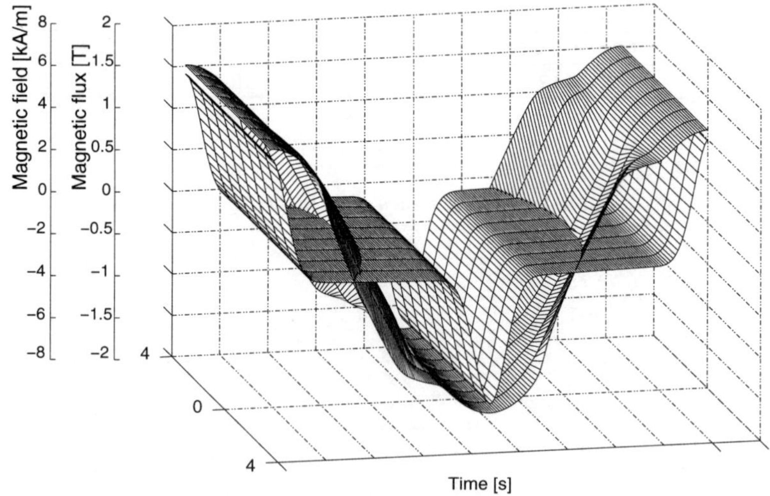

Figure 11.12 Waveforms of B and H

11.5.2 *Modelling of hysteresis*

In equation (11.23) H and B are periodic functions of time and a, b and c are real coefficients. Moreover, the derivative function of H with respect to B is

$$\frac{\partial H}{\partial B} = a + bpB^{p-1} + cqB^{q-1} \tag{11.32}$$

In the harmonic domain, this expression becomes a vector of complex harmonic terms, and so are the terms $bp\mathbf{B}^{p-1}$ and $cq\mathbf{B}^{q-1}$. The DC-like term, a, becomes a vector with only a DC entry or, alternatively, a diagonal matrix with a as the only entry, i.e.

$$\mathbf{a} = \begin{bmatrix} 0 \\ 0 \\ a \\ 0 \\ 0 \end{bmatrix} \Rightarrow \begin{bmatrix} a & & & & \\ & a & & & \\ & & a & & \\ & & & a & \\ & & & & a \end{bmatrix} \tag{11.33}$$

It should be remarked that this is a standard harmonic domain operation, which corresponds to the case when a complex conjugate vector is expressed as a Toeplitz matrix, but this of course is the limiting case.

It has been found that for harmonic domain purposes, a suitable way to model hysteresis is to supplement the fundamental frequency terms, which are real, with an imaginary component [2]. This is done in such a way that the new fundamental frequency terms are complex conjugates, e.g. $a_1 = a + \mathrm{j}a_{\mathrm{hyst}}$ and $a_{-1} = a_1^*$. The new diagonal matrix takes the following form:

$$\begin{bmatrix} a & & & & \\ & a_{-1} & & & \\ & & a & & \\ & & & a_1 & \\ & & & & a \end{bmatrix} \tag{11.34}$$

This is a simple way to represent hysteresis in harmonic domain calculations. Its main effect is to produce the hysteresis loop at flux densities below saturation levels, independently of the magnitude of the base frequency, which is largely the main characteristic of the hysteresis phenomenon. The magnitude and phase angle of a_1 may be estimated by examining the experimental hysteresis curve: a is approximately equal to the slope H/B at the origin and a_{hyst} is the intercept on the H axis of the hysteresis loop at $B_{\max} = 1$. The exact, final values may be obtained by further experimentation.

By way of example, Figure 11.13 shows the hysteresis curves generated with the method just described. The magnetising characteristic in equation (11.30) was used with the following complex term: $a_1 = 51.8025 - \mathrm{j}30$.

11.5.3 *Calculation of iron losses*

A simple method to calculate the iron losses due to eddy currents is based on the observation that the profile of E is largely linear. Using such an assumption, the instantaneous power responsible for causing losses is

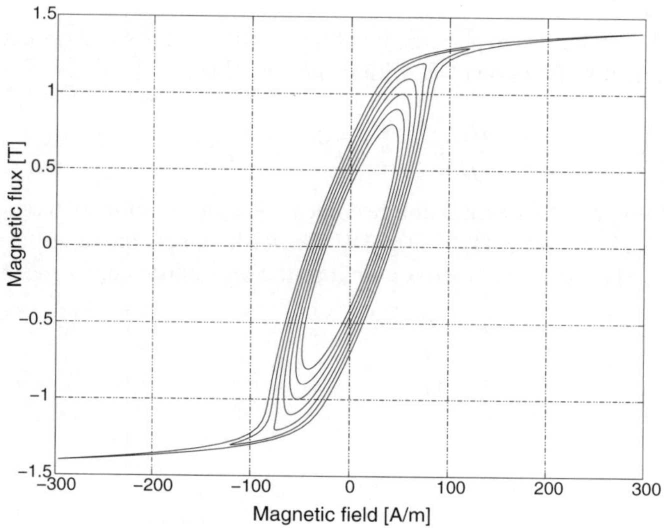

Figure 11.13 Magnetising curves with hysteresis

$$p(t) = 2\int_0^d \sigma E^2(x,t)\mathrm{d}x = \frac{2\sigma d}{3}E_n^2(t) \tag{11.35}$$

The average (active) power for a single lamination of 1 m^2 surface area is

$$P = \frac{2\sigma d}{3}E_{n,\mathrm{rms}}^2 \tag{11.36}$$

where $E_{n,\mathrm{rms}}$ is the true RMS value of E at the surface of the lamination.

11.5.4 External system

Either a Norton or a Thévenin equivalent circuit may be used to represent the external circuit in Figure 11.14 in the harmonic domain. For instance, if the latter representation is used we have

$$\mathbf{V} = \mathbf{V}_\mathrm{T} - \mathbf{Z}_\mathrm{T}\mathbf{I} \tag{11.37}$$

where the interfacing variables between the lamination and the external circuit are

$$\mathbf{V} = 2\mathbf{E}_n wNn \tag{11.38}$$

$$\mathbf{I} = \mathbf{H}_n h_l/n \tag{11.39}$$

where:

n is the number of turns,
N is the number of laminations,
w is the width of laminations,

h_l is the height of the winding,

$\mathbf{E}_n$ and $\mathbf{H}_n$ are the electric and magnetic surface fields.

Figure 11.14 shows the circuit representation of the lamination and the external electric circuit, represented by a Thévenin equivalent.

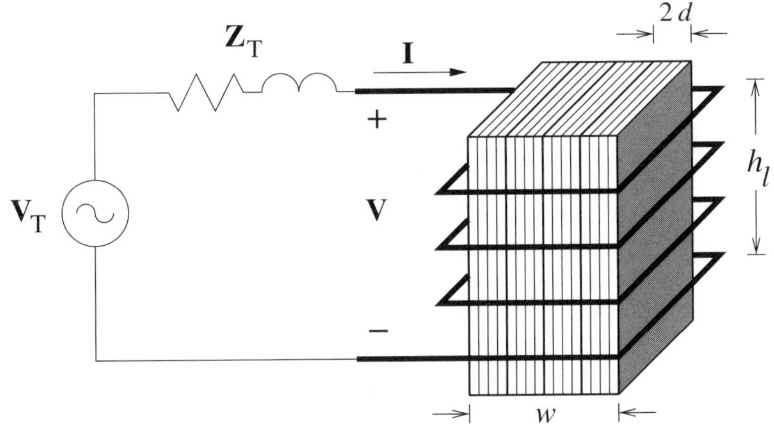

Figure 11.14 Circuit representation

11.6 Summary

This chapter has presented comprehensive harmonic domain information on many of the key issues that directly relate to magnetising harmonics, namely single-phase transformers, three-phase banks of transformers, inrush currents, laminated iron cores and core losses.

A single-phase transformer model that includes tapping positions on both the primary and the secondary windings was derived from first principles. The basic model was then extended to incorporate saturating characteristics in single-phase transformers and three-phase banks of transformers, using harmonic domain techniques. Several three-phase transformer connections were considered. The periodic, steady-state solution of laminated iron cores was studied by using a discrete representation of Maxwell's equations, which are amenable to harmonic domain representations. The method was used very effectively to gain an insight of the waveforms of B and H inside the lamination, which may exhibit significant distortion. A simple and yet effective way to model hysteresis in the harmonic domain for more accurate calculations was discussed. The inrush current phenomenon arising during transformer energisation, and containing a slowly decaying transient response, was also discussed and a connection with harmonic domain techniques was made.

11.7 Bibliography

1. R. Yacamini, A. Abu-Nasser, "Numerical Calculation of Inrush Current in Single-Phase Transformers", *Proceedings of the IEE*, Part B, Vol. 128, No. 6, November 1981, pp. 327–334.
2. N. Rajakovic, A. Semlyen, "Harmonic Domain Analysis of Field Variables Related to Eddy

Current and Hysteresis Losses in Saturated Laminations", *IEEE Transactions on Power Delivery*, Vol. 4, No. 2, April 1989, pp. 1111–1116.

3. N. Rajakovic, A. Semlyen, "Harmonic Domain Modelling of Laminated Iron Core", *IEEE Transactions on Power Delivery*, Vol. 4, No. 1, January 1989, pp. 382–390.
4. J.J. Rico, E. Acha, M. Madrigal, "The Study of Inrush Current Phenomenon Using Operational Matrices", *IEEE Transactions on Power Delivery*, paper PE-004PRD(09-2000).
5. N. Rajakovic, A. Semlyen, "Investigation of the Inrush Phenomenon: A Quasi-Stationary Approach in the Harmonic Domain", *IEEE Transactions on Power Delivery*, Vol. 4, No. 4, October 1989, pp. 2114–2120.
6. H.W. Dommel, A. Yan, S. Wei, "Harmonics from Transformer Saturation", *IEEE Transactions on Power Systems*, Vol. PWRD-1, No. 2, April 1986, pp. 209–215.
7. A. Semlyen, E. Acha, J. Arrillaga, "Newton-Type Algorithms for the Harmonic Analysis of Nonlinear Power Circuits in Periodical Steady State with Special Reference to Magnetic Nonlinearities", *IEEE Transactions on Power Delivery*, Vol. 3, No. 3, July 1988, pp. 1090–1098.
8. H.W. Dommel, "Transformer Models in Simulation of Electromagnetic Transients", *5th Power Systems Computation Conference*, Paper No. 3.1/4, Cambridge, England, 1–5 September 1975.
9. CIGRE Working Group, 36-05, "Harmonics, Characteristic Parameters, Methods of Study, Estimates of Existing Values in the Networks", *Electra*, No. 77, July 1981, pp. 35–54.
10. M.S. Chen, W.E. Dillo, "Power System Modelling", *Proceeding of the IEEE*, Vol. 62, 1974, p. 901.
11. L.F. Blue, G. Camilli, S.B. Farnham, H.A. Peterson, "Transformer Magnetizing Inrush Currents and Influence on System Operation", *AIEE Transactions*, Vol. 63, 1944, pp. 366–375.
12. D. Povh, W. Schultz, "Analysis of Overvoltages Caused by Transformer Magnetizing Inrush Current", *IEEE Transactions on Power Apparatus and Systems*, Vol. PAS-97, No. 4, July/August 1978, pp. 1355–1365.
13. J.F. Witte, F.P. DeCesaro, S.R. Mendis, "Damaging Long-Term Overvoltages on Industrial Capacitor Banks due to Transformer Energization Inrush Currents", *IEEE Transactions on Industry Applications*, Vol. 30, No. 4, July/August 1994, pp. 1107–1115.
14. J. Arrillaga, N.R. Watson, S. Chen, *Power System Quality Assessment*, John Wiley & Sons, Chichester, 2000.

12

Electric Arcs

12.1 Introduction

Careful examination of the electric arc that occurs between two identical electrodes leads to the development of qualitative and quantitative tools with which to study the behaviour of electric arcs in both electric arc furnaces and arc discharge lamps. The arc is a thermoelectric phenomenon which is a function of the medium in which the arc discharge takes place and the voltage level applied across the electrodes. At the point of voltage inception, the plasma may contain insufficient ions to prevent the arc from extinguishing and it becomes necessary to increase the voltage applied across the electrodes, up to a minimum level, to generate a strong enough ionisation to maintain a stable arc.

The electric arc established between the anode and the cathode may be assumed to take place along three different zones, with different voltage distributions, as defined below and shown in Figure 12.1:

1. The arc column. This takes the form of a tube and exhibits a progressive voltage drop, which is a function of the current intensity through the arc and the arc length.
2. The cathode zone. This has a small voltage drop, which is a function of plasma and electrode characteristics. It is largely independent of the current intensity through the arc.
3. The anode zone. The voltage drop in this zone is two to three times larger than in the cathode, and is a function of the current intensity through the arc.

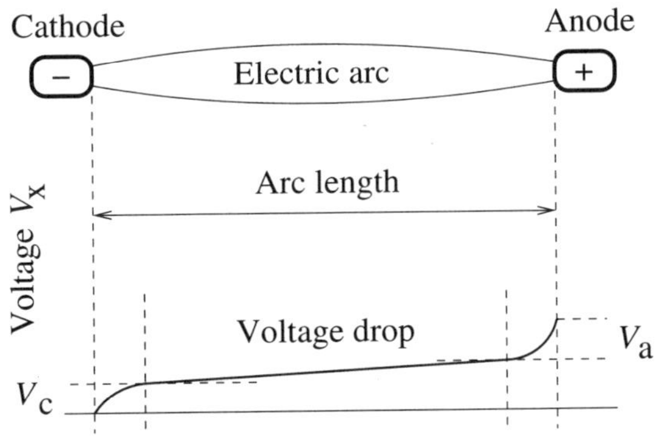

Figure 12.1 Voltage drops across the arc

For a given arc length, it is necessary to have a minimum voltage applied to maintain a

stable arc. This is given by the sum of the voltage drops across the three zones mentioned above.

An important characteristic exhibited by AC electric arcs is the extinction of the arc at each zero crossing of the current. Once the arc vanishes, it is necessary to increase the voltage applied across the electrodes until it reaches the minimum value in order to re-establish the arc. This intrinsic behaviour of the arc may lead to discontinuous arc operation between the instant of current extinction and the point at which sufficient ions are generated to re-establish the arc.

The electric arc model presented in this chapter, and used to study the most relevant electrical parameters, comes in the form of a non-linear differential equation, which appeals to the fundamental concept of power balance. The differential equation is found to represent very well the periodic behaviour of arcs found in electric arc furnaces and electric discharge lamps. As expected, owing to the very different media in which the two plasmas occur, their respective characteristics are quite different and each application is presented separately. The arc in the electric furnace contains large, random perturbations for long periods of the melting cycles whereas the arc in electric discharge lamps is a very stable mechanism, following an initial, relatively large transient.

12.2 Electric Arc Furnaces: Overview

The electric arc furnace is a time-varying, non-linear load that has the ability to generate large, random perturbations at the point of common coupling with the network of the electricity supply company. This is particularly the case for large-size units, some of which are rated at 100 MW and above. Among the most common adverse power quality effects introduced by electric arc furnaces are: voltage and current harmonics, voltage and current imbalances, low power factors, and voltage "flicker".

It has been reported [1] that voltage stabilisation significantly improves the operation of electric furnaces in two ways: (1) by increasing the maximum power and hence the rate of steel production, and (2) by enabling the furnace to work at the same maximum power but with a shorter arc, which reduces refractory wear and helps to reduce voltage flicker. Voltage stabilisation is undoubtedly of great direct benefit to the steel-maker but to the utility company the benefit may be in the form of reduced voltage disturbances, i.e. flicker, caused by the rapid, erratic variations in furnace currents. Similarly to harmonic voltages and currents, flicker is also known to propagate upstream in the utility company network, adversely affecting its ability to supply high-quality electricity to its customers. This would be particularly the case in the neighbourhood of a steel plant.

Over the years, several methods have been used for voltage stabilisation, such as connection to higher network voltage and synchronous condensers with buffer reactors, but a more up-to-date alternative relies on the use of thyristor-controlled compensators. More advanced VAR compensators based on the use of GTO and IGBT technology may soon become economic enough to be used in this very demanding application, enabling better voltage stabilisation and flicker correction than is possible today with thyristor-based technology.

Electric furnace global capacity has continued to increase unabated since the 1950s. The so-called ultra-high-power furnaces, with capacities of 50–150 tonnes, first appeared in the mid 1960s, enabling up to 12 melting cycles per day as opposed to the 4 or 5 achieved with the previous generation of furnaces. Such large gains in productivity, coupled with higher efficiencies in the melting process itself, saw the electric furnace become competitive with the classic steel-making processes. Some of the largest furnaces are more than 30 ft (9 m) in diameter and require up to 165 MVA of electric power, with capacities of around 400 tonnes. The highly efficient steel-making processes achieved with electric furnaces led to their widespread deployment, and, together with the large increases in their individual ratings, enabled the electric furnace to become one of the most important non-linear loads in the electrical power network.

The melting process may be analysed purely from a metallurgical viewpoint or from an electrical one. We shall follow the latter approach, to present a harmonic model for the electric furnace and study its associated parameters such as voltage, current, arc radius, and active and reactive powers. Naturally, the electric arc model should follow the metallurgical process very closely for the model to have practical meaning.

From the electrical viewpoint, it is important to look closely at the power consumed by the furnace, and the arc prevailing conditions as the melting process progresses. It is in this respect that the energy consumed in each basket of the melting cycle may be divided into three main steps:

1. Drilling period. This period is characterised by the use of reduced voltages and powers. It is of short duration, lasting between 3 and 4 minutes. The material to be reduced, i.e. scrap, is highly heterogeneous, causing very irregular variations in the electric arc.
2. Melting period. In this period, the electric arc is surrounded by melting scrap, which at this stage forms a more homogeneous material than in the drilling period. Full voltage and power is used in this process, lasting between 21 and 23 minutes. Typical values for medium-power electric furnaces are: 610 V, 42 kA, 37 MW and PF=0.83.
3. End of melting period and reheating. This period uses lower voltage and higher current than in the melting period. It involves short arcs and lasts for about 5 minutes. Typical values in this process are: 560 V, 45 kA, 34 MW and PF=0.78.

Figure 12.2 shows the calculated active power consumed by an electric furnace as a function of time. In this case, the melting cycle involves three baskets and a final refining period [2]. The values shown correspond to an actual furnace where the voltages and currents were measured on the secondary side of the transformer. Measurements were taken every 10 seconds. These results reflect the highly stochastic nature of arc voltage over long periods.

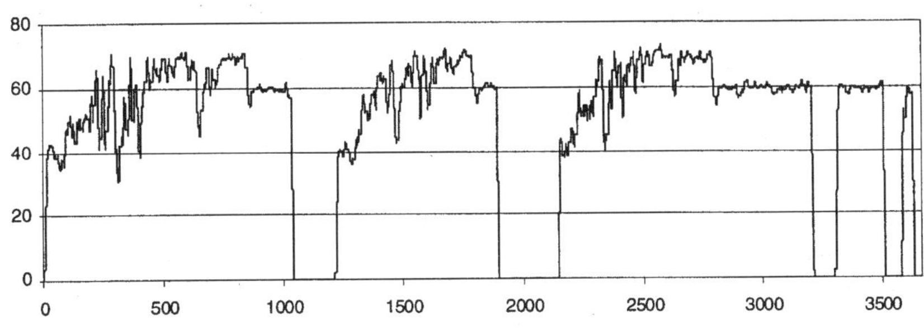

Figure 12.2 Active power, in MW, as a function of time, consumed in a melting cycle, with three baskets and one refining period

12.3 Electric Furnace Installations

Owing to the very large power drawn by AC electric furnaces from the utility network, they are invariably connected to network points that have high short-circuit levels. Practice will vary from country to country but suitable connection points are normally found in substations operating in the range of 10–40 kV. For instance, in Spain [2] it is quite common to consider 30 kV as the preferred voltage level, but some of the larger installations are directly connected at 130 and 220 kV points using step-down transformers of ratios 130/30 kV and 220/30 kV, respectively.

Figure 12.3 gives the one-line equivalent diagram of an electric furnace installation showing the main components and typical voltage levels associated with these kinds of installations. From the figure, it can be appreciated that the electric arc is the main non-linearity in the system, but not the only one. The magnetic cores of the two transformers may become important non-linearities, particularly if pushed into asymmetric over-excitation by the presence of direct currents generated by the electric arcs. Similarly, the reactive power compensator, such as an SVC, is likely to generate harmonic current distortion, which in most cases will be removed by filtering equipment.

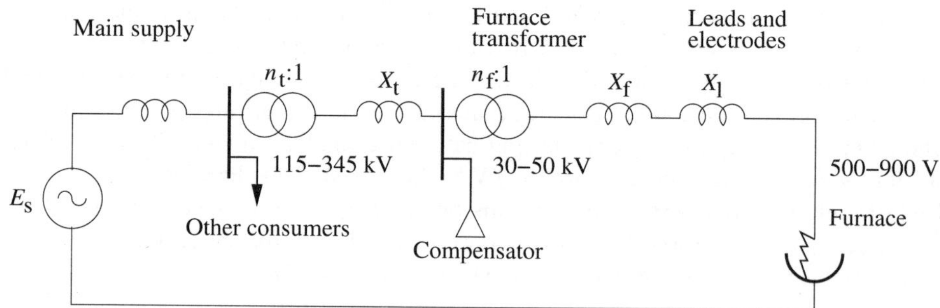

Figure 12.3 One-line equivalent diagram of an electric furnace installation

The electric furnace transformer deserves special attention. These are transformers of very sturdy construction as they have to withstand large over-voltages and over-currents for long periods of time, and with rapidly fluctuating operating requirements. The electrodynamic forces the transformer windings are subjected to are formidable. Short circuits caused by the electrodes touching each other during furnace operation are frequent. The primary winding may be connected in either delta or star configuration, whereas the secondary winding is always delta connected. Voltage regulation on the secondary side of the transformer is achieved by using the tapping mechanism found on the primary winding. The currents are smaller on the primary winding than on the secondary winding and this reduces the possibility of arcing in the winding during on-load tapping operation. Sufficient tap steps are provided to allow for 12–16 voltage points on the secondary side of the transformer. The tapping mechanism is normally designed to achieve maximum power using the top four steps, leaving the remaining lower steps to work with reduced powers.

The transformer reactance alone, which is of the order of 4–6%, is sufficient to stabilise the arc when the operation involves short- and medium-length arcs. These are achieved at voltage levels of 400–800 V. However, when long arcs are involved, achieved at voltage levels of around 1000 V, additional reactances must be connected in series to enable voltage stabilisation and to prevent the electrodes from breaking.

From the delta-connected secondary winding, three flexible cables are derived which connect to the electrode arm busbar, i.e. rounded conductors supported by the electrode arm support structure, and from there to the graphite electrodes. On the primary side of the electric furnace transformer of large installations, it is common practice to provide fast reactive power control to aid voltage stabilisation and flicker reduction.

The main system asymmetries appear in the impedances of the flexible cables and in the electrode arm busbar but they are normally quite small compared to the asymmetries introduced by uneven electric arcs in the three phases. This has motivated most researchers and practising engineers working with electric furnace problems to assume that the electric furnace installation is perfectly balanced with the only asymmetry being in the electric arcs. This may have been a reasonable assumption to make when computing resources were limited but imposing such constraints in the solution algorithm may no longer be necessary with present-day hardware and software capabilities.

Comprehensive power quality studies of electric furnace installations would involve the interaction of the various non-linearities in the neighbourhood of the electric arcs, which by necessity would be represented in the phase domain since harmonic and inter-harmonic terms are highly influenced by network and operation imbalances. It is in this respect that it would be more appropriate to look at the frame of reference of the phases as the natural coordinate system in which to conduct such studies, with restrictions being imposed only by the available data.

12.4 Electric Arc Furnace Modelling

In this section, a comprehensive arc model is developed for harmonic studies of electric furnaces [3]. It should be mentioned that the model of the complete furnace is not addressed here but rather the emphasis is placed on the arc. The other non-linear components in the furnace installation, namely the furnace transformer and the VAR compensator, are addressed separately in other chapters of the book.

We start by developing a general dynamic arc model in the form of a differential equation, based on the fundamental principle of conservation of energy. The approach is different from those methods where the electric arc is represented by some empirical relation. In the dynamic model such relations are implicit for steady-state conditions but are not pre-defined and will result in different conditions, depending on both frequency and current magnitude. The power balance equation for the arc is

$$p_1 + p_2 = p_3 \tag{12.1}$$

where:

p_1 represents the power transmitted in the form of heat to the external environment,

p_2 represents the power which increases the internal energy in the arc and which therefore affects its radius,

p_3 represents the total power developed in the arc and converted into heat.

In equation (12.1) it is assumed that the cooling effect is a function of the arc radius r only. Thus

$$p_1 = k_1 r^n \tag{12.2}$$

While, in fact, it is also a function of the arc temperature, this dependence is assumed to be less significant and is therefore ignored, in order to keep the model simple. Thus, only the arc radius r appears as a state variable. If the environment around the arc is hot the cooling of the arc may not depend on its radius at all, so that in this case $n = 0$. If this is not the case and the arc is long, then the cooling area is mainly its lateral surface, so that $n = 1$. If the arc is short, then the cooling is proportional to its cross-section at the electrodes, so that $n = 2$.

The term p_2 is proportional to the derivative of the energy inside the arc which is proportional to r^2,

$$p_2 = k_2 r \frac{\mathrm{d}r}{\mathrm{d}t} \tag{12.3}$$

Finally,

$$p_3 = vi = \frac{k_3/r^m}{r^2} i^2 \tag{12.4}$$

where the arc voltage is given by

$$v = \frac{k_3}{r^{m+2}} i \tag{12.5}$$

In this expression, the resistivity of the arc column is assumed to be inversely proportional to r^m , where $m = 0 \ldots 2$, to reflect the fact that the arc may be hotter in the interior if it has a larger radius.

Substitution of equations (12.2)–(12.4) into (12.1) gives the differential equation of the arc:

$$k_1 r^n + k_2 r \frac{\mathrm{d}r}{\mathrm{d}t} = \frac{k_3}{r^{m+2}} i^2 \tag{12.6}$$

12.4.1 Static electric arc model

For steady-state purposes, equation (12.6) may be solved for r,

$$r = \frac{k_3^{1/(m+n+2)}}{k_1^{1/(m+n+2)}} i^{2/(m+n+2)} \tag{12.7}$$

and equation (12.5) may be expressed by

$$v = \frac{k}{i^q} \tag{12.8}$$

where

$$q = \frac{m+2-n}{m+2+n} \tag{12.9}$$

and

$$k = k_3^{n/(m+n+2)} k_1^{(m+2)/(m+n+2)} \tag{12.10}$$

For $m = 0,\ 1,\ 2$, the values of q are given below:

$n\backslash m$	0	1	2
0	1	1	1
1	1/3	1/2	3/5
2	0	1/5	1/3

The values of q do not vary significantly with m. Their variation as a function of n indicates the following:

If $n = 0,\ q = 1$, then $v = k/i$ which corresponds to a hyperbolic v-i characteristic.

If $n = 1,\ q \approx 1/2$, then $v = k/\sqrt{i}$ which corresponds to a square-root–hyperbolic characteristic.

If $n = 2,\ q = 0$ then $v \approx$ constant.

This analysis provides enough evidence to suggest that the static arc characteristic depends mainly on the conditions of cooling and only a little on the variation of the arc resistivity with temperature.

12.5 Harmonic Domain Equation of the Electric Arc

The differential equation (12.6) may be rewritten as

$$f(r) = k_1 r^{m+n+2} + k_2 r^{m+3}\frac{\mathrm{d}r}{\mathrm{d}t} - k_3 i^2 = 0 \tag{12.11}$$

Equation (12.11) in the harmonic domain is given by

$$\mathbf{F} = k_1\mathbf{R}^{m+n+2} + k_2\mathbf{R}^{m+3}\mathbf{D}(\mathrm{j}h\omega_0)\mathbf{R} - k_3\mathbf{I}^2 = 0 \tag{12.12}$$

where the exponents mean convolutions. Also, using (12.5) the voltage is given by

$$\mathbf{V} = k_3\left(\mathbf{R}^{m+2}\right)^{-1}\mathbf{I} \tag{12.13}$$

Equation (12.12) is a non-linear equation, and Newton's method may be used very effectively to solve it,

$$\begin{aligned} \Delta\mathbf{R}^{(\mathrm{k}+1)} &= -\left[\frac{\partial\mathbf{F}}{\partial\mathbf{R}}\right]^{-1}\mathbf{F}^{(\mathrm{k})} && (12.14)\\ \mathbf{R}^{(\mathrm{k}+1)} &= \mathbf{R}^{(\mathrm{k})} + \Delta\mathbf{R}^{(\mathrm{k}+1)} && (12.15) \end{aligned}$$

where k is the iteration counter, and

$$\frac{\partial\mathbf{F}}{\partial\mathbf{R}} = (m+n+2)k_1\mathbf{R}^{m+n+1} + (m+3)k_2\mathbf{R}^{m+2}\mathbf{D}(\mathrm{j}h\omega_0)\mathbf{R} + k_2\mathbf{R}^{m+3}\mathbf{D}(\mathrm{j}h\omega_0)\frac{\mathrm{d}\mathbf{R}}{\mathrm{d}\mathbf{R}} \tag{12.16}$$

The Jacobian in (12.14) is expressed in matrix form. It must be noted that equation (12.16) involves many self- and mutual convolutions. In this equation $\mathbf{R}^{m+n+1}$, $\mathbf{R}^{m+2}$ and $\mathbf{R}^{m+3}$ are matrices, and only $\mathbf{R}$ is a vector. Also, the product $\mathbf{R}^{m+2}\mathbf{D}(\mathrm{j}h\omega_0)\mathbf{R}$ should be represented in matrix form.

12.5.1 Dynamic characteristic

To illustrate how the v-i characteristic of the electric arc varies with time, Figure 12.4 shows measured dynamic v-i characteristics in a 250 V, 70 kA AC arc [4]. Figure 12.4(a) corresponds to the drilling period, (b) to the melting period and (c) to the end of the melting period and reheating.

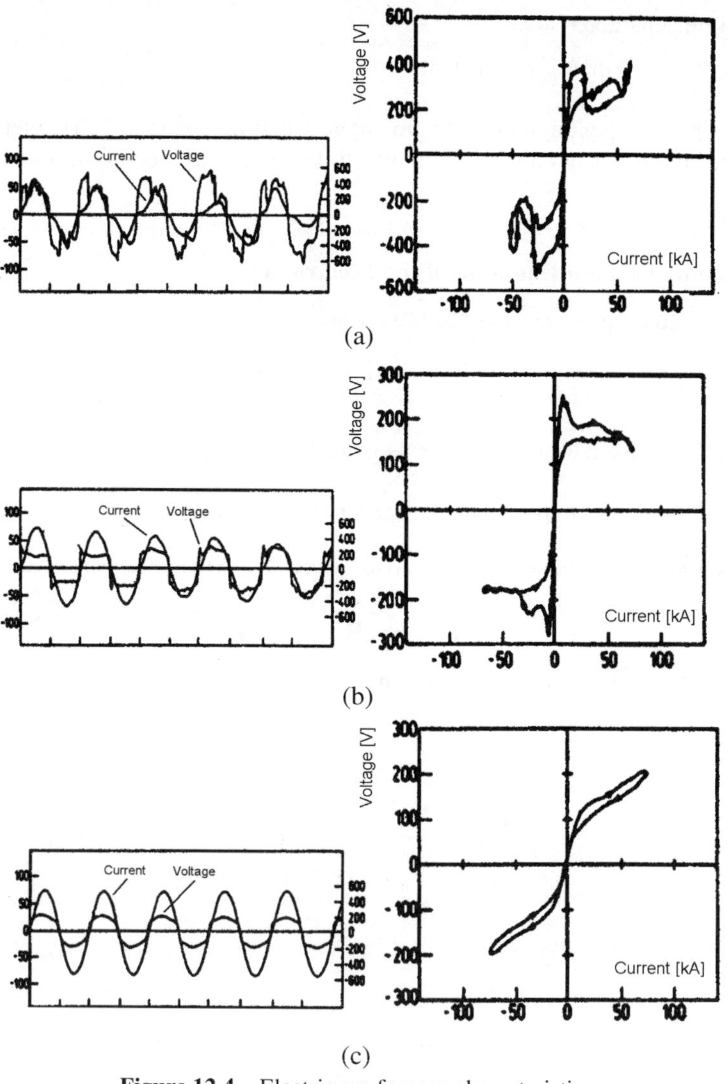

Figure 12.4 Electric arc furnace characteristics

Large variations are observed in the voltage and current measured during the drilling period, giving rise to a badly distorted v-i characteristic, which resembles more a transient operation than the steady-state operation. During the melting-down process the voltage and current waveforms are more regular but still the voltage is highly non-sinusoidal. At the end of the melting-down period, the voltage waveform becomes more sinusoidal and the v-i characteristic exhibits a high degree of linearity.

Figure 12.5 shows a v-i characteristic corresponding to the melting period obtained by harmonic domain techniques, by solving equation (12.12) using Newton's method. The following parameters were used: $k_1 = 3000$, $k_2 = 1$, $k_3 = 12.5$, $m = 0$, $n = 2$ and a sinusoidal current excitation at 60 Hz. These values were used to reproduce the characteristic in Figure 12.4(b).

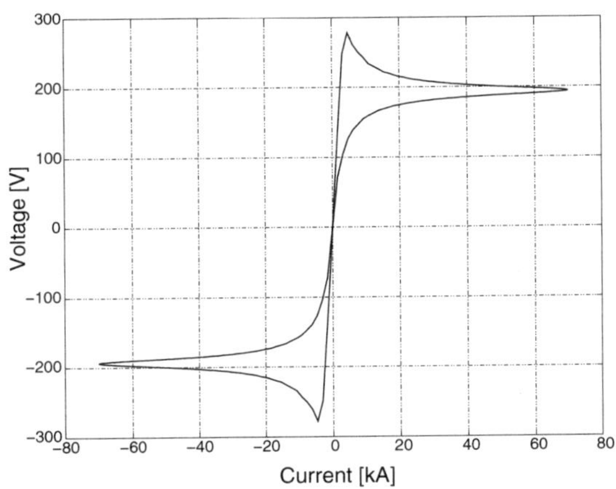

Figure 12.5 Simulated v-i characteristic of a 250 V, 70 kA AC arc

The two characteristics, measured and calculated, compare well with each other: they follow the same basic pattern. The measured arc characteristic has a deterministic and a stochastic part, with the latter becoming less and less significant as the process of meltdown progresses.

The simulated arc voltage, expressed in the time domain, is shown in Figure 12.6. The square form of the arc voltage is typical of the waveshapes associated with electric arcs. The variation of the arc radius for a full cycle is shown in Figure 12.7.

The results of the simulation must satisfy the power balance equation (12.1). A full period of the variation of p_1, p_2 and p_3 is shown in Figure 12.8.

The next MATLAB™ program was used to solve equation (12.12) by Newton's method:

```
% example Electric Arc Furnace
clear all
h = 100;
k1= 3000;
k2= 1;
k3= 12.5;
m = 0;
n = 2;
f = 60;
```

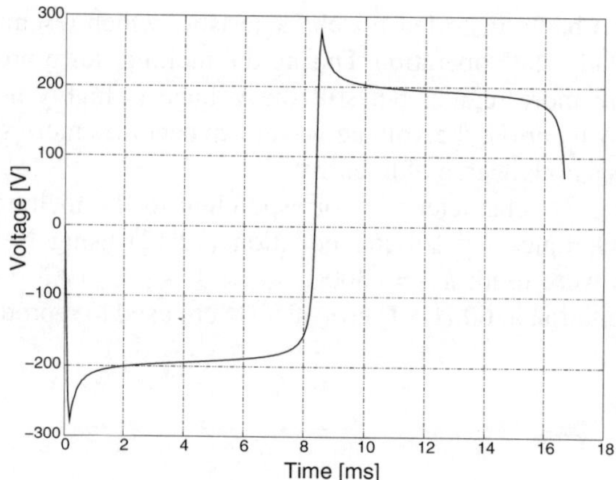

Figure 12.6 One period of arc voltage

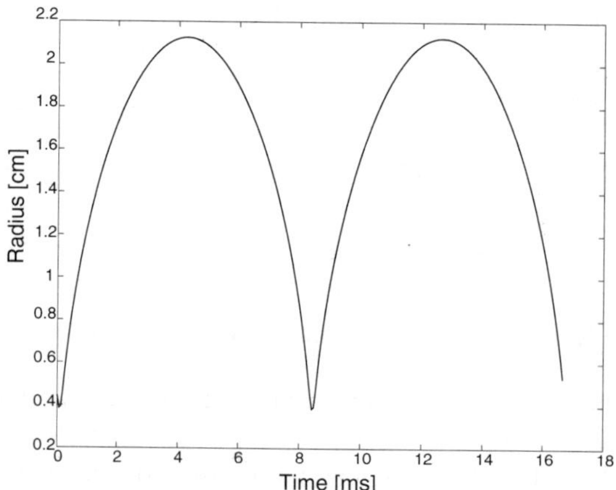

Figure 12.7 Variation of the arc radius with time for one period

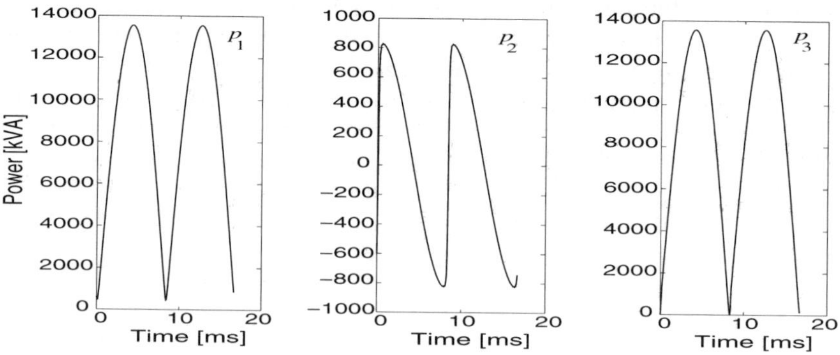

Figure 12.8 One period of the variation of p_1, p_2, and p_3

```
w = 2*pi*f;
Imax=i*70;    % kA
r0  =1.6;     % cm
              % voltage in V
D=form_Zm(0,w,h);
R=zeros(2*h+1,1); R(h+1)=r0;
I=zeros(2*h+1,1); I(h)=Imax'/2;   I(h+2)=Imax/2;
error=1;
iter=0;
while error>1e-6
 Rmn2  = self_conv(R,h,m+n+2);
 Rm3   = self_conv(R,h,m+3);
 Rm3   = calc_Fm(Rm3,h);
 F     = k1*Rmn2+k2*Rm3*D*R-k3*self_conv(I,h,2);
 Rmn1  = self_conv(R,h,m+n+1);
 Rmn1  = calc_Fm(Rmn1,h);
 Rm2   = self_conv(R,h,m+2);
 Rm2   = calc_Fm(Rm2,h);
 Rm2DR= Rm2*D*R;
 Rm2DR= calc_Fm(Rm2DR,h);
 Rm3D  = Rm3*D;
 dF_R  =  (m+n+2)*k1*Rmn1+(m+3)*k2*Rm2DR+k2*Rm3D;
 dR    = -inv(dF_R)*F;
 R     = R+dR;
 error= norm(dR)
 iter  = iter+1
end
Rm2  = self_conv(R,h,m+2);
Rm2  = calc_Fm(Rm2,h);
V    = k3*inv(Rm2)*I;
```

12.5.2 Static characteristic

Conditions of cooling (parameter n) impose bigger changes on the shape of static arc characteristics than the variation of the arc resistivity with temperature (parameter m). Figure 12.9 shows static characteristics corresponding to different values of n (0,1,2) with m being kept equal to 1, $k_1 = 3000$ and $k_3 = 12.5$. The voltages are obtained from equation (12.8) for different current values, i.e. 0–100 kA. The simple calculation of (12.8) is far more efficient than frequency domain and time domain computations but it is restricted to the case of static characteristics. The constants associated with the non-linear element are determined by experimentation.

The following MATLAB™ steps were used to obtain the results in Figure 12.9:

```
>> m=1;
>> n=2;
>> k1=3000;
>> k3=12.5;
>> Imax=100;
>> I=1:Imax;
>> p1=n/(m+n+2);
```

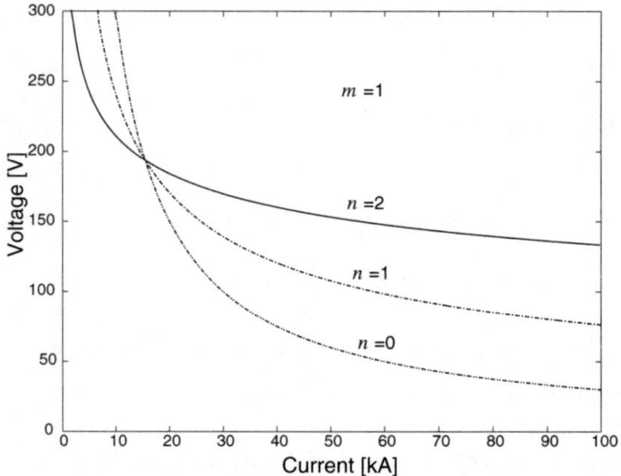

Figure 12.9 Static characteristic of an electric arc

```
>> p2=(m+2)/(m+n+2);
>> q=(m-n+2)/(m+n+2);
>> k=k3^p1*k1^p2;
>> V=k./(I.^q);
>> plot(I,V);
>> axis([0 100 0 300]);
```

12.6 Electric Arc Lamps: Overview

Similar to the arc in electric furnaces, the arc in discharge lamps is a time-varying, non-linear load but operating at a much lower current and power rating. Also, it takes place in a benign environment, i.e. a sealed, gas-filled tube, and it is not subject to extreme variations in operating conditions experienced by the arc in an electric arc furnace. Nevertheless, owing to its non-linear nature, an arc discharge lamp also has the ability to generate voltage and current waveform distortion but because of its low rating, it is highly unlikely that a single lamp will be able to have much impact in a building installation, let alone in a distribution company network. The issue may take on an entirely different dimension if the requirement is to look at the aggregated contribution of thousands of arc discharge lamps and their impact on a building's transformer. By the same token, the adverse effects contributed by the hundreds of thousands of arc discharge lamps that are likely to be in operation at any one time in a busy city may be compelling enough to attract the interest of distribution systems engineers to look closely at the characteristics of such devices [12]. This is all the more relevant and timely because of the widespread adoption of a new generation of arc discharge lamps, termed energy saving lamps, which use solid-state ballasts as opposed to reactor-type ballasts. These operate at very low power factors and have a remarkably rich harmonic current spectrum [11].

It will come as no surprise that over the past 10 years or so, power systems researchers and practising distribution engineers have turned their attention to the study of arc discharge lamp as a potential source of waveform distortion, and more recently to assess the impact of large concentrations of arc discharge lamps on electricity distribution networks.

Arc discharge lamps have been in use for more than 60 years. They have been well studied by illuminating engineers and physicists, who have introduced continuous improvements on both circuit and ballast design, leading to higher efficiencies and significant reductions in both size and cost. A ballast, in the form either of a reactor or an electronic circuit, must be present in discharge lamps to prevent the lamps from being damaged by excessive current due to their negative resistance.

Over the years, theoretical and practical approaches have been taken to study the arc in discharge lamps, with some of the early work relying on the use of measured, instantaneous, voltage–current characteristics to gain an understanding of arc behaviour. For instance, in the discussion of an early paper on the subject [6], an analogy is drawn between incandescent and fluorescent lamps to explain the intrinsic behaviour of the electric arc at power and higher frequencies.

It is pointed out there that the dynamic volt–ampere characteristic of an incandescent lamp depends on the instantaneous temperature of the filament and on its thermal history. At low frequencies, say 25 Hz, the filament cools off during the cycle and the resistance of the lamp is not constant. As a result, the volt–ampere characteristic has a looped shape, i.e. it exhibits a hysteretic behaviour. At higher frequencies, the current changes so quickly that there is no time for the lamp filament to cool appreciably during each cycle. The resistance is effectively constant and the volt-ampere characteristic becomes a straight line, i.e. it contains no hysteresis. It is argued that thermal equilibrium is attained at power frequencies, e.g. 60 Hz, and that this is an entirely satisfactory frequency for the operation of the incandescent lamp. This analysis is used to advance an explanation of the behaviour of the electric arc. It is mentioned that the volt–ampere characteristic of a fluorescent lamp depends primarily on the state of ionisation in the electric discharge. At power frequencies there is time for the column of gas to de-ionise. As the frequency is raised, the discharge goes into a state of dynamic equilibrium in which the state of ionisation is essentially constant. The difference is that with the fluorescent lamp equilibrium of the discharge column is not obtained at 60 but at 3000 Hz. Thus the frequencies that were entirely suitable for incandescent lamps are not the most suitable for fluorescent lighting. The difference occurs merely because the time constant for de-ionisation of the fluorescent lamp discharge is much less than the time constant for thermal cooling of the incandescent lamp filament.

Qualitative and experimental work on arc discharge lamps has gone hand in hand with the development of differential equations for the analysis and design of operating circuits. Francis [7] and Peek [8] use differential equations that possess the electron density as a function of time. These equations have been criticised for being approximations that involve two undetermined constants that must be extracted from the measured current and voltage characteristics of each lamp analysed [9], and also because the lamp current cannot be easily calculated unless the instantaneous lamp voltage is known [10]. This provided beforehand the motivation for Lowke and Zollweg [9] to develop an alternative differential equation that appeals to the fundamental concept of energy balance. This equation has no undetermined constants but several parameters of the plasma must be known as a function of temperature, e.g. density, specific heat, thermal conductivity, electrical conductivity and the radiation emission coefficient. This fact renders the differential equation too cumbersome for most applications but perhaps more importantly, the authors show in their work that the two undetermined constants in the Francis equation [9] may in fact be calculated by integrating terms that are functions of the plasma parameters, as in their equation. The new coefficients in the differential equation are not constant with time as in the Francis equation, enabling more accurate electric arc analysis. It is argued that Francis equations are approximations because they involve integrations of the energy balance equation over the whole temperature profile.

An alternative differential equation was put forward by Bo and Masumi [10], where, from the outset, the arc conductance of the discharge lamp as opposed to the electron density was used as the state variable. There are clear indications that this is the first model actually intended for solution on a digital computer as opposed to solution by hand calculation or using an analogue computer. Also, this mathematical model was clearly developed to facilitate interfacing with the external electric circuit.

To derive the mathematical model of the discharge lamp, the following postulates were used:

1. The time-rate of increase of the electron density is proportional to the square of the lamp current.
2. The time-rate of decrease of the electron density is proportional to the square of the electron density.
3. The equivalent instantaneous conductance of the lamp is proportional to the electron density.

From the above postulates, the following equations are assembled:

$$\frac{\mathrm{d}g}{\mathrm{d}t} = Ai^2 - Bg^2 \tag{12.17}$$

and

$$g = \frac{i}{v} \tag{12.18}$$

where v is the voltage across the lamp and i is the lamp current. A and B are constant coefficients derived by suitable substitution of equation (12.18) into (12.17), and measurements of the lamp voltage and current at a time when $\mathrm{d}v/\mathrm{d}t = 0$ and $\mathrm{d}i/\mathrm{d}t = 0$ on the respective waveforms.

These constants are related to the RMS value of the voltage across the lamp, as indicated by the following equation:

$$V_{\mathrm{rms}} = \sqrt{\frac{B}{A}} \tag{12.19}$$

Similar to the coefficients found in the Francis equation, coefficients A and B are time-independent parameters and they would therefore be approximations. Such approximations would be fully equivalent to carrying out integrations of the energy balance equation over the whole temperature profile, as argued by Lowke and Zollweg [9] when referring to the Francis equation.

Nevertheless, applications of these equations have yielded very good results when compared to measurements carried out in actual circuits. This is significant in cases when the object is neither the study of the electric arc itself nor the design of arc discharge lamps but, rather, to assess the impact of arc discharge lamps on the external circuit. Examples of these are the interaction of the arc discharge lamp with the circuit ballast, or the assessment of adverse power quality effects introduced by a cluster of arc discharge lamps on a building installation.

12.7 Arc Discharge Lamps

This section contains the derivation of an alternative differential equation representing the dynamic behaviour of arc discharge lamps. The starting point is the differential equation used for modelling the electric arc in Section 12.4. In order to obtain a model similar to that of Bo and Masumi [10], the arc conductance g, rather than the arc radius r, is used as the state variable. Values of $n = 2$ and $m = 0$ are chosen in equation (12.6) for this purpose, yielding

$$k_1 r^2 + k_2 r \frac{\mathrm{d}r}{\mathrm{d}t} = \frac{k_3}{r^2} i^2 \tag{12.20}$$

If we use in (12.20) the expression of the arc conductance,

$$g = \frac{r^2}{k_3}$$

the resultant differential equation becomes

$$g \frac{\mathrm{d}g}{\mathrm{d}t} = a i^2 - b g^2 \tag{12.21}$$

where $a = 2/k_2 k_3$ and $b = 2k_1/k_2$. Equation (12.21) can be represented by

$$f(g) = g \frac{\mathrm{d}g}{\mathrm{d}t} + b g^2 - a i^2 = 0 \tag{12.22}$$

and the voltage given by

$$v = \frac{i}{g} \tag{12.23}$$

For steady-state purposes, using harmonic domain techniques, equations (12.22) and (12.23) are represented by

$$\mathbf{F} = \mathbf{G}\mathbf{D}(\mathrm{j}h\omega_0)\mathbf{G} + b\mathbf{G}^2 - a\mathbf{I}^2 = 0 \tag{12.24}$$

and

$$\mathbf{V} = \mathbf{G}^{-1}\mathbf{I} \tag{12.25}$$

Equation (12.24) is non-linear, and it may be solved using Newton's method,

$$\Delta\mathbf{G}^{(\mathrm{k}+1)} = -\left[\frac{\partial \mathbf{F}}{\partial \mathbf{G}}\right]^{-1} \mathbf{F}^{(\mathrm{k})} \tag{12.26}$$

$$\mathbf{G}^{(\mathrm{k}+1)} = \mathbf{G}^{(\mathrm{k})} + \Delta\mathbf{G}^{(\mathrm{k}+1)} \tag{12.27}$$

where k is the iteration counter, and

$$\frac{\partial \mathbf{F}}{\partial \mathbf{G}} = \mathbf{D}(\mathrm{j}h\omega_0)\mathbf{G} + \mathbf{G}\mathbf{D}(\mathrm{j}h\omega_0)\frac{\mathrm{d}\mathbf{G}}{\mathrm{d}\mathbf{G}} + 2b\mathbf{G} \tag{12.28}$$

It should be noted that $\mathbf{G}$ in the second term on the rhs of (12.28) is expressed in matrix form in order to carry out the operation $\mathbf{GD}(\mathrm{j}h\omega_0)$. Also, the three products should be expressed in matrix form to allow evaluation of matrix $\frac{\partial \mathbf{F}}{\partial \mathbf{G}}$.

Simulated characteristics using equation (12.24), with $a = 0.0625$ and $b = 2500$, correspond to a 30 W fluorescent lamp. The results are shown in Figure 12.10.

The characteristics of Figure 12.10.(a) are similar to the measured characteristic [6] shown in Figure 12.10.(b) for the same frequencies. The computations show that, according to expectation, the loop becomes narrower as frequency increases. Discrepancies are due mainly to the fact that the measurements also include the effect of external components.

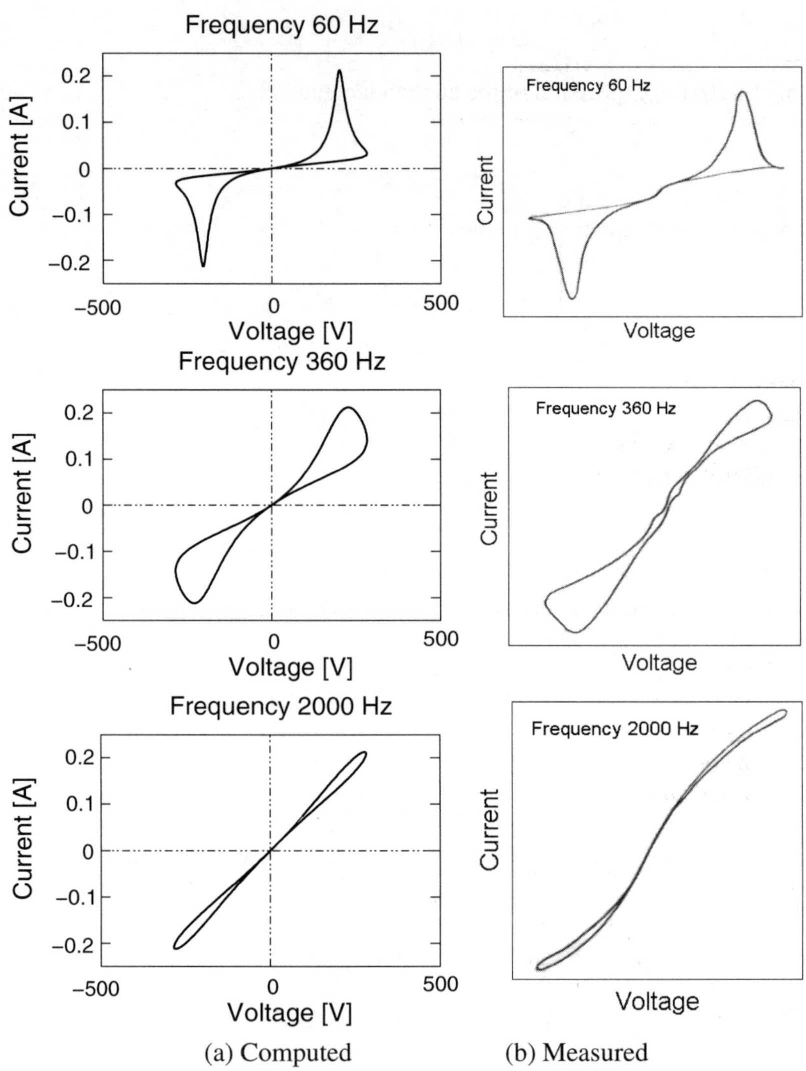

(a) Computed (b) Measured

Figure 12.10 Characteristics of a fluorescent lamp at 60, 360 and 2000 Hz

The following MATLAB™ program can be used to solve equation (12.26):

```
% example Electric Discharge Lamps
clear all
h = 100;
a = 0.0625;
b = 2500;
f = 60; % 60, 360 2000 Hz.
w = 2*pi*f;
Imax=i*0.2;   % A
g0  =0.2;     % 1/ohms
              % voltage in V
D=form_Zm(0,w,h);
G=zeros(2*h+1,1); G(h+1)=g0;
I=zeros(2*h+1,1); I(h)=Imax'/2;   I(h+2)=Imax/2;
error=1;
iter=0;
while error>1e-6
 Gm    = calc_Fm(G,h);
 G2    = self_conv(G,h,2);
 F     = Gm*D*G+b*self_conv(G,h,2)-a*self_conv(I,h,2);
 DG    = D*G;
 DG    = calc_Fm(DG,h);
 GD    = Gm*D;
 dF_G  = DG+GD+2*b*Gm;
 dG    = -inv(dF_G)*F;
 G     = G+dG;
 error= norm(dG)
 iter = iter+1
end
V    = inv(Gm)*I;
```

12.8 Summary

This chapter has presented an overview of electric arcs, where a distinction has been drawn between arcs present in electric furnaces and in discharge lamps. Nevertheless, it was shown that a single differential equation, with quite different constant coefficients, could be used to represent the electric arc in both applications. The differential equation is derived first with the arc furnace in mind, and appeals to the basic principle of power balance and emphasises the interdependency between arc radius and cooling. Reference is made to short and long radius, both of which are utilised at different stages of the steel reduction process. The great variations in the active power consumed by the arc furnace throughout the melting cycle are emphasised. The random behaviour of the arc during the initial stages of the meltdown process is contrasted with the smooth variations that take place at the end of the cycle, during the refining period. The arc model used here, in the form of a differential equation, is not intended for the study of the arc during the initial stage of the meltdown process, i.e. random, but rather, it is intended for the intermediate and final stages. Simulation results are compared against available measurements and it is concluded that they compare well with each other.

A model for the discharge lamp is derived as a particular case of the generic arc model developed above, but a change of variable is carried out to obtain a model that is more in line with models available in the open literature. In the discharge lamp the arc conductance is the state variable as opposed to the arc radius. In this case too, simulation results are compared against measurements, achieving very close agreement between them. In this application, the hysteresis loop narrows as frequency increases.

12.9 Bibliography

1. T.J.E. Miller, *Reactive Power Control in Electric Systems*, Wiley-Interscience, New York, 1982.
2. L. Fernández, "Modelado en Dominio de la Frequencia de Hornos de Arco de Corriente Alterna para la Estimación de Armónicos, Fluctuaciones y Desequilibrios", *PhD Thesis* (in Spanish), Universidad Politécnica de Madrid, 1999.
3. E. Acha, A. Semlyen, N. Rajakovic, "A Harmonic Domain Computational Package for Nonlinear Problems and Its Application to Electric Arcs", *IEEE Transactions on Power Delivery*, Vol. 5, No. 3, July 1990, pp. 1390–1397.
4. E. Acha, J.J. Rico, "Harmonic Domain Modelling of Non-linear Power Plant Component", *Proceedings of IEEE ICHPS VI*, Bologna, Italy, 1994, pp. 206–213.
5. V.G. Schönfelder, "Elektrische Lichtbogenöfen und ihr Einsatz in der Eisenschaffenden Industrie", *Electrowärme International*, No. 5, October 1983, pp. B214–B221.
6. J.H. Campbell, H.E. Schultz, D.D. Kershaw, "Characteristics and Applications of High-Frequency Fluorescent Lighting", *Illuminating Engineering* (North America), Vol. 48, February 1953, pp. 95–103.
7. V.J. Francis, *Fundamentals of Discharge Tube Circuits*, John Wiley & Sons, New York, 1948.
8. S.C. Peek, D.E. Spencer, "A Differential Equation for the Fluorescent Lamp", *Illuminating Engineering* (North America), Vol. 63, No. 4, April 1968, p. 157.
9. J.J. Lowke, R.J. Zollweg, "Theoretical Predictions of AC Characteristics of Mercury Arc Lamps", *Journal of the Illuminating Engineering Society*, July 1975, pp. 253–259.
10. H. Bo, K. Masumi, "Analysis and Operating Circuits for Discharge Lamps by the Simulation Method", *Journal of the Illuminating Engineering Society*, Vol. 5, No. 2, January 1976, pp. 92–98.
11. E. Acha, N.P. Johnson, L.L. Loh, P. Miller, "The Impact of Energy Saving Apparatus on Power Quality", *Proceedings of 8th ICHQP*, Vol. 1, 14–16 October, Athens, Greece, 1998, pp. 36–41.
12. W.M. Grady, G.T. Heydt, "Prediction of Power System Harmonics due to Gaseous Discharge Lighting", *IEEE Transactions on Power Apparatus and Systems*, Vol. PAS-104, No. 3, March 1985, pp. 554–562.

Part III

POWER ELECTRONICS EQUIPMENT

13

Static VAR Compensator

13.1 Introduction

The use of power-electronics-based devices in bulk power transmission has increased steadily over the last 30 years. With it has come the increased risk of harmonic distortion in the power network because several of these devices achieve their main operating state at the expense of generating harmonic currents. In the early days, most applications of this technology were in the area of HVDC transmission. However, the static VAR compensator (SVC), which is a more recent development, is finding widespread use in the area of reactive power management and control. Over the last 15 years a substantial number of SVCs have been incorporated into existing AC transmission systems.

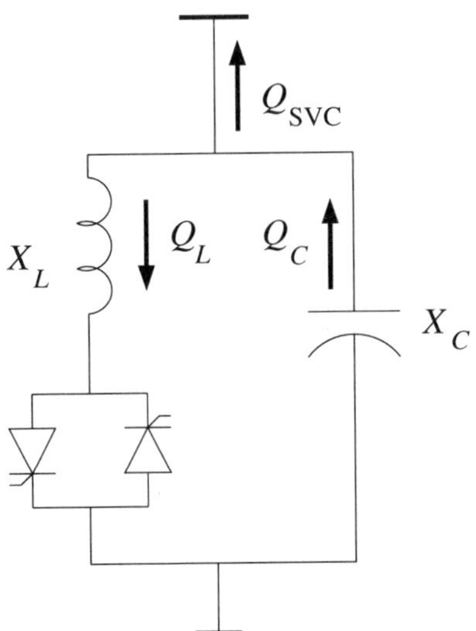

Figure 13.1 Static VAR compensator arrangement

The SVC is a plant component which is used to supply a variable amount of reactive power at a specific location in the electrical power network. The SVC comprises a parallel combination of a fixed capacitor bank and a thyristor-controlled reactor (TCR), as shown in Figure 13.1. The reactive power control capability of the TCR makes the SVC operate as a variable reactance which supplies/absorbs reactive power in an adaptive fashion.

The reactive power in the SVC can be varied from a maximum absorbed value (thyristor valves bypassed) to a maximum injected value (thyristor valves open), by varying the thyristor's firing angle α within the range $90° < \alpha < 180°$. On the other hand, the reactive power absorbed by the TCR can be varied from a maximum value when the thyristor valves are bypassed, i.e. $\alpha = 90°$, to a zero value when the thyristor valves are open, i.e. $\alpha = 180°$. The former case corresponds to the TCR in a full conducting state whereas the latter corresponds to a non-conducting state.

The relevant voltage and current waveforms in the TCR operation are shown in Figure 13.2, where the TCR current, $i_{\mathrm{TCR}}(t)$, is shown as a waveform that varies in time in response to increased firing angle delays, which is amenable to shorter conducting periods and decreased amplitudes of the current waveform. The current $i_{\mathrm{TCR}}(t)$ is rich in harmonics due to the dead bands created by the thyristor's gating action. Increasing the thyristor's gating is amenable to a reduced fundamental frequency current. It should be mentioned that the fundamental frequency component $i_{\mathrm{f}}(t)$ is the only component that contributes to the SVC reactive power supply. The magnitude of $i_{\mathrm{f}}(t)$ can be expressed as a function of the firing angle by

$$I_{\mathrm{f}}(\alpha) = \frac{V}{\pi X_L}(2\pi - 2\alpha + \sin 2\alpha) \tag{13.1}$$

where V is the voltage amplitude of $v_{\mathrm{TCR}}(t)$, and X_L is the reactance of the linear inductor at fundamental frequency.

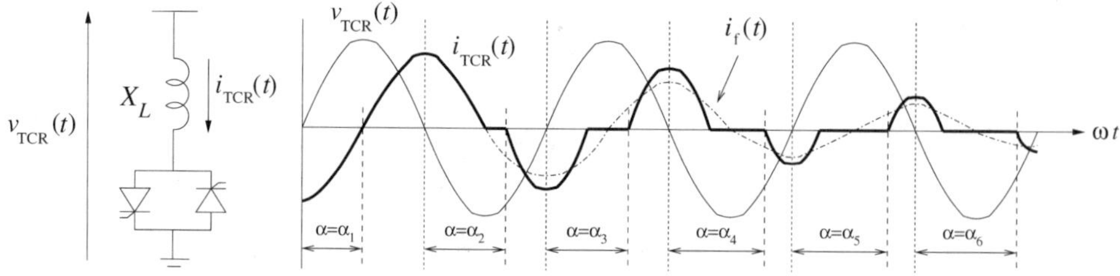

Figure 13.2 TCR current response

The fundamental frequency equivalent impedance of the TCR is readily available from (13.1),

$$X_{\mathrm{TCR}} = \frac{\pi X_L}{2\pi - 2\alpha + \sin 2\alpha} \tag{13.2}$$

The parallel combination of X_{TCR} with the capacitive reactance X_C gives the SVC equivalent impedance at fundamental frequency, X_{SVC},

$$X_{\mathrm{SVC}} = \frac{\pi X_C X_L}{X_C(2\pi - 2\alpha + \sin 2\alpha) - \pi X_L} \tag{13.3}$$

The reactive power injected by the SVC is given by

$$Q_{\mathrm{SVC}} = \frac{V^2}{X_{\mathrm{SVC}}} \tag{13.4}$$

The reactive power and impedance characteristics of the SVC are given in Figure 13.3. A well-designed SVC should have two operating regions, one inductive and one capacitive, in order to enable both the absorption and the injection of reactive power at the point of connection. An intrinsic property of the impedance characteristic is the resonant point that appears when the SVC changes from the inductive to the capacitive region as the SVC firing angle increases. In this example the reactive power increases from maximum power absorbed when the SVC operates in the inductive region to maximum power generated when it operates in the capacitive region. As also illustrated in the figure, the variation of reactive power with firing angle is non-linear.

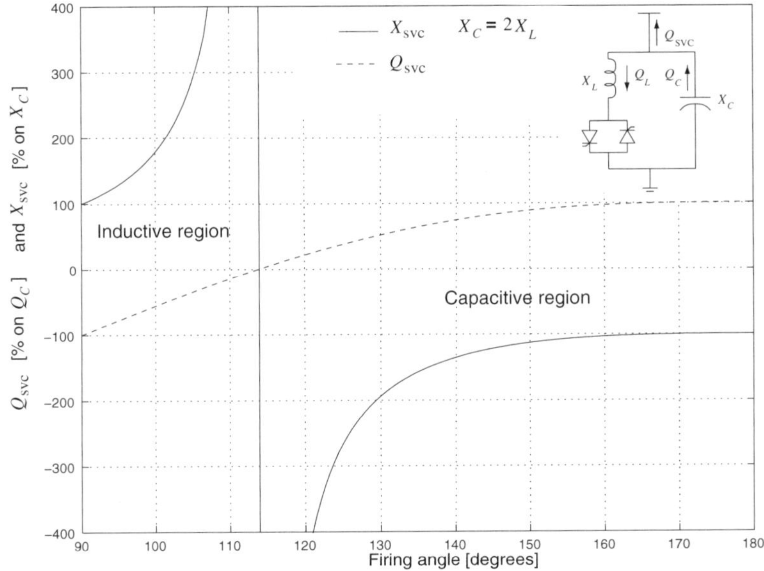

Figure 13.3 SVC reactive power and impedance characteristics as a function of firing angle

The harmonic magnitudes of the current $i_{\mathrm{TCR}}(t)$ may be calculated as a function of the firing angle by using the following relation,

$$I_h(\alpha) = \frac{4V}{\pi X_L}\left(\frac{h\sin\alpha\cos h\alpha - \cos\alpha\sin h\alpha}{h\left(h^2-1\right)}\right) \tag{13.5}$$

for $h = 2k+1$, with $k = 1, 2, 3 \ldots$

This expression shows that the TCR generates the third, fifth, seventh, ninth . . . harmonics. Figure 13.4 shows the magnitudes of the harmonic currents generated by a single-phase TCR. Normally, three-phase TCRs are delta connected to prevent the triplen harmonics from reaching the electric network.

13.2 Single-phase TCR Basic Operation

Figure 13.5 shows the main variables associated with the operation of the TCR $v_R(t)$, and the current through the reactor, $i_{\mathrm{TCR}}(t)$. The firing angle, α, and the conduction angle, σ, are related by the following relation:

$$\sigma = 2(\pi - \alpha) \tag{13.6}$$

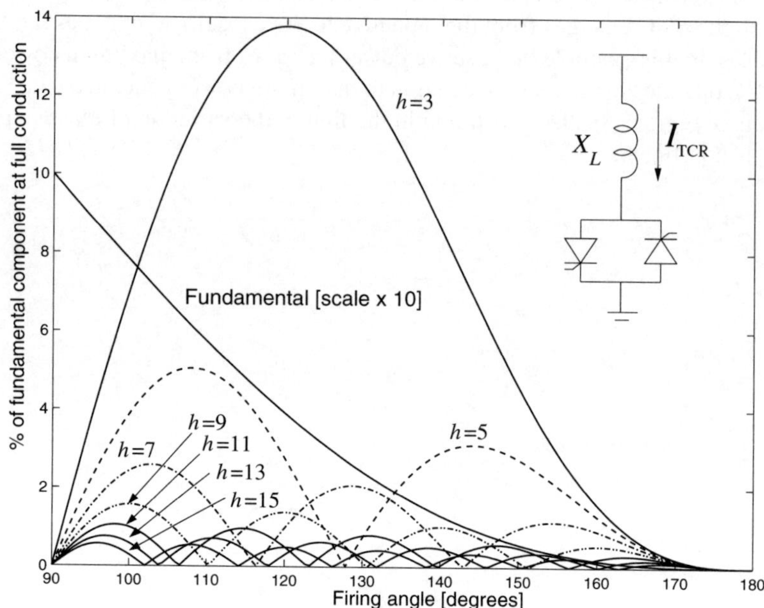

Figure 13.4 TCR harmonics content

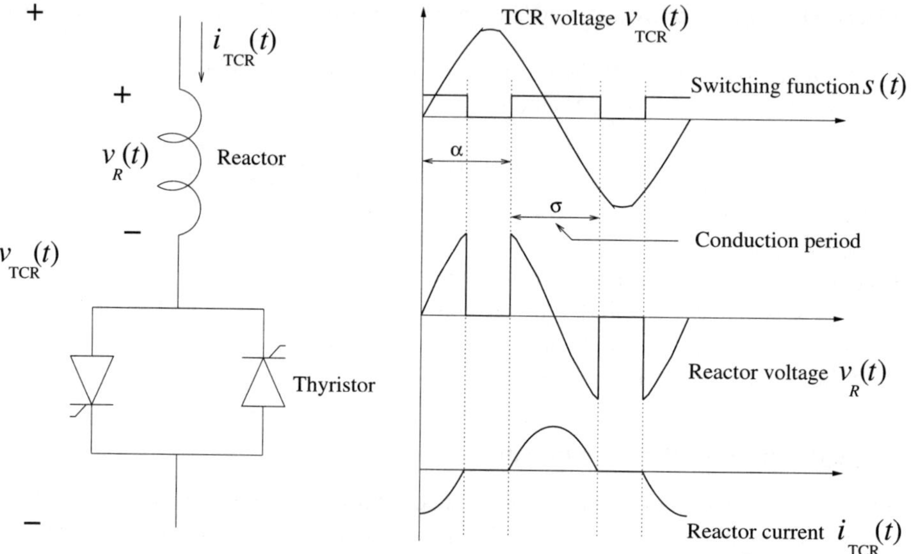

Figure 13.5 TCR ideal operation

This figure illustrates that the switching function, $s(t)$, governs the behaviour of both the voltage across the reactor,

$$v_R(t) = s(t)v_{\mathrm{TCR}}(t) \tag{13.7}$$

and

$$v_R(t) = L\frac{\mathrm{d}i_{\mathrm{TCR}}(t)}{\mathrm{d}t} \tag{13.8}$$

Equations (13.7) and (13.8) may be expressed in the harmonic domain as follows:

$$\mathbf{V}_R = \mathbf{S}\mathbf{V}_{\mathrm{TCR}} \tag{13.9}$$

$$\mathbf{V}_R = L\mathbf{D}(\mathrm{j}h\omega_0)\mathbf{I}_{\mathrm{TCR}} \tag{13.10}$$

Combining (13.9) and (13.10) we obtain

$$\mathbf{I}_{\mathrm{TCR}} = \mathbf{Y}_{\mathrm{TCR}}\mathbf{V}_{\mathrm{TCR}} \tag{13.11}$$

where the TCR admittance matrix $\mathbf{Y}_{\mathrm{TCR}}$ is

$$\mathbf{Y}_{\mathrm{TCR}} = \frac{1}{L}\mathbf{D}^{-1}(\mathrm{j}h\omega_0)\mathbf{S} \tag{13.12}$$

The admittance matrix $\mathbf{Y}_{\mathrm{TCR}}$ includes the cross-couplings between frequencies, giving a complete representation of the TCR during steady-state operation.

13.2.1 Voltage zero-crossing reference

The operation of the thyristor is a function of the voltage zero crossing; it starts conduction just after the voltage zero crossing is reached. In cases when the TCR is connected to a strong system, the voltage zero crossing from one period of the voltage waveform to the next is constant and so is the matrix $\mathbf{S}$. However, when the TCR is connected to a weak system, the voltage zero crossing becomes a function of system conditions; in particular, it becomes a function of the voltage waveform at the point of connection. Figure 13.6 illustrates this point, where the zero crossing in a voltage waveform containing distortion is shown. A sinusoidal waveform is superimposed to emphasise the change in zero crossing due to distortion. In this situation the matrix $\mathbf{S}$ is a function of the voltage $\mathbf{V}_{\mathrm{TCR}}$.

If we assume that the TCR voltage zero crossing occurs at a time t when

$$v_{\mathrm{TCR}} = \sum_{n=-\infty}^{\infty} V_n \mathrm{e}^{\mathrm{j}n\omega_0 t} = 0 \tag{13.13}$$

or at an angle θ_0 when

$$\sum_{n=-\infty}^{\infty} V_n \mathrm{e}^{\mathrm{j}n\theta_0} = 0 \tag{13.14}$$

then it is possible to find the voltage zero-crossing reference by solving (13.14) by iteration using the Newton–Raphson method,

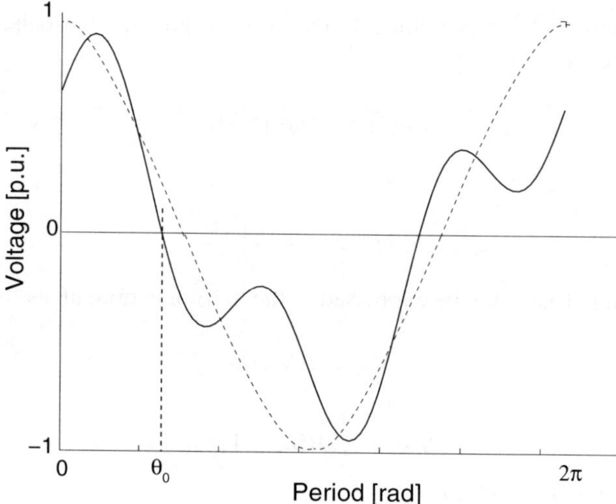

Figure 13.6 Voltage zero crossing

$$\theta^{(k+1)} = \theta^{(k)} - \frac{v_{TCR}}{v'_{TCR}} \tag{13.15}$$

where k is an iteration counter.

$$v_{TCR} = \sum_{n=-\infty}^{\infty} V_n e^{jn\theta_0^{(k)}} \tag{13.16}$$

and

$$v'_{TCR} = \sum_{n=-\infty}^{\infty} jnV_n e^{jn\theta_0^{(k)}} \tag{13.17}$$

The iterative solution of equation (13.15) requires that a suitable initial condition be selected, and a sensible practice is to assume a sinusoidal waveform as the starting condition. In this case the angle $\theta_0^{(0)}$ is given by

$$\theta_0^{(0)} = \frac{\pi}{2} - \arctan\left(\frac{\Im\{V_1\}}{\Re\{V_1\}}\right) \tag{13.18}$$

In voltage waveforms containing a high level of distortion, and where unsuitable initial conditions are made, the Newton–Raphson method may fail to reach convergence or it may converge to a solution different from the one intended. This is illustrated in Figure 13.7 where the zero crossing of a sinusoidal waveform is used as the initial condition, which in this case is not good enough to reach the target, and the solution moves instead towards the second zero-crossing point. It should be mentioned that in such situations the Gauss–Seidel method, whose behaviour towards the solution is monotonic as opposed to quadratic, provides an alternative iterative method in which a solution may be attempted.

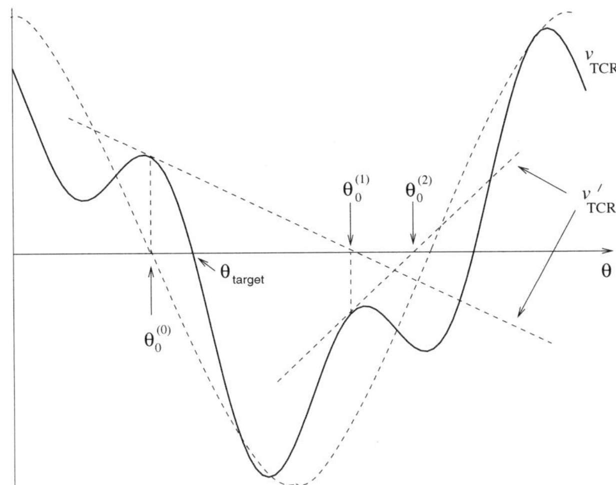

Figure 13.7 Newton–Raphson method convergence

Example 13-1: The voltage waveform in Figure 13.6 contains a 50% third harmonic distortion, where the fundamental term is 2 p.u. and the third harmonic is 1 p.u. Using the algorithm outlined above, the voltage zero crossing is found to be: $\theta_0 = 1.2637$ rad or $\theta_0 = 72.4°$.

The following MATLAB™ function was used to obtain the first zero crossing of the harmonic vector, `V=[i*0.5;0;1;0;1;0;-i*0.5]` with `h=3`:

```
function theta=zero_cross(V,h)
% V : harmonic vector V=[-h ... -1 0 1 ... h]
% h : number of harmonics
% theta : angle in radians where the signal v(t) crosses zero.
%---- initial conditions
theta0=pi/2-angle(V(h+2));
while abs(theta0)>pi
  theta0=theta0-sign(theta0)*pi;
end
error=1;
while error>1e-12
  Vtcr=0;
  dVtcr=0;
  for n=-h:h
    Vtcr =V(n+h+1)*exp(i*n*theta0)+Vtcr;
    dVtcr=i*n*V(n+h+1)*exp(i*n*theta0)+dVtcr;
  end
  theta1=theta0-Vtcr/dVtcr;
  error=abs(theta1-theta0);
  theta0=theta1;
end
theta=real(theta1);
```

13.2.2 Thyristor turn-on and turn-off

A realistic representation of the TCR should include the thyristor's turn-on and turn-off actions. Figure 13.8 illustrates the various parameters associated with the TCR voltage waveform, where turn-on and turn-off instants are represented. The TCR voltage zero crossing is taken as the reference for issuing the firing signal α. If the first thyristor fires at a time θ_{a1}, the thyristor turns on and conducts for a period σ_1. It turns off with the zero crossing of the thyristor's current, at a time θ_{b1}. An equidistant firing scheme is assumed, i.e. $\theta_{a2} - \theta_{a1} = \pi$. In this figure, the switching function $s(\theta)$ is also shown.

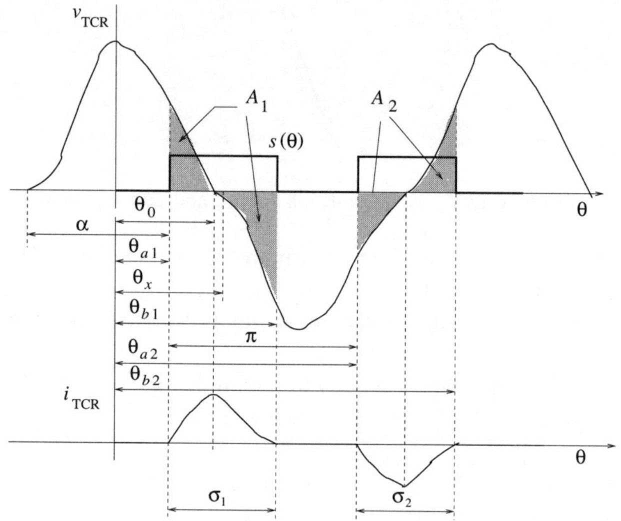

Figure 13.8 TCR operation

The analysis starts at a time of voltage zero crossing θ_0, which may be expressed in terms of complex Fourier harmonic coefficients,

$$v_{\mathrm{TCR}} = \sum_{n=-\infty}^{\infty} V_n \mathrm{e}^{\mathrm{j}n\theta_0} = 0 \tag{13.19}$$

The end of the conduction period, given by the zero crossing of the thyristor's current, occurs when the areas A_1 and A_2 are both zero, i.e.

$$\begin{aligned} A_1 &= \int_{\theta_{a1}}^{\theta_{b1}} v_{\mathrm{TCR}} \mathrm{d}\theta &= \sum_{n=-\infty}^{\infty} \frac{V_n}{\mathrm{j}n} \left(\mathrm{e}^{\mathrm{j}n\theta_{b1}} - \mathrm{e}^{\mathrm{j}n\theta_{a1}} \right) = 0 \\ A_2 &= \int_{\theta_{a2}}^{\theta_{b2}} v_{\mathrm{TCR}} \mathrm{d}\theta &= \sum_{n=-\infty}^{\infty} \frac{V_n}{\mathrm{j}n} \left(\mathrm{e}^{\mathrm{j}n\theta_{b2}} - \mathrm{e}^{\mathrm{j}n\theta_{a2}} \right) = 0 \end{aligned} \tag{13.20}$$

where

$$\begin{aligned} \theta_{a1} &= \theta_0 - (\pi - \alpha) \\ \theta_{a2} &= \theta_{a1} + \pi \end{aligned} \tag{13.21}$$

Solving (13.19) and (13.20) for θ_0, θ_{b1} and θ_{b2} using Newton's method, the following equation is solved by iteration:

$$\begin{bmatrix} \theta_0 \\ \theta_{b1} \\ \theta_{b2} \end{bmatrix}^{(k+1)} = \begin{bmatrix} \theta_0 \\ \theta_{b1} \\ \theta_{b2} \end{bmatrix}^{(k)} - \begin{bmatrix} J_{11} & J_{12} & J_{13} \\ J_{21} & J_{22} & J_{23} \\ J_{31} & J_{32} & J_{33} \end{bmatrix}^{-1} \begin{bmatrix} v_{\mathrm{TCR}} \\ A_1 \\ A_2 \end{bmatrix}^{(k)} \tag{13.22}$$

k is an iteration counter and the Jacobian elements are:

$$\begin{aligned}
J_{11} &= \frac{\partial v_{\mathrm{TCR}}}{\partial \theta_0} = \sum_{n=-\infty}^{\infty} \mathrm{j} n V_n \mathrm{e}^{\mathrm{j} n \theta_0} \\
J_{12} &= \frac{\partial v_{\mathrm{TCR}}}{\partial \theta_{b1}} = 0 \\
J_{13} &= \frac{\partial v_{\mathrm{TCR}}}{\partial \theta_{b2}} = 0 \\
J_{21} &= \frac{\partial A_1}{\partial \theta_0} = -\sum_{n=-\infty}^{\infty} V_n \mathrm{e}^{\mathrm{j} n (\theta_0 - (\pi - \alpha))} \\
J_{22} &= \frac{\partial A_1}{\partial \theta_{b1}} = \sum_{n=-\infty}^{\infty} V_n \mathrm{e}^{\mathrm{j} n \theta_{b1}} \\
J_{23} &= \frac{\partial A_1}{\partial \theta_{b2}} = 0 \\
J_{31} &= \frac{\partial A_2}{\partial \theta_0} = -\sum_{n=-\infty}^{\infty} V_n \mathrm{e}^{\mathrm{j} n (\theta_0 + \alpha)} \\
J_{32} &= \frac{\partial A_2}{\partial \theta_{b1}} = 0 \\
J_{33} &= \frac{\partial A_2}{\partial \theta_{b2}} = \sum_{n=-\infty}^{\infty} V_n \mathrm{e}^{\mathrm{j} n \theta_{b2}}
\end{aligned} \tag{13.23}$$

The initial conditions required for the solution of equation (13.22) may be taken to correspond to a sinusoidal waveform, which is given by the zero crossing of the fundamental component of v_{TCR}. The initial value for θ_0 is given by

$$\theta_0^{(0)} = \frac{\pi}{2} - \arctan\left(\frac{\Im\{V_1\}}{\Re\{V_1\}}\right) \tag{13.24}$$

and the initial conditions for θ_{b1} and θ_{b2} can given by assuming quarter-wave symmetry,

$$\begin{aligned} \theta_{b1}^{(0)} &= \theta_0^{(0)} + \left(\theta_0^{(0)} - \theta_{a1}^{(0)}\right) \\ \theta_{b2}^{(0)} &= \theta_{b1}^{(0)} + \pi \end{aligned} \tag{13.25}$$

After convergence the centre of the switching function $s(\theta)$ is taken to be at

$$\theta_x = \theta_{a1} + \frac{\sigma_1}{2} \tag{13.26}$$

where the conduction angles are given as

$$\begin{aligned} \sigma_1 &= \theta_{b1} - \theta_{a1} \\ \sigma_2 &= \theta_{b2} - \theta_{a2} \end{aligned} \tag{13.27}$$

Then, θ_x, σ_1 and σ_2 are used to calculate the harmonic content of the switching function given by

$$\begin{aligned} S_0 &= \frac{\sigma_1 + \sigma_2}{2\pi} \\ S_n &= \frac{1}{n\pi}\left(\sin\frac{n\sigma_2}{2}\cos n\pi + \sin\frac{n\sigma_1}{2}\right)\mathrm{e}^{-\mathrm{j}n\theta_x} \end{aligned} \tag{13.28}$$

It should be noted that these equations are general. They cater for the case when the conduction periods of thyristors 1 and 2 differ owing to the existence of voltage waveforms with loss of quarter-wave symmetry.

The following MATLAB™ function is used to obtain values for the angles σ_1, σ_2 and θ_x, which give the conduction periods of both thyristors:

```
function [sigma1,sigma2,thetax]=Thy_turn_on_off(V,alpha,h)
% V       : harmonic vector V=[-h ... -1 0 1 ... h]
% alpha   : firing angle in radians
% h       : number of harmonics
% sigma1  : first thyristor conduction angle in radians
% sigma2  : second thyristor conduction angle in radians
% thetax  : angle in sigma1/2;
% ------ initial conditions
theta00  = pi/2-angle(V(h+2));
thetab10 = theta00+pi-alpha;
thetab20 = thetab10+pi;
error=1;
while error>1e-12
  Vtcr = 0; A1 = 0; A2 = 0;
  J11 = 0; J12 = 0; J13 = 0;
  J21 = 0; J22 = 0; J23 = 0;
  J31 = 0; J32 = 0; J33 = 0;
  for n=-h:h
    if n~=0
      Vtcr = V(n+h+1)*exp(i*n*theta00)+Vtcr;
      thetaa1=theta00-(pi-alpha);
      thetaa2=thetaa1+pi;
      A1 = 1/(i*n)*V(n+h+1)*(exp(i*n*thetab10)-exp(i*n*thetaa1))+A1;
      A2 = 1/(i*n)*V(n+h+1)*(exp(i*n*thetab20)-exp(i*n*thetaa2))+A2;
      J11 = i*n*V(n+h+1)*exp(i*n*theta00)+J11;
      J12 = 0;
      J13 = 0;
      J21 = V(n+h+1)*exp(i*n*(theta00-pi+alpha))+J21;
      J22 = V(n+h+1)*exp(i*n*thetab10)+J22;
      J23 = 0;
      J31 = V(n+h+1)*exp(i*n*(theta00+alpha))+J31;
```

```
      J32 = 0;
      J33 = V(n+h+1)*exp(i*n*thetab20)+J33;
    end
  end
  fun =[Vtcr;A1;A2];
  Jac =[J11 J12 J13;-J21 J22 J23;-J31 J32 J33];
  theta0=[theta00;thetab10;thetab20];
  theta1=theta0-inv(Jac)*fun;
  error=norm(theta1-theta0);
  theta00  = theta1(1);
  thetab10 = theta1(2);
  thetab20 = theta1(3);
end
theta01   = theta1(1);
thetab11  = theta1(2);
thetab21  = theta1(3);
ta1       = real(theta01-pi+alpha);
ta2       = real(theta01+alpha);
sigma1    = real(thetab11-ta1);
sigma2    = real(thetab21-ta2);
thetax    = ta1+sigma1/2;
```

The next MATLAB™ function is used to obtain the harmonic content of the switching function corresponding to the conduction angles σ_1 and σ_2:

```
function S=calc_S(sigma1,sigma2,thetax,h)
% S : Switching harmonic vector S=[-h ... -1 0 1 ... h]
S=zeros(2*h+1,1);
c=1;
for n=-h:h
  if n~=0
    S(c)=1/(n*pi)*(sin(n*sigma2/2)*cos(n*pi)+
         sin(n*sigma1/2))*exp(-i*n*thetax);
  else
    S(c)=(sigma1+sigma2)/(2*pi);
  end
c=c+1;
end
```

This MATLAB™ function is used to generate the TCR admittance as given by equation (13.12):

```
function Ytcr=calc_TCR(V,alpha,h,w,L)
  [sigma1,sigma2,thetax]=Thy_turn_on_off(V,alpha,h);
  Sv=calc_S(sigma1,sigma2,thetax,h);
  S=calc_Fm(Sv,h);
  D=form_Zm(0,w,h);
  Ytcr=1/L*inv(D)*S;
```

Figure 13.9 shows the impact of the voltage waveform on the thyristor current. Figure 13.9(a) corresponds to a TCR operating under sinusoidal voltage conditions. In cases of non-sinusoidal voltage operation, as shown in Figure 13.9(b), it is possible for the current to deviate appreciably from the more idealised form shown in Figure 13.9(a). In the former case, the conduction period of the thyristor is 120°, with a current amplitude of approximately 0.32 p.u. In the latter condition, the same firing angle is used but the voltage is non-sinusoidal, giving rise to a conduction period longer than 120°, with a current amplitude of approximately

0.4 p.u.

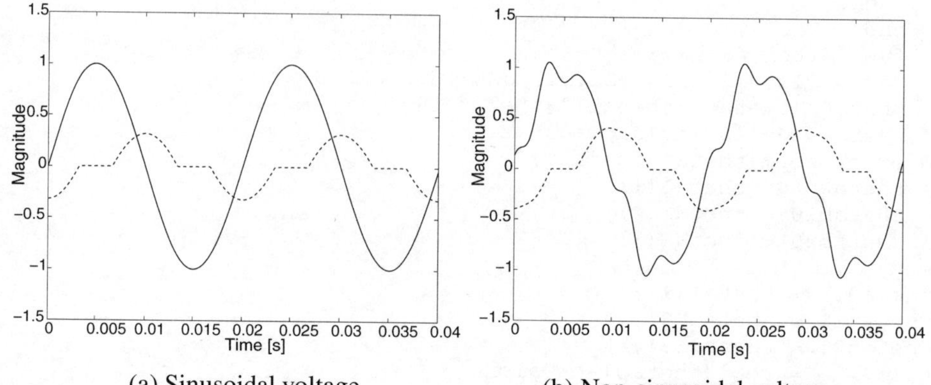

(a) Sinusoidal voltage (b) Non-sinusoidal voltage

Figure 13.9 Effect of voltage waveform in the thyristor current for a firing angle of 120°

The MATLAB™ program used to obtain the waveforms of Figure 13.9(b) and Figure 13.9(a) is:

```
clear all
h=50;
f=50;
w=2*pi*f;
L=0.005;
alpha=120*pi/180;
V=zeros(2*h+1,1);
V(h-6) = 0.0260+i*0.0150; % zero for sinusoidal voltage
V(h-4) = 0.04-i*0.05;     %               ''
V(h-2) = i*0.01;          %               ''
V(h)   = i*0.5;
V(h+2) = -i*0.5;
V(h+4) = -i*0.01;         %               ''
V(h+6) = 0.04+i*0.05;     %               ''
V(h+8) = 0.0260-i*0.0150; %               ''
Ytcr=calc_TCR(V,alpha,h,w,L);
Itcr=Ytcr*V;
v=plot_f(V,h,2);
itcr=plot_f(Itcr,h,2);
t=0:1:363;
t=t/363*2/f;
plot(t,v,t,itcr)
```

13.2.3 Single-phase SVC model

The SVC is well described by the parallel combination of the TCR admittance and the capacitor bank admittance,

$$\mathbf{Y}_{\mathrm{SVC}} = \mathbf{Y}_{\mathrm{TCR}} + \mathbf{Y}_C \tag{13.29}$$

where $\mathbf{Y}_C = C\mathbf{D}(\mathrm{j}h\omega_0)$ is the admittance matrix of the shunt capacitor and C is its capacitance.

13.3 Three-phase TCR Modelling

Normally, a three-phase TCR consists of three single-phase TCRs delta connected order to confine the 3rd, 9th and 15th harmonic currents within the delta connection. Accordingly, equation (13.12) may be used as the basic building block for modelling a three-phase TCR, as shown in Figure 13.10. The equation that represents the invariant power, three-phase TCR equivalent is given by

$$\begin{bmatrix} \mathbf{I}_a \\ \mathbf{I}_b \\ \mathbf{I}_c \end{bmatrix} = \frac{1}{3} \begin{bmatrix} \mathbf{Y}^{ab}_{\mathrm{TCR}} + \mathbf{Y}^{ca}_{\mathrm{TCR}} & -\mathbf{Y}^{ab}_{\mathrm{TCR}} & -\mathbf{Y}^{ca}_{\mathrm{TCR}} \\ -\mathbf{Y}^{ab}_{\mathrm{TCR}} & \mathbf{Y}^{bc}_{\mathrm{TCR}} + \mathbf{Y}^{ab}_{\mathrm{TCR}} & -\mathbf{Y}^{bc}_{\mathrm{TCR}} \\ -\mathbf{Y}^{ca}_{\mathrm{TCR}} & -\mathbf{Y}^{bc}_{\mathrm{TCR}} & \mathbf{Y}^{ca}_{\mathrm{TCR}} + \mathbf{Y}^{bc}_{\mathrm{TCR}} \end{bmatrix} \begin{bmatrix} \mathbf{V}_a \\ \mathbf{V}_b \\ \mathbf{V}_c \end{bmatrix} \tag{13.30}$$

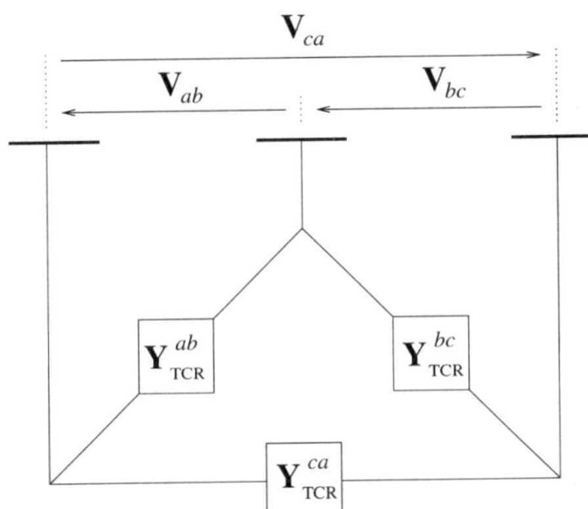

Figure 13.10 Delta-connected three-phase TCR

The MATLAB™ function used to calculate the equivalent admittance matrix for delta- and star-connected TCRs is:

```
function Ytcr=calc_TCR_ThreePhase(Va,Vb,Vc,alpha_a,alpha_b,alpha_c,w,
                                  Ltcr,h,connection)
% alpha's in rad
% connection 1 : star
% connection 0 : delta
if connection==1
 Ytcr_a= calc_TCR(Va,alpha_a,h,w,Ltcr);
 Ytcr_b= calc_TCR(Vb,alpha_b,h,w,Ltcr);
 Ytcr_c= calc_TCR(Vc,alpha_c,h,w,Ltcr);
 Ytcr=[ Ytcr_a Ytcr_a*0 Ytcr_a*0
      Ytcr_a*0    Ytcr_b Ytcr_a*0
      Ytcr_a*0 Ytcr_a*0    Ytcr_c];
else
 Ytcr_ab= calc_TCR(Va-Vb,alpha_a,h,w,Ltcr);
 Ytcr_bc= calc_TCR(Vb-Vc,alpha_b,h,w,Ltcr);
 Ytcr_ca= calc_TCR(Vc-Va,alpha_c,h,w,Ltcr);
 Ytcr=[Ytcr_ab+Ytcr_ca          -Ytcr_ab          -Ytcr_ca
             -Ytcr_ab Ytcr_bc+Ytcr_ab          -Ytcr_bc
             -Ytcr_ca          -Ytcr_bc Ytcr_ca+Ytcr_bc]/3;
end
```

Example 13-2: The compensation system shown in Figure 13.11 consists of a 100 MVA, 230/24 kV power transformer with a star–delta connection, and a 0–80 MVAR static VAR compensator. The equivalent circuit used for harmonic domain calculations is also shown in this figure.

The electric parameters of the various plant components are expressed in Ω:

$$\text{Transformer} \quad : \quad X_{\text{trans}} = \frac{(Z\%)\,(\text{kV})^2}{(100)(\text{MVA})} = \frac{(10)(24)^2}{(100)(100)} = 0.5760\,\Omega$$

$$R_{\text{trans}} = X_{\text{trans}}/(X/R) = 0.5760/10 = 0.05760\,\Omega$$

$$\text{Capacitor} \quad : \quad X_{\text{cap}} = \frac{(\text{kV})^2}{\text{MVAR}} = \frac{(24)^2}{80} = 7.2\,\Omega$$

$$\text{Reactor} \quad : \quad X_{\text{reac}} = \frac{(\text{kV})^2}{\text{MVAR}} = \frac{(24)^2}{80} = 7.2\,\Omega$$

From the equivalent circuit the system of equations to be solved is

$$\begin{bmatrix} \mathbf{I}_1 \\ 0 \end{bmatrix} = \begin{bmatrix} \mathbf{Y}_{t,11} & \mathbf{Y}_{t,12} \\ \mathbf{Y}_{t,21} & \mathbf{Y}_{t,22} + \mathbf{Y}_{\text{SVC}} \end{bmatrix} \begin{bmatrix} \mathbf{V}_1 \\ \mathbf{V}_2 \end{bmatrix}$$

from which the following equation is derived

$$\mathbf{V}_2 = -\left(\mathbf{Y}_{t,22} + \mathbf{Y}_{\text{SVC}}\right)^{-1} \mathbf{Y}_{t,21} \mathbf{V}_1$$

It should be noted that the matrix $\mathbf{Y}_{\text{SVC}}$ is a function of the voltage $\mathbf{V}_2$ which is not known at first, so an iterative process is required to obtain the correct result.

This example is well suited to study the effects of the TCR connection and imbalances.

Four cases are considered in this example: (i) A delta-connected TCR. (ii) A star-connected TCR. (iii) SVC subjected to unbalanced voltage excitation. (iv) SVC with unequal firing angles.

(i) Delta-connected TCR with a firing angle of $\alpha = 110°$. Figure 13.12 shows the voltage and current waveforms and their harmonic contents, at node 2. As expected, the voltages and currents are well balanced and only the characteristic harmonics appear, i.e. 5th, 7th, 11th and 13th. The voltage contains 2% of the fifth harmonic, with the rest of the harmonics having a small enough value to be of no concern. The currents exhibit a larger amount of distortion where the fifth harmonic current reaches a value of more than 10% in the line current and around 9% in the phase current. The seventh harmonic current has a value of 2% and 3%, respectively. As shown in the figure, the 11th and 13th harmonic currents have a value of around 1%. It should be noted that although a single-phase TCR generates the third harmonic current at $\alpha = 110°$, as shown in Figure 13.4, the TCR delta connection prevents this current from reaching node 2. On the other hand, a single-phase TCR does not generate a ninth harmonic current at $\alpha = 110°$, as shown in Figure 13.4.

(ii) Star-connected TCR with a firing angle of $\alpha = 110°$. Figure 13.13 shows similar results to case (i). The only major difference is that the third harmonic currents generated by the three individual TCRs are injected directly into node 2. In this case, a 13% line current and 22%

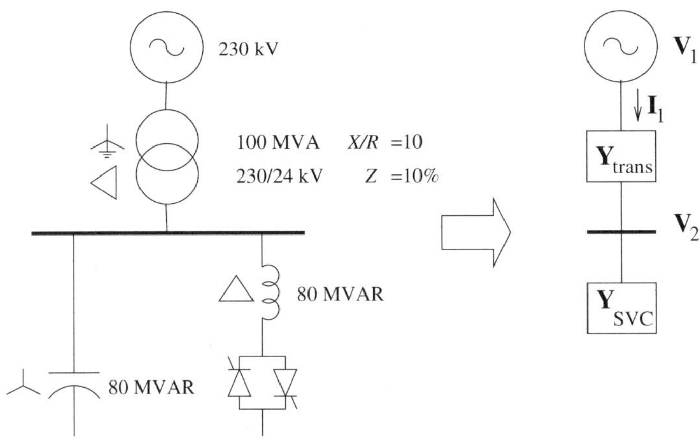

Figure 13.11 Electric system with SVC

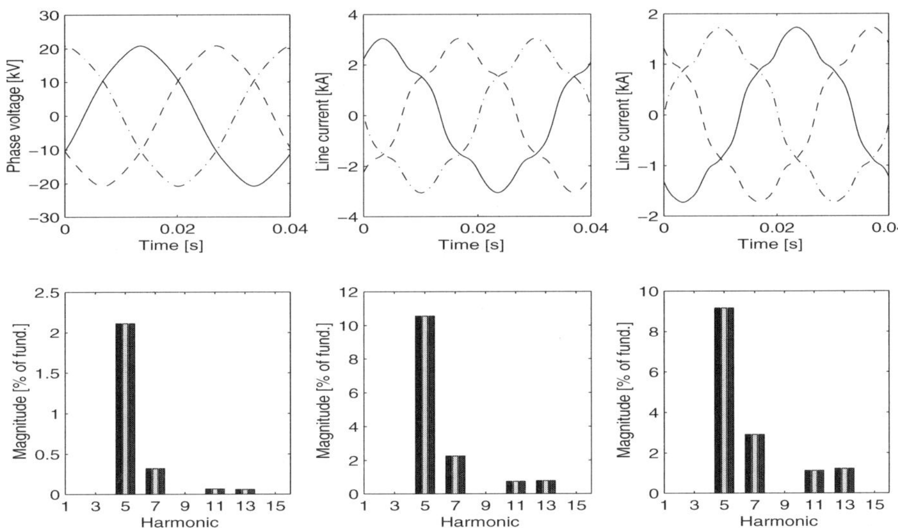

Figure 13.12 Three-phase voltages at the 24 kV bus, and three-phase currents in the capacitor and TCR, respectively. Top half shows the information in time domain form and bottom half shows the respective harmonic spectra

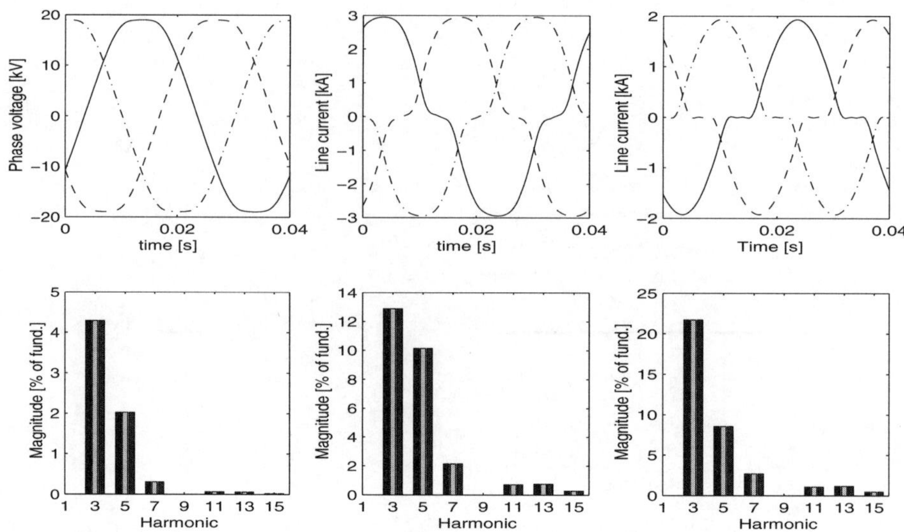

Figure 13.13 Case (ii): Voltage at the 24 kV bus, currents in the capacitors and TCR, respectively

phase current are injected. This, in turn, will give rise to more than 4% third harmonic voltage distortion at node 2.

(iii) Delta-connected TCR operating under unbalanced voltage magnitude excitation and a firing angle of $\alpha = 110°$. The following voltage magnitudes are used in this example: $V_a = 1$ p.u., $V_b = 0.98$ p.u. and $V_c = 1.1$ p.u. The results in Figure 13.14 show that unbalanced voltage excitation causes the harmonic currents generated by the TCR to be unbalanced, with the following effects: a significant amount of the third harmonic appears in the voltage waveform at node 2, and all harmonic voltages are unbalanced. The reason that third harmonic voltages appear in node 2 is that a certain amount of third harmonic current will escape the delta connection towards the network, because of the unequal third harmonic currents being generated by the TCR.

(iv) Delta-connected TCR with unbalanced firing angle operation. The following firing angles are used in this example: $\alpha_a = 110°$, $\alpha_b = 100°$ and $\alpha_c = 112°$. The results in Figure 13.15 show that for the values of firing angle selected, the harmonic currents generated by the TCR produce important imbalances. In particular, the third harmonic currents are very unbalanced giving rise to equally unbalanced third harmonic voltages in node 2. Once again, third harmonic voltages appear in node 2 because a certain amount of third harmonic current will escape the delta connection towards the network, because of the unequal third harmonic currents being generated by the TCR.

The following MATLAB™ program was used to generate the results of this example:

```
clear all
% Transformer
  MVAT=100;
  Z   =10;
  X_R =10;
  KV  =24;
% Capacitor bank
  MVARC=80;
% Reactor bank of the TCR
```

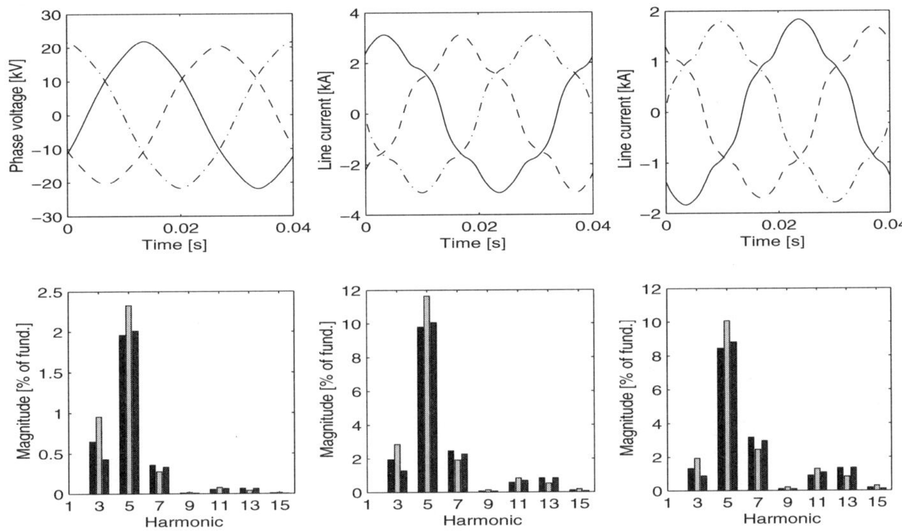

Figure 13.14 Case (iii): Voltage at the 24 kV bus, currents in the capacitors and TCR, respectively

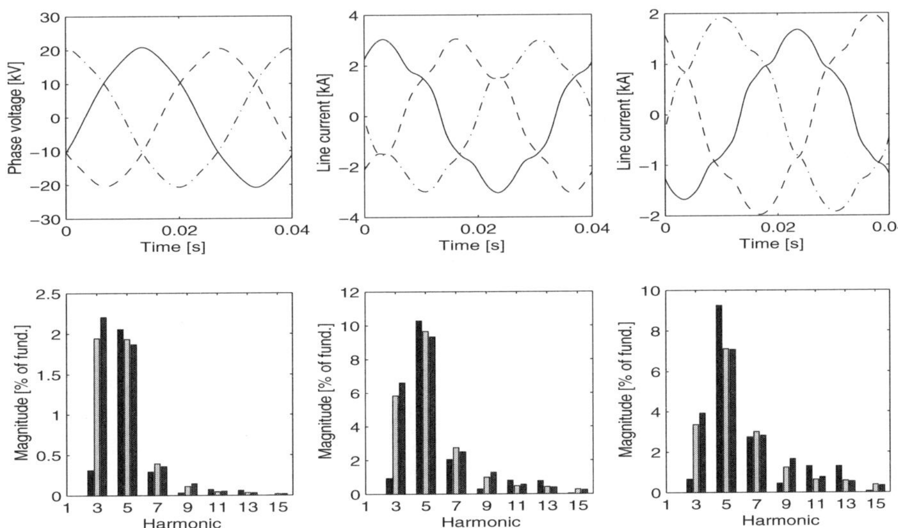

Figure 13.15 Case (iv): Voltage at the 24 kV bus, currents in the capacitors and TCR, respectively

```
  MVARR=80;
  alpha_a=110*pi/180;
  alpha_b=110*pi/180;
  alpha_c=110*pi/180;
% Impedances
  Xtrans=Z*KV^2/(100*MVAT);
  Rtrans=Xtrans/X_R;
  Xcap  =KV^2/MVARC;
  Xrea  =KV^2/MVARR;
% Three-phase balanced voltage source at 24 kV
h=15;
Vap=KV*sqrt(2/3); Vbp=KV*sqrt(2/3); Vcp=KV*sqrt(2/3);
[Va1,Vb1,Vc1]=source_h(Vap,Vbp,Vcp,0,-120,120,h);
V1=[Va1;Vb1;Vc1];
% Initial three-phase balanced voltage in the SVC
  [Va2,Vb2,Vc2]=source_h(Vap,Vbp,Vcp,0-30,-120-30,120-30,h);
  V20=[Va2;Vb2;Vc2];
  [Yt11,Yt12,Yt21,Yt22]=transf_bank(Rtrans,Xtrans,h,'Ws-D');
  Ycap=inv(form_Zm(0,-Xcap,h));
  Ycap_wye=[Ycap Ycap*0 Ycap*0 Ycap*0 Ycap Ycap*0 Ycap*0 Ycap*0 Ycap];
error=1; iter=0;
while error>0.0001
  Ytcr_delta=calc_TCR_ThreePhase(Va2,Vb2,Vc2,alpha_a,alpha_b,alpha_c,1,
                                  Xrea,h,0);
  Ysvc=Ycap_wye+Ytcr_delta;
  V2=-inv(Yt22+Ysvc)*Yt21*V1;
  error=norm(V2-V20)
  iter=iter+1
  Va2=V2(1:2*h+1);
  Vb2=V2(2*h+2:4*h+2);
  Vc2=V2(4*h+3:6*h+3);
  V20=V2;
end
```

Example 13-3: The reduced transmission system shown in Figure 13.16 is a part of the 220 kV New Zealand power system for which complete harmonic information exists in the open literature [11]. The system exhibits a parallel resonance between the fourth and fifth harmonic frequencies, i.e. 200–250 Hz at Tiwai [12].

The transmission lines have been modelled with full frequency dependence, geometric imbalances and long-line effects. Generators, transformers and loads have been assumed to behave linearly.

By way of example, Figure 13.17 gives the driving point impedance for phase a, as seen from Tiwai.

At Tiwai a delta-connected three-phase TCR is assumed to be connected. The TCR reactance is $X_{\text{react}} = 0.2339$ p.u. and the firing angle $\alpha = 110°$. As shown in Figure 13.4, TCRs will inject maximum fifth harmonic current when the firing angle is close to $110°$. In the case being analysed, this current excites the parallel resonance that exist near the fifth harmonic frequency, giving rise to a badly distorted voltage waveform at Tiwai. Figure 13.18 shows the voltage waveforms for the three phases. Large harmonic voltage imbalances are shown in this result where the percentage of the fifth harmonic reaches just over 8% for phase a, 7.5% for phase b and 10.8% for phase c, (Table 13.1).

Under perfectly balanced conditions the fifth and seventh TCR harmonic currents can be

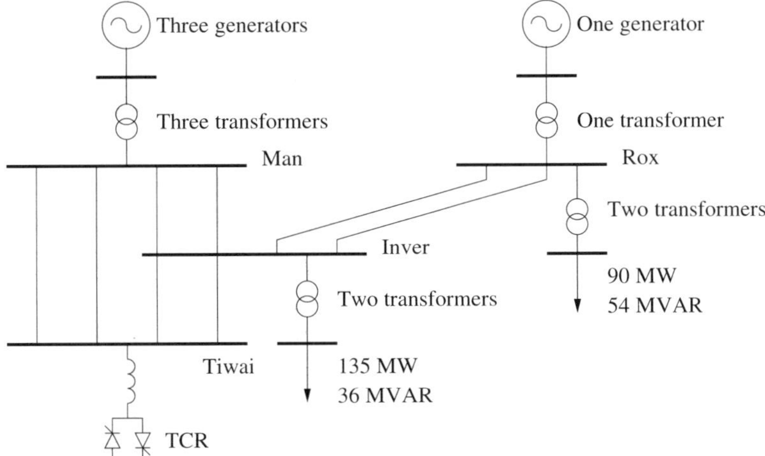

Figure 13.16 South Island reduced test system

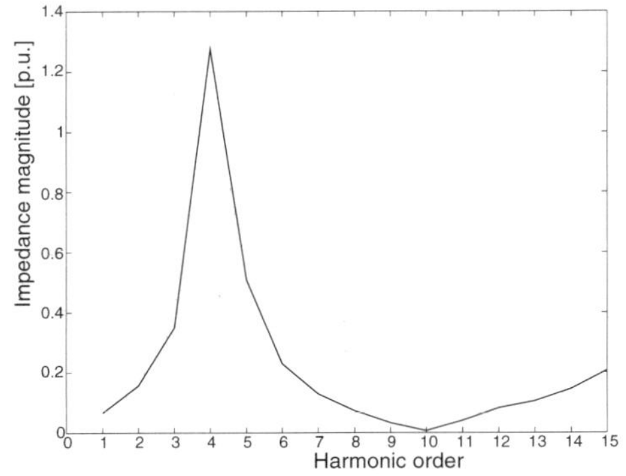

Figure 13.17 Driving point impedance at Tiwai (phase a)

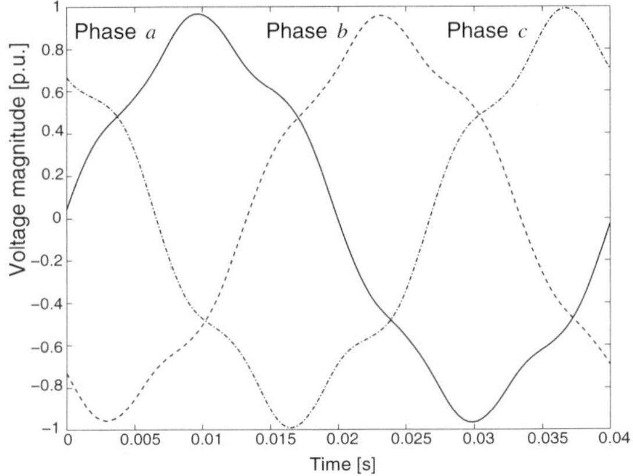

Figure 13.18 Harmonic voltage waveform at Tiwai for the case of a six-pulse TCR

Table 13.1 Harmonic voltage magnitude in % of the fundamental for the case of a six-pulse TCR

h	Phase a	Phase b	Phase c
3	0.0614	0.2937	0.3146
5	8.0397	7.5128	10.8665
7	0.2762	0.3978	0.3066
9	0.0190	0.0159	0.0257
11	0.1226	0.1017	0.1236
13	0.1951	0.2103	0.1898
15	0.0264	0.0212	0.0478

removed effectively from the high-voltage side of the network by employing two identical half-sized TCR units and a three-phase transformer with two secondary windings. The primary winding is grounded star connected. One secondary winding is star connected whilst the other secondary winding is delta connected. One half-sized unit is connected to each secondary winding. In this situation the fifth and seventh harmonic currents of both TCRs will be equal in magnitude and in phase opposition at the primary side of the transformer.

However, the effectiveness of the arrangement may diminish under the presence of network or TCR asymmetries, because even a reduced amount of fifth or seventh harmonic currents escaping towards the primary side of the transformer may be sufficient to distort the voltage waveform if conditions for harmonic voltage magnification exist.

Example 13-4: The case of an electronically compensated transmission system is presented below. It is a 500 kV, 500 km, single-circuit line that carries maximum and minimum powers of 750 and 375 MW, respectively. The transmission line is of flat configuration and has four bundled conductors per phase, as shown in Figure 10.3. It contains a single compensation point located half-way along the line. This compensating station consists of a series capacitor of 70 Ω reactance to provide 50% series compensation and a shunt inductor of 950 Ω reactance to provide 50% shunt compensation. The series capacitor is fixed and will remain connected regardless of the loading condition.

(i) From a three-phase load flow solution it can be verified that the case of minimum power transmission, with a reasonably flat voltage profile, can be achieved with 50% shunt and 50% series compensation. In this situation the TCR will produce no harmonic current distortion. The fully loaded condition is achieved with 25% shunt compensation and 50% series compensation. The load flow solution at the fundamental frequency determines the amount of shunt compensation which is obtained with a 120° delay angle in the TCR firing control. This operating condition will give rise to harmonic currents being generated by the TCR. Figure 13.19 shows the resulting harmonic currents just after the TCR delta connection. Filtering equipment is normally provided for the 5th, 7th, 11th and 13th harmonics and they normally do not reach the high-voltage side of the transmission system. However, no filters are normally connected for the third and ninth harmonic currents because it is hoped that these will be confined to the delta connection of the TCRs. However, owing to geometric transmission imbalances the amount of third harmonic current flowing towards the TCR transformer is 1.5% in phase a.

(ii) Next, the TCR is operated in an unbalanced manner in order to restore geometric balance at the point of compensation. A three-phase load flow solution gives the firing angle

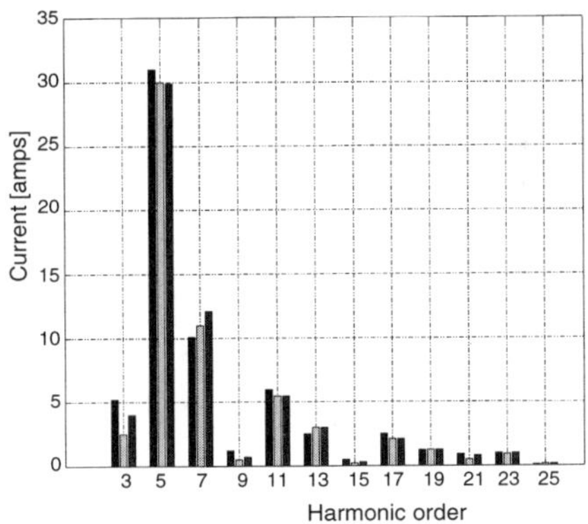

Figure 13.19 Harmonic currents drawn by the TCR operating under unbalanced voltage excitation conditions

delays that will be necessary to keep the voltage level at the values shown in Table 13.2, whilst Figure 13.20 shows the harmonic currents generated by the TCR under such operating conditions. It can be seen that harmonic currents in the three phases are very unbalanced, but perhaps more importantly, the amount of third harmonic has increased to an unacceptably high level, i.e. 11% in phase b.

Table 13.2 Load flow voltage solution for the case of unbalanced TCR firing angles

Node	Phase a	Phase b	Phase c
Source end	$1.020\angle 0.00°$	$1.020\angle 240.00°$	$1.020\angle 120.00°$
Sending end of compensation	$1.038\angle -5.91°$	$1.038\angle 236.49°$	$1.038\angle 116.96°$
Receiving end of compensation	$0.993\angle -1.55°$	$0.993\angle 240.48°$	$0.991\angle 120.53°$
Load end	$0.989\angle -6.43°$	$1.011\angle 236.10°$	$1.018\angle 116.98°$

13.4 Summary

SVC models for the analysis of harmonics have been presented in this chapter. The SVC is formed by the parallel combination of a TCR with a bank of capacitors. A complete three-phase TCR model which is based on the use of harmonic switching functions has been presented. This harmonic model is completely general and can be interpreted as a harmonic admittance equivalent. It is suitable for direct incorporation into the multi-phase harmonic domain frame of reference where it combines easily with the frequency-dependent admittances of the transmission network, and with other linearised components such as saturated transformers, rotating machinery, electric arcs and power converters.

The use of harmonic switching functions has proved to be a simple and yet powerful tool in the solution of TCR harmonics. All operations are conducted entirely in the frequency domain, and the use of alternate time domain and frequency domain representations is no longer required. This approach leads to efficient iterative solutions of power networks containing TCRs. The general operating condition corresponds to a case when the TCR is connected

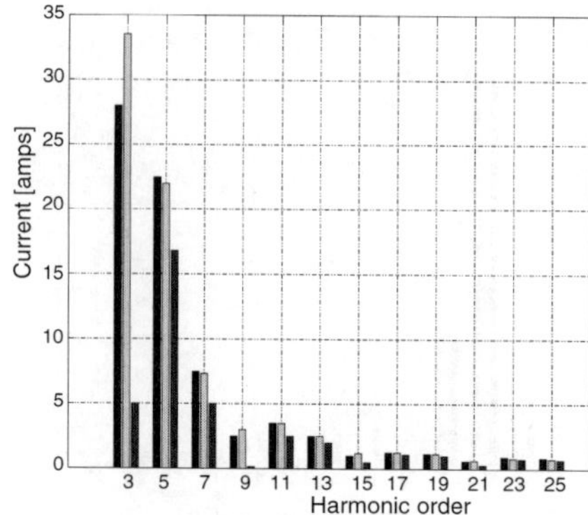

Figure 13.20 Harmonic currents drawn by a TCR operating under unbalanced firing angle conditions

to a weak point of the network, where the switching instants of the thyristor valves are a function of network nodal voltage conditions. This applies to both thyristor turn-on and turn-off instants. In such a situation, the voltage zero-crossing points need to be calculated with good accuracy at each iterative step, and these values are used to update the TCR switching function. Newton–Raphson techniques are used to determine the zero-crossing points.

The impact of the three-phase TCR connection on harmonic generation was investigated by using numeric examples. Voltage excitation imbalances and thyristor firing angle imbalances are responsible for third harmonic voltages and currents appearing outside the TCR circuit, and such imbalances should be avoided as much as practicable.

13.5 Bibliography

1. T.J.E. Miller, *Reactive Power Control in Electric Systems*, Wiley-Interscience, New York, 1982.
2. N. Mohan, T.M. Undeland, W.P. Robbins, *Power Electronics: Converters, Applications and Design*, John Wiley & Sons, New York, 1989.
3. M.H. Rashid, *Power Electronics: Circuits, Devices, and Applications*, Prentice–Hall, Englewood Cliffs, NJ, 1993.
4. N.H. Hingorani, "Flexible AC Transmission", *IEEE Spectrum*, April 1993, pp. 40–45.
5. L.J. Bohmann, R.H. Lasseter, "Harmonic Interactions in Thyristor Controlled Reactor Circuits", *IEEE Transactions on Power Delivery*, Vol. 4, No. 3, July 1989, pp. 1919–1926.
6. J.J. Rico, E. Acha, T.J.E. Miller, "Harmonic Domain Modelling of Three Phase Thyristor-Controlled Reactors by Means of Switching Vectors and Discrete Convolutions", *IEEE Transactions on Power Delivery*, Vol. 11, No. 3, July 1996, pp. 1678–1684.
7. W. Xu, J.R. Marti, H.W. Dommel, "Harmonic Analysis of Systems with Static Compensators", *IEEE Transactions on Power Systems*, Vol. 6, No. 1, February 1991, pp. 183–190.
8. W. Xu, J.R. Marti, H.W. Dommel, "A Multiphase Harmonic Load Flow Solution Technique", *IEEE Transactions on Power Systems*, Vol. 6, No. 1, February 1991, pp. 174–182.
9. L. Gyugyi, "Power Electronics in Electric Utilities: Static Var Compensators", *Proceeding of the IEEE*, Vol. 76, April 1988, pp. 483–494.

10. R. Yacamini, J.W. Resende, "Thyristor Controlled Reactors as Harmonic Sources in HVDC Converter Stations and AC Systems", *IEE Proceedings*, Part B, Vol. 133, July 1986, pp. 263–269.
11. E. Acha, J. Arrillaga, A. Medina, A. Semlyen, "General Frame of Reference for Analysis of Harmonic Distortion in Systems with Multiple Transformer Nonlinearities", *IEE Proceedings*, Part C, Vol. 136, No. 5, September 1989, pp. 271–278.
12. G.H. Robinson, "Harmonic Phenomena Associated with Benmore-Haywards HVDC Transmission Scheme", *New Zealand Engineering*, Vol. 21, No. 1, January 1966, pp. 16–29.

14

Thyristor-controlled Series Compensator

14.1 Introduction

The use of series capacitors in long-distance transmission to enhance power transfer capabilities has been a popular resource among power engineers [1]. Very recently, with the availability of high-current, high-power electronic switches the series compensation technology has taken a major step forwards. Thyristor-controlled series compensators (TCSCs) provide active power flow control with very little delay, leading to much improved system damping and stability margins for the compensated transmission circuit.

The TCSC is the electronically controlled counterpart of the long-enduring, mechanically controlled series compensator. A particular case of the TCSC is the advanced series compensator (ASC). It consists of a number of compensating units of small rating connected in tandem as shown in Figure 14.1.

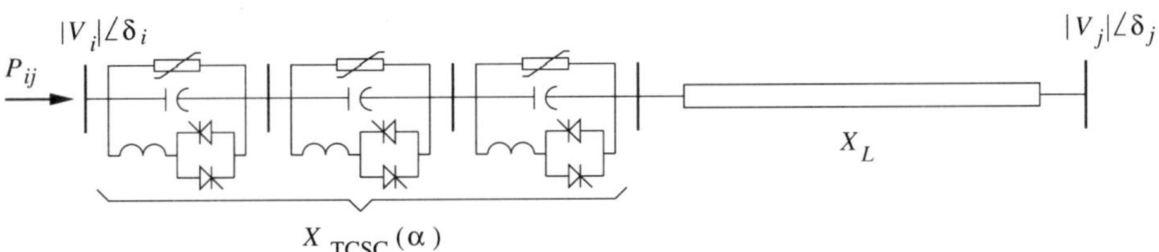

Figure 14.1 TCSC main circuit

Each compensating unit is made up of the parallel combination of a thyristor-controlled reactor (TCR), a fixed bank of capacitors, and a metal–oxide varistor (MOV) to prevent capacitor over-voltages. In this application, the TCR's firing angle, α, is used to provide the TCSC regulating capabilities. With reference to Figure 14.1 the following simplified equation may be used to gain a basic understanding of the role that the TCSC plays in varying the overall reactance of the compensated transmission line, hence affecting active power transmission between nodes i and j:

$$P_{ij} = \frac{|V_i||V_j|}{X_{\text{TCSC}}(\alpha) + X_L} \sin \delta_{ij}$$

The equivalent reactance of the TCSC at the fundamental frequency is given by [4]

$$X_{\mathrm{TCSC}}(\alpha) = -X_C + (X_C + X_{LC})\frac{2(\pi-\alpha)+\sin 2(\pi-\alpha)}{\pi} - \frac{4X_{LC}^2\cos^2(\pi-\alpha)}{X_L}\frac{k\tan k(\pi-\alpha)-\tan(\pi-\alpha)}{\pi} \tag{14.1}$$

where

$$k = \frac{1}{\sqrt{LC}\omega_0}, \qquad X_{LC} = \frac{X_C X_L}{X_C - X_L} \tag{14.2}$$

and X_L and X_C are the rated reactances of the inductor and the capacitor, respectively. This equation accounts for the harmonic interactions between the TCR and the capacitor, giving an exact solution at the fundamental frequency.

Owing to the principle of operation of the TCSC and its series connection with the transmission system, TCSCs are exposed to disturbances resulting from normal operation of the transmission system. Hence, it becomes a matter of great importance to predict possible over-voltages and resonant impedance conditions well in advance.

Figure 14.2 shows the equivalent circuit representation of the basic TCSC scheme where a TCR is placed in parallel with a fixed capacitor, with the TCSC equivalent reactance being a function of the TCR firing angle. TCRs achieve their fundamental frequency operating point at the expense of generating harmonic currents, which are a function of the thyristor's conduction angle. It should be noted that only two TCR operating conditions are harmonic-generation free: when the TCR is conducting fully and when it is not conducting at all.

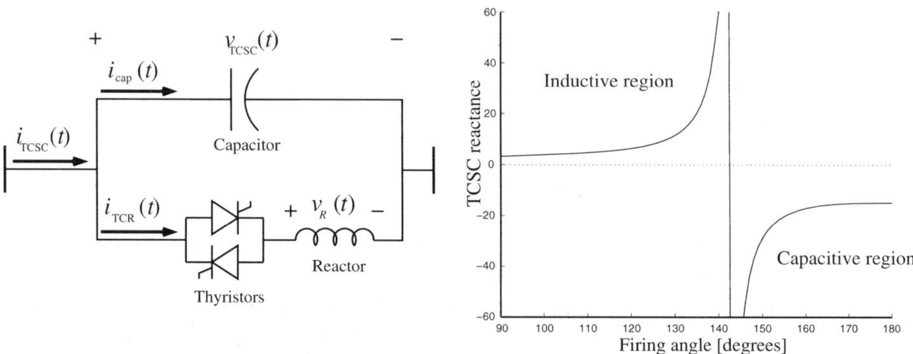

Figure 14.2 TCSC scheme and impedance characteristic at fundamental frequency

In practice, the harmonic currents generated by the TCR are trapped inside the TCSC owing to the low impedance of the capacitor compared to the network equivalent impedance, producing considerable voltage distortion in the TCSC capacitor.

Under normal conditions, the TCSC operates with sinusoidal line currents, as illustrated in Figures 14.3 and 14.4. The former figure refers to the case when the TCSC operates in the inductive region whereas the latter refers to operation in the capacitive region. Inside the

TCSC, a typical TCR current and voltage across the inductor are shown in Figure 14.3, where the conduction period of the TCR current is sufficiently long to enable the TCSC to operate in the inductive region. The current in the capacitive branch and the voltage across the capacitor are both non-sinusoidal.

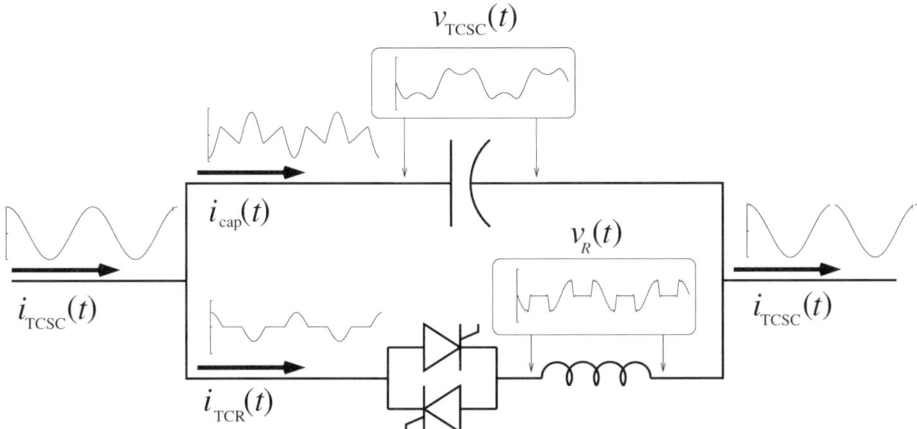

Figure 14.3 TCSC operation in the inductive region

Figure 14.4 refers to the case when the conduction period of the TCR current is decreased sufficiently to enable the TCSC to operate in the capacitive region. The voltage across the inductor for this operating condition is shown in the figure. The voltage across the capacitor is more sinusoidal when the TCSC operates in the capacitive region than when it operates in the inductive region. However, the current through the capacitor is still highly non-sinusoidal and changes in such a way that when combined with the TCR current it produces a largely sinusoidal TCSC line current.

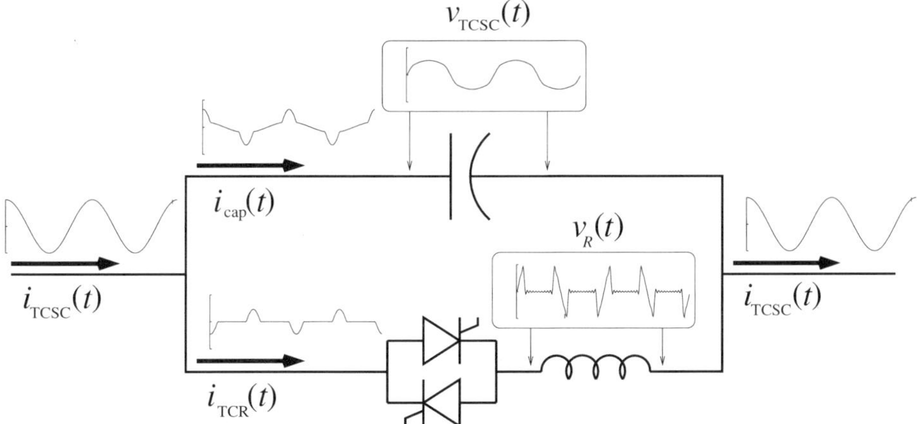

Figure 14.4 TCSC operation in the capacitive region

Measurements taken in the 500 kV Slatt TCSC system [8] support this observation, where for capacitive operation mode, the line current THD had a maximum value of 1.4%. Similarly for the 230 kV Kayenta TCSC system [4], in capacitive operation, the maximum THD voltage in the system was less than 1.5%.

However, when the electrical network undergoes dynamic conditions in the vicinity of the TCSC, its line current shows a slowly decaying transient that can last for up to several cycles [8]. In this situation, the line current loses quarter-wave symmetry and exhibits a rich harmonic spectrum for the duration of the transient period. Such network behaviour is undesirable for it may result in TCSC resonant conditions leading to either capacitor over-voltages or very large currents across the thyristors [8]. Some of the phenomena which may trigger such undesirable TCSC behaviour are faulted conditions in neighbouring locations, switching of heavy loads, line and transformer energisation. These are very practical operating conditions that very much resemble a quasi-steady-state, non-sinusoidal behaviour, which may be investigated by studying the TCSC response using harmonic domain techniques.

14.2 TCSC Operation

The TCSC has three basic modes of operation: blocked thyristors (zero conduction), bypassed thyristors (full conduction) and vernier mode operation (partial conduction). In vernier mode, and depending on firing angle value, the fundamental frequency operating point lies in either the capacitive or the inductive region.

The TCSC steady-state response may be calculated by using both time domain and frequency domain solution techniques. The latter approach, however, is only accurate if the solution incorporates the fundamental and harmonic phasors of the various quantities involved. The frame of reference afforded by the harmonic domain, where all the harmonics and cross-couplings between harmonics are explicitly shown, is used below to represent the TCSC equivalent impedance.

14.2.1 TCSC model

The TCSC is well described by the parallel combination of the TCR and capacitor admittances,

$$\mathbf{Y}_{\mathrm{TCSC}} = \mathbf{Y}_C + \mathbf{Y}_{\mathrm{TCR}} \tag{14.3}$$

where $\mathbf{Y}_C = C\mathbf{D}(\mathrm{j}h\omega_0)$ is the admittance matrix of the series capacitor and C is its capacitance.

The MATLAB™ function used to calculate the TCSC equivalent admittance is:

```
function Ytcsc=calc_TCSC(V,alpha,h,w,L,C)
  Xc=1/(w*C);
  Yc=inv(form_Zm(0,-Xc,h));
  Ytcr=calc_TCR(V,alpha,h,w,L);
  Ytcsc=Yc+Ytcr;
```

The transfer admittance matrix $\mathbf{Y}_{\mathrm{TCSC}}$ represents a complete model for the TCSC at steady state, where all the harmonics and cross-couplings between harmonics are explicitly represented.

It should be noted that equation (14.3) represents a single-phase TCSC. However, owing to the de coupled nature of the three phases, the mathematical model for the three-phase TCSC is straightforward to derive:

$$\mathbf{Y}^{abc}_{\text{TCSC}} = \begin{bmatrix} \mathbf{Y}^{a}_{\text{TCSC}} & 0 & 0 \\ 0 & \mathbf{Y}^{b}_{\text{TCSC}} & 0 \\ 0 & 0 & \mathbf{Y}^{c}_{\text{TCSC}} \end{bmatrix} \tag{14.4}$$

Example 14-1: Figure 14.5 shows a power system which includes a TCSC and a TCR. The relevant parameter information is given in the diagram, per unit. The system of equations that represent the power network is given by

$$\begin{bmatrix} \mathbf{I}_1 \\ 0 \\ 0 \\ 0 \end{bmatrix} = \begin{bmatrix} \mathbf{Y}_{11} & -\mathbf{Y}_{\text{TCSC}} & -\mathbf{Y}_{L2} & 0 \\ -\mathbf{Y}_{\text{TCSC}} & \mathbf{Y}_{22} & 0 & -\mathbf{Y}_{L1} \\ -\mathbf{Y}_{L2} & 0 & \mathbf{Y}_{33} & -\mathbf{Y}_{L3} \\ 0 & -\mathbf{Y}_{L1} & -\mathbf{Y}_{L3} & \mathbf{Y}_{44} \end{bmatrix} \begin{bmatrix} \mathbf{V}_1 \\ \mathbf{V}_2 \\ \mathbf{V}_3 \\ \mathbf{V}_4 \end{bmatrix}$$

where

$$\begin{aligned} \mathbf{Y}_{11} &= \mathbf{Y}_{\text{TCR}} + \mathbf{Y}_{L2} \\ \mathbf{Y}_{22} &= \mathbf{Y}_{\text{TCSC}} + \mathbf{Y}_{L1} \\ \mathbf{Y}_{33} &= \mathbf{Y}_{L2} + \mathbf{Y}_{L3} + \mathbf{Y}_{\text{load2}} \\ \mathbf{Y}_{44} &= \mathbf{Y}_{L1} + \mathbf{Y}_{L3} + \mathbf{Y}_{\text{load1}} + \mathbf{Y}_{\text{TCR}} \end{aligned}$$

$\mathbf{I}_1$ is the current injection in node 1, $\mathbf{Y}_{L1}$ is the admittance between nodes 2 and 4, $\mathbf{Y}_{L2}$ is the admittance between nodes 1 and 3, $\mathbf{Y}_{L3}$ is the admittance between nodes 3 and 4, $\mathbf{Y}_{\text{load1}}$ is the admittance of the load in node 4, and $\mathbf{Y}_{\text{load2}}$ is the admittance of the load in node 3.

The solution for the unknown voltages is given by

$$\begin{bmatrix} \mathbf{V}_2 \\ \mathbf{V}_3 \\ \mathbf{V}_4 \end{bmatrix} = \begin{bmatrix} \mathbf{Y}_{22} & 0 & -\mathbf{Y}_{L1} \\ 0 & \mathbf{Y}_{33} & -\mathbf{Y}_{L3} \\ -\mathbf{Y}_{L1} & -\mathbf{Y}_{L3} & \mathbf{Y}_{44} \end{bmatrix}^{-1} \begin{bmatrix} \mathbf{Y}_{\text{TCSC}} \\ \mathbf{Y}_{L2} \\ 0 \end{bmatrix} \mathbf{V}_1 \tag{14.5}$$

Equation (14.5) is solved by iteration because the admittances $\mathbf{Y}_{\text{TCSC}}$ and $\mathbf{Y}_{\text{TCR}}$ are a function of the zero crossings of voltages $\mathbf{V}_1 - \mathbf{V}_2$ and $\mathbf{V}_3$, respectively.

The initial conditions are taken to be the nodal voltages given by the fundamental frequency solution of (14.5).

The TCSC impedance at the fundamental frequency, calculated using (14.1), is $X_{\text{TCSC}} = -\text{j}0.0288$ p.u. For the TCR, using (13.2), $X_{\text{TCR}} = \text{j}0.2558$ p.u. With these impedances, the nodal voltages solution at fundamental frequency are: $V_{2\text{f}} = 1.0869\angle 0.3399°$, $V_{3\text{f}} = 0.9571\angle 0.3636°$ and $V_{4\text{f}} = 0.9360\angle - 0.2906°$.

Table 14.1 gives the harmonic nodal voltages as calculated by the MATLAB ™ program given below. It can be seen bacause of the harmonic distortion present in the system, the fundamental voltages are slightly different from those obtained with no harmonics considered. As expected, the nodal voltage at node 2 contains very little harmonic distortion because this is the receiving node of the TCSC. The SVC is connected at node 4 and it injects maximum third harmonic current since the conduction angle is $\alpha = 120°$. In turn this current injection will give rise to 4.62% third harmonic voltage.

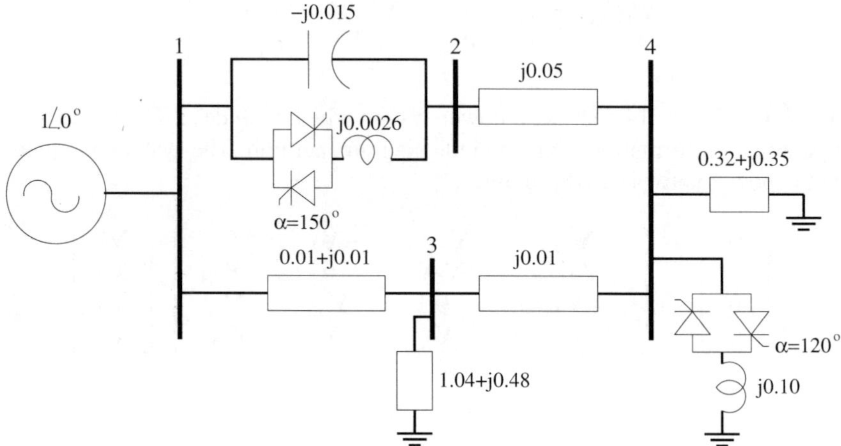

Figure 14.5 Power systems with TCSC and TCR

Table 14.1 Nodal voltages harmonic content

h	V_2	V_3	V_4
1	$1.0900\angle 0.3623°$	$0.9591\angle 0.2735°$	$0.9398\angle -0.3651°$
3	$0.0140\angle 10.0391°$	$0.0238\angle 162.2561°$	$0.0462\angle 171.4542°$
5	$0.0043\angle 21.9919°$	$0.0091\angle -10.9788°$	$0.0181\angle -5.2109°$
7	$0.0012\angle 30.1546°$	$0.0038\angle -7.0894°$	$0.0077\angle -2.9154°$
9	$0.0000\angle -58.2294°$	$0.0073\angle 171.3665°$	$0.0147\angle 174.6308°$
11	$0.0003\angle 223.4329°$	$0.0036\angle -9.6731°$	$0.0073\angle -6.9948°$
13	$0.0002\angle 232.2763°$	$0.0019\angle -7.5381°$	$0.0038\angle -5.2681°$
15	$0.0000\angle 11.1989°$	$0.0043\angle 171.4474°$	$0.0087\angle 173.4167°$
17	$0.0001\angle 74.6021°$	$0.0024\angle -9.4481°$	$0.0049\angle -7.7092°$
19	$0.0001\angle 81.2239°$	$0.0012\angle -7.4114°$	$0.0024\angle -5.8548°$
21	$0.0000\angle -14.6850°$	$0.0031\angle 170.3276°$	$0.0062\angle 171.7364°$
23	$0.0000\angle -87.6074°$	$0.0019\angle -11.0363°$	$0.0038\angle -9.7496°$
25	$0.0000\angle -77.2906°$	$0.0008\angle -9.1563°$	$0.0016\angle -7.9723°$
27	$0.0000\angle -27.1545°$	$0.0024\angle 168.8058°$	$0.0048\angle 169.9023°$
29	$0.0000\angle 124.5027°$	$0.0017\angle -12.2076°$	$0.0034\angle -11.1866°$
THD%	1.3458	2.9175	5.8261

The steady-state periodic solution of the circuit in Figure 14.5 is given in the form of voltage and current waveforms in Figure 14.6. The nodal voltages in nodes 1 and 2 are sinusoidal but the nodal voltage in node 4 is clearly non-sinusoidal, and so is the voltage across the TCSC, i.e. voltage $V_{1,2}$. All currents in this circuit are non-sinusoidal except for the load currents. Of particular interest are the currents through the capacitive and the inductive branches of the TCSC. In this case the line current at the terminals of the TCSC are non-sinusoidal but it is likely that these are due to the currents injected by the TCR as opposed to the TCSC. The harmonic currents injected by the TCR travel to the supply (sink) via transmission lines 4–3 and 3–1.

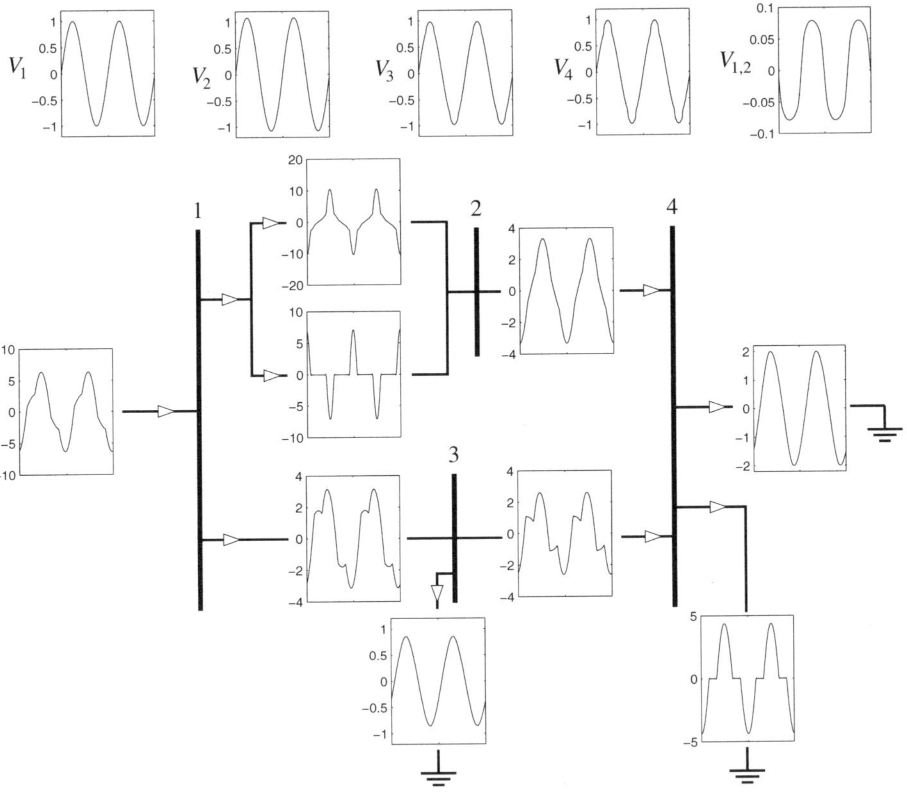

Figure 14.6 Current and voltage waveform response of the power system

The MATLAB™ program used to solve this example is given below:

```
clear all
h      =30;
C      =1/0.015;                          % TCSC
L      =0.0026;
alpha =150*pi/180;
alpha2=120*pi/180;                        % TCR
Ltcr  =0.10;
RL1=0;      XL1=0.05;                     % Transmission system
RL2=0.01; XL2=0.01;
```

```
RL3=0;    XL3=0.01;
Rload1=0.32; Xload1=0.35;                  % Loads
Rload2=1.04; Xload2=0.48;
% ---------------------------------- initial conditions
V1=zeros(2*h+1,1); V2=zeros(2*h+1,1);
V3=zeros(2*h+1,1); V4=zeros(2*h+1,1);
O=zeros(2*h+1,2*h+1);
V1f=1;              V2f=1.0869+i*0.0064;
V3f=0.9571+i*0.0061; V4f=0.9360-i*0.0047;
V1(h)= i*V1f'/2; V1(h+2)= -i*V1f/2;
V2(h)= i*V2f'/2; V2(h+2)= -i*V2f/2;
V3(h)= i*V3f'/2; V3(h+2)= -i*V3f/2;
V4(h)= i*V4f'/2; V4(h+2)= -i*V4f/2;
V2340=[V2;V3;V4];
% ---------------------------------- admittance matrices
YL1   =inv(form_Zm(RL1,XL1,h));
YL2   =inv(form_Zm(RL2,XL2,h));
YL3   =inv(form_Zm(RL3,XL3,h));
Yload1=inv(form_Zm(Rload1,Xload1,h));
Yload2=inv(form_Zm(Rload2,Xload2,h));
Ytcsc=calc_TCSC(V1-V2,alpha,h,1,L,C);
Ytcr =calc_TCR(V4,alpha2,h,1,Ltcr);
error=1;
%---------------------------------- iterative process
while error>0.0001
  Y21=[-Ytcsc
       -YL2
          O];
  Y22=[ Ytcsc+YL1             O                 -YL1
                O YL2+YL3+Yload2                -YL3
            -YL1            -YL3 YL1+YL3+Yload1+Ytcr];
  V234=-inv(Y22)*Y21*V1;
  V2=V234(1:2*h+1);
  V3=V234(2*h+2:4*h+2);
  V4=V234(4*h+3:6*h+3);
  Ytcsc=calc_TCSC(V1-V2,alpha,h,1,L,C);
  Ytcr =calc_TCR(V4,alpha2,h,1,Ltcr);
  error=norm(V234-V2340)
  V2340=V234;
end
```

14.3 TCSC Impedance Response

From equation (14.3) the impedance matrix is

$$\mathbf{Z}_{\mathrm{TCSC}} = \mathbf{Y}_{\mathrm{TCSC}}^{-1} \tag{14.6}$$

and for a given current flowing through the TCSC, the voltage across it is given by

$$\Delta\mathbf{V}_{\mathrm{TCSC}} = \mathbf{Z}_{\mathrm{TCSC}}\mathbf{I}_{\mathrm{TCSC}} \tag{14.7}$$

and, in an expanded form,

$$\begin{bmatrix} \vdots \\ \Delta V_{-m} \\ \vdots \\ \Delta V_m \\ \vdots \end{bmatrix} = \begin{bmatrix} \ddots & & & & \\ & X_{-m,-n} & \cdots & X_{-m,n} & \\ & \vdots & \ddots & \vdots & \\ & X_{m,-n} & \cdots & X_{m,n} & \\ & & & & \ddots \end{bmatrix} \begin{bmatrix} \vdots \\ I_{-n} \\ \vdots \\ I_n \\ \vdots \end{bmatrix} \tag{14.8}$$

From equation (14.8) and assuming that the line current is a periodic even signal, in the Fourier sense, $I_{-n} = I_n$, then the harmonic voltage ΔV_m produced by the harmonic line current I_n is given as

$$\Delta V_m = (X_{m,-n} + X_{m,n})\, I_n = X_{\mathrm{TCSC}_{mn}} I_n \tag{14.9}$$

where

$$X_{\mathrm{TCSC}_{mn}} = (X_{m,-n} + X_{m,n}) \tag{14.10}$$

This impedance is a function of both frequency and firing angle. It should be noted that, for a given firing angle value, the analytical solution given by (14.1) would be a particular case of the solution given by (14.10) when $n = m = 1$.

The equivalent reactance of the TCSC installed in the Kayenta substation is calculated with the harmonic domain model. The following parameters are taken for the TCSC plant: $X_L = 2.56\,\Omega$ and $X_C = 15\,\Omega$ at 60 Hz [4]. Any number of harmonics can be included in the calculations but for this example only seven harmonic terms are used.

By way of comparison, the equivalent reactance of the Kayenta TCSC is also calculated using equation (14.1), which is only valid for fundamental frequency operation. This result is plotted in Figure 14.7, alongside the result given by the harmonic domain solution for the case of $n = m = 1$. A good correspondence exists for the full range of firing angle variation, i.e. 90°-180°. The results show the well-known resonant point for this TCSC scheme at around 143^0.

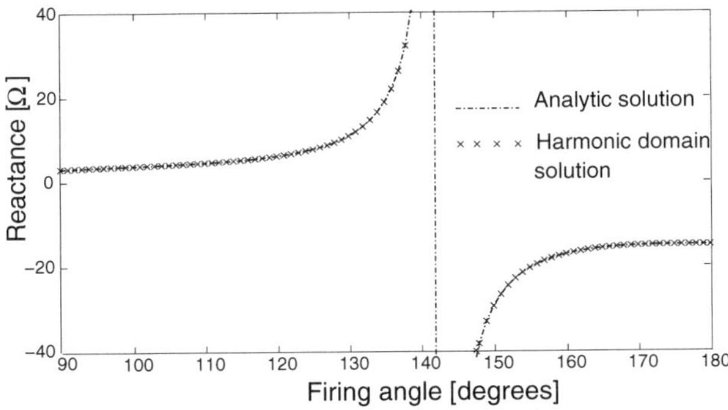

Figure 14.7 Comparison of TCSC equivalent reactances

The TCSC equivalent reactances in Figure 14.8 show that, in addition to the fundamental frequency resonant impedance point, i.e. X_{11}, the TCSC also exhibits resonant impedance points at other harmonic frequencies, e.g. X_{22}, X_{33}, X_{44}. Arguably, this is an expected result; however, only the use of harmonic domain techniques will yield accurate harmonic impedances, since equation (14.1) cannot be used for impedances other than the fundamental. Moreover, harmonic domain techniques also give important information corresponding to resonant impedance points that may exist as a result of cross-couplings between harmonics.

These results show that in the vernier capacitive region associated with the fundamental frequency voltage there are several resonant impedance points that have links to harmonic currents of various frequencies via X_{13}, X_{14}, X_{15}, X_{16}, X_{17}. Similarly, it is shown that the fundamental frequency current has the potential to give rise to excessive third harmonic voltage distortion via X_{13}. In this particular example, however, this is not a problem since this resonant point coincides with the fundamental frequency resonant point and this should already be a non-operational firing angle region.

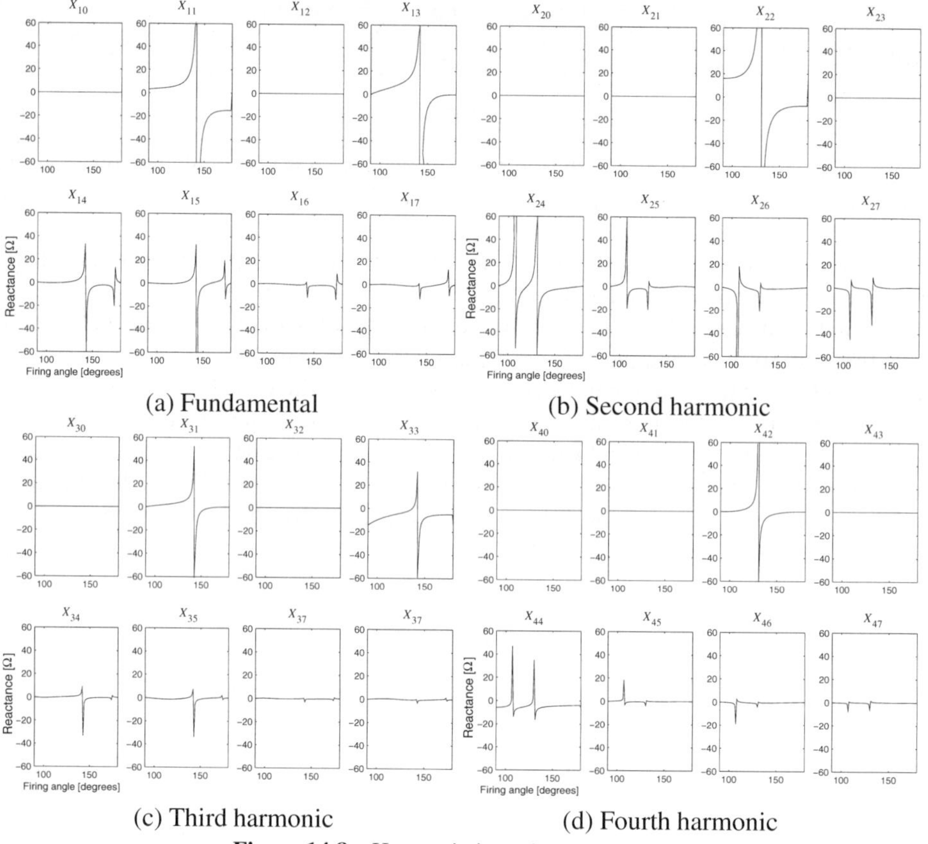

Figure 14.8 Harmonic impedance responses

The vernier inductive region proves to be a problematic region to operate under the presence of odd harmonic currents due to the existence of high resonant impedance points. This is better appreciated in Figure 14.9, where the various TCSC harmonic reactances are shown per unit, and resonant points can be observed at 107°, 130°, 143°, and 173°. The first two resonant points are in the inductive region, with respect to the fundamental resonant point, and are associated with the second and fourth harmonic impedances. If operation were to take place in the vicinity of any of these resonant points and the TCSC line current were to contain even a small amount of second or fourth harmonic, this would be sufficient to trigger voltages and currents exhibiting large harmonic distortion inside the TCSC.

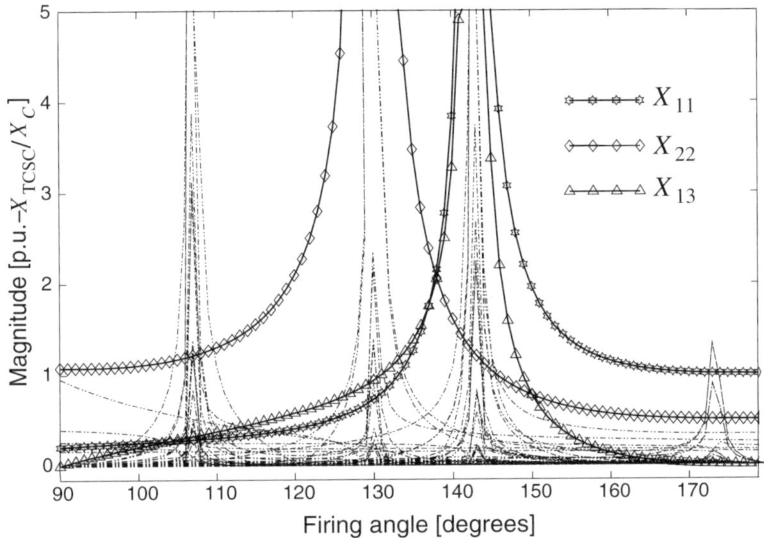

Figure 14.9 Impedance magnitude responses for the first seven harmonics and their interactions for the Kayenta system

The effect of the second resonant point is shown in Figure 14.10 where a firing angle of 129° is used and the line current is (a) purely sinusoidal and (b) contains a 2% second harmonic with respect to the fundamental. Results are presented for the voltage across the capacitor and for the line currents in the TCSC, TCR and capacitor. It can be observed, by comparing the results in Figures 14.10(a) and 14.10(b), that the 2% second harmonic superimposed on the TCSC line current has a very large distorting effect on the voltages and currents inside the TCSC owing to the nearby resonant point existing at 130 °. A similar distortion effect, but in the capacitive vernier region, has been reported [8] where, due to a small offset in the line current, the quarter symmetry of the capacitor voltage was lost, causing the thyristor to conduct uneven currents at each half cycle.

Figure 14.11 shows the harmonic content of the various waveforms in Figure 14.10. By comparing Figures 14.11(a) and 14.11(b) it can be seen that the effect of the second harmonic in the line current is to produce even harmonics in the TCSC voltage, with the odd harmonics remaining almost the same. This effect is due to the fact that the cross-couplings between the line current second harmonic and even harmonics in the TCSC voltage are strongly coupled via reactances X_{22} and X_{42}. In contrast, X_{12} and X_{32} present zero harmonic impedances at this firing angle value, as evidenced by the results in Figure 14.8. Is should be noticed that the harmonic conversion process taking place between the line current second harmonic and the TCSC voltage fourth harmonic via X_{42} generates a fourth harmonic circulating current inside the TCSC. A similar situation arises between the line current second harmonic and TCSC voltage sixth harmonic.

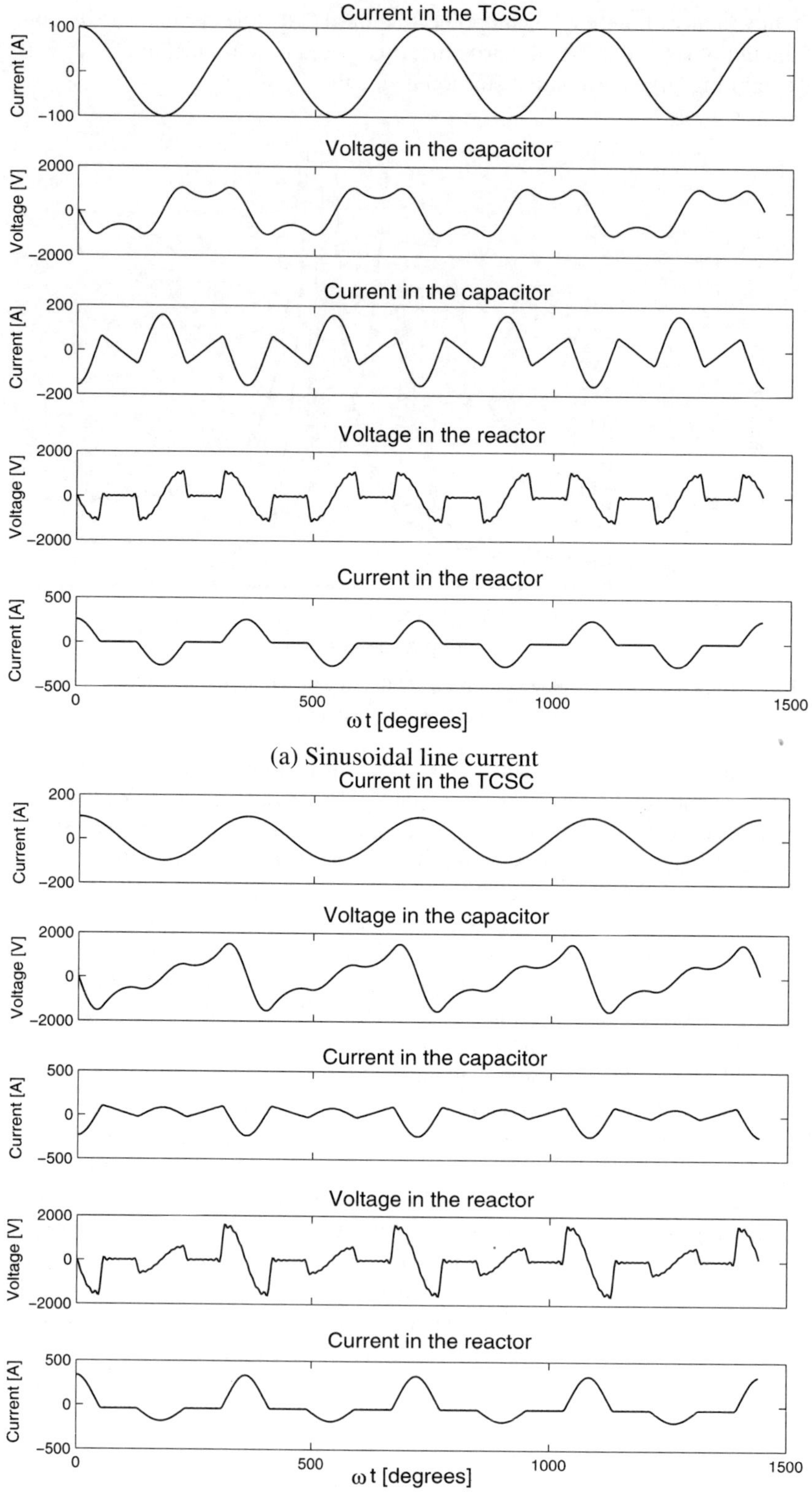

Figure 14.10 TCSC response with firing angle of 129°

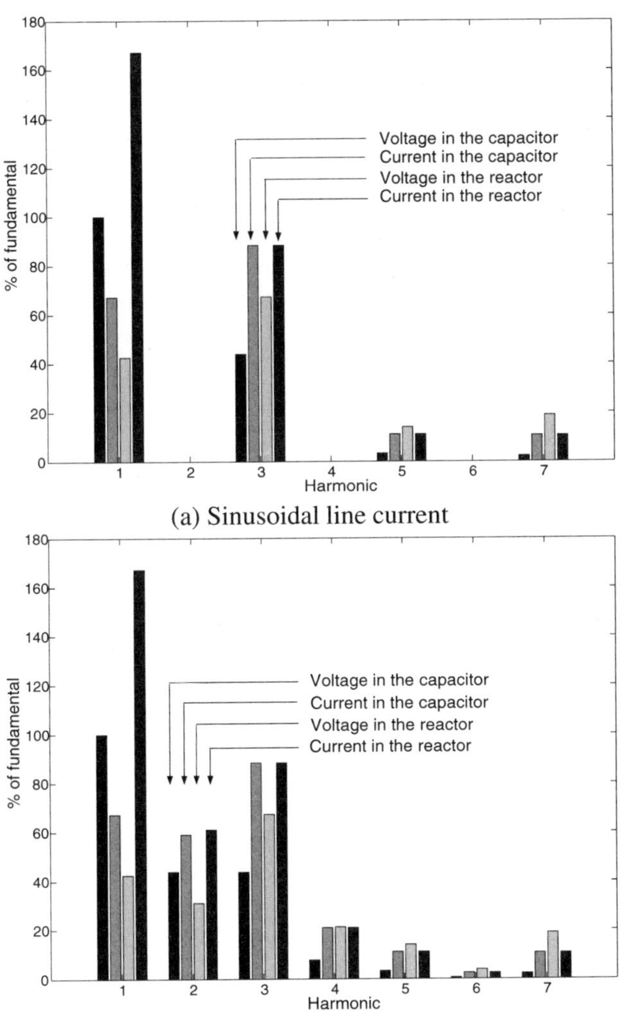

(a) Sinusoidal line current

(b) Line current with 2% of second harmonic

Figure 14.11 Harmonic content of TCSC response with firing angle of 129°

14.4 Summary

A model for the thyristor controlled series compensator has been presented. The model is derived from first principles using harmonic domain theory. The TCSC impedance matrix includes all the harmonics and cross-couplings between harmonics, giving a full representation of the TCSC in the steady state. Hence, the model is ideal for conducting fundamental analysis of TCSCs operating under periodic, non-sinusoidal conditions. The model has been used to study the TCSC circuit installed in the Kayenta substation. In addition to the well-known fundamental frequency resonant point reported in the open literature, many other resonant points, at other harmonic frequencies, are shown to exist. Resonant points due to cross-couplings between harmonics are cases of particular interest because they represent conditions of high impedance, responsible for the TCSC frequency conversion effect. It should be remarked that time domain simulations do not yield the level of detail necessary to study the TCSC's frequency conversion effect since they only provide the aggregated final response. The information given by harmonic domain techniques could be used to achieve better informed TCSC rating specifications and to gather more realistic firing angle constraints for TCSC models used in various power systems application studies, e.g. load flows, optimal power flows, small-signal stability studies.

14.5 Bibliography

1. T.J.E. Miller, *Reactive Power Control in Electric Systems*, Wiley-Interscience, New York, 1982.
2. E. Larsen, C. Bowler, B. Damsky, S. Nilsson, "Benefits of Thyristor Controlled Series Compensator", *International Conference on Large High Voltage Electric Systems (CIGRE)*, paper 14/37/38-04, Paris, 1992.
3. X. Zhou, J. Liang, "Overview of Control Scheme for TCSC to enhance the Stability of Power Systems", *IEE Proceedings General Transmission and Distribution*, Vol. 146, No. 2, March 1999, pp. 125–130.
4. N. Christl, R. Hedin, K. Sadek, P. Lutzelberger, P.E. Krause, S.M. McKenna, A.H. Montoya, D. Togerson, "Advanced Series Compensation (ASC) with Thyristor Controlled Impedance", *International Conference on Large High Voltage Electric Systems (CIGRE)*, paper 14/37/38-05, Paris, 1992.
5. E. Larsen, K. Clark, S.A. Miske, J. Urbanek, "Characteristics and Rating Considerations of Thyristor Controlled Series Compensation", *IEEE Transactions on Power Delivery*, Vol. 9, No. 2, April 1994, pp. 992–1000.
6. C. Fuerte, E. Acha, H. Ambriz, "A Thyristor Controlled Series Compensator Model for the Power Flow Solution of Practical Power Networks", *IEEE Transactions on Power Systems*, paper PE019PRS(05-99).
7. D.N. Kosterev, W.A. Mittelstadt, R.R. Mohler, W.J. Kolodsziej, "An Application Study for Sizing and Rating Controlled and Conventional Series Compensation", *IEEE Transactions on Power Delivery*, Vol. 11, No. 2, April 1996, pp. 1105–1111.
8. S.J. Kinney, M.A. Reynolds, J.F. Hauer, R.J. Piwko, B.L. Damsky, J.D. Eden, "Slatt Thyristor Controlled Series Capacitor System Test Results", *International Conference on Large High Voltage Electric Systems (CIGRE)*, paper 530-02, Tokyo, 1995.
9. S.G. Helbing, G.G. Karady, "Investigations of an Advanced Form of Series Compensator", *IEEE Transactions on Power Delivery*, Vol. 9, No. 2, April 1994, pp. 939–947.

10. S.G. Jalali, R.H. Lasseter, I. Dobson, "Dynamic Response of a Thyristor Controlled Switched Capacitor", *IEEE Transactions on Power Delivery*, Vol. 9, No. 3, July 1994, pp. 1609–1615.

15

Three-phase, Six-pulse Rectifier

15.1 Introduction

The six-pulse rectifier is the most common three-phase converter used in the power range 2–1000 kVA and in the voltage range 220 V to 13.8 kV. As the power level rises, it becomes increasingly attractive to use 12-pulse rectifiers. These rectifiers are made up of two six-pulse units, with one unit connected to the secondary side of a star–star transformer. The second unit is connected to the secondary side of a delta–star transformer. The two rectifiers are connected in series on the DC side in order to achieve the high voltages required in DC transmission. On the primary side, the two rectifiers are connected in parallel in order to achieve cancellation of mainly the fifth and seventh harmonic currents. However, this is only possible if the two star connections in both secondary windings are shifted by 30 ° from each other. This is in addition to the 30° that naturally exists in the delta–star connection of the second unit, see page 388 of reference [1].

The classic literature gives equations (15.1) and (15.2) as the means to determine the order and magnitude of the AC harmonic currents drawn by a q-pulse rectifier [2,3]:

$$h = kq \pm 1 \qquad k = 1, 2, 3, \ldots \tag{15.1}$$

$$\frac{I_h}{I_1} = 1/h \tag{15.2}$$

These equations indicate that a six-pulse rectifier generates the characteristic harmonics of orders 5th, 7th, 11th, 13th, ... The magnitude of the harmonic I_h per unit of the fundamental is the reciprocal of the harmonic order. Such results are based on the following oversimplified assumptions: (1) There is sufficient DC load inductance to produce constant direct current. (2) There is no commutating inductance on the AC side.

Figure 15.1(a) shows the AC current waveform drawn by an idealised six-pulse converter connected to a star–star transformer. Figure 15.1(b) shows a similar result but for a six-pulse converter connected to a delta–star transformer.

The harmonic content of the AC waveform of Figure 15.1(a) can be extracted using Fourier analysis, resulting in the following harmonic series:

$$i(t) = I_1 \left[\sin \omega_0 t - \frac{1}{5} \sin 5\omega_0 t - \frac{1}{7} \sin 7\omega_0 t + \frac{1}{11} \sin 11\omega_0 t + \frac{1}{13} \sin 13\omega_0 t + \cdots \right] \tag{15.3}$$

Likewise, for the AC current waveform given by the solid line in Figure 15.1(b) the harmonic series is

$$\begin{aligned} i(t) &= I_1 \left[\sin(\omega_0 t + 30^\circ) + \frac{1}{5}\sin 5(\omega_0 t + 30^\circ) + \frac{1}{7}\sin 7(\omega_0 t + 30^\circ) \right. \\ &\quad \left. + \frac{1}{11}\sin 11(\omega_0 t + 30^\circ) + \frac{1}{13}\sin 13(\omega_0 t + 30^\circ) + \cdots \right] \end{aligned} \tag{15.4}$$

It should be noted that in Figure 15.1(b), a second waveform appears which is shifted by 30 ° to the right with respect to the solid line waveform. The harmonic series of the dashed line waveform would be

$$i(t) = I_1 \left[\sin\omega_0 t + \frac{1}{5}\sin 5\omega_0 t + \frac{1}{7}\sin 7\omega_0 t + \frac{1}{11}\sin 11\omega_0 t + \frac{1}{13}\sin 13\omega_0 t + \cdots \right] \tag{15.5}$$

By comparing (15.3)–(15.5), it may be seen that the three series contain identical harmonic magnitudes but their phase shifts differ. The harmonic series in (15.4) includes the effect of the delta–star transformer connection. In addition, the harmonic series in (15.5) includes the effect of the 30° phase shift introduced by the fact that the two secondary windings, connected in star, are phase shifted by 30° from each other. This effect is only relevant if the two six-pulse rectifiers are operated in parallel, forming a 12-pulse rectifier. In such a case, the alternating pair of harmonics will cancel out at the common source path, giving the current waveform shown in Figure 15.1(c).

The harmonic content of the AC current waveform of Figure 15.1(c) will be given by the following harmonic series:

$$i(t) = I_1 \left[\sin\omega_0 t + \frac{1}{11}\sin 11\omega_0 t + \frac{1}{13}\sin 13\omega_0 t + \frac{1}{23}\sin 23\omega_0 t + \frac{1}{25}\sin 25\omega_0 t + \cdots \right] \tag{15.6}$$

where the characteristic harmonics in a 12-pulse rectifier are the 11th, 13th, 23th, 25th. It should be noted that this harmonic series may be derived by adding up the harmonic series (15.3) and (15.5).

Figure 15.2(a) shows the effect of a finite commutating reactance on the AC line current drawn by the converter. It causes a gradual rise time and fall time governed by the angle μ. The angle μ is termed the commutation angle or overlap angle. The length of this commutation period is a function of the magnitude of the AC system inductance and the thyristor's firing angle, α, where larger values of the firing angle result in lower commutation angles. Figure 15.2(b) shows the AC current waveform where the instant of the thyristor switching is delayed by a firing angle α.

When the rectifier is working under non-ideal conditions such as unbalanced AC excitation, unbalanced firing angle control, unbalanced AC impedance, as well as pre-existing harmonic distortion in the AC supply, a wide range of non-characteristic harmonics will be generated on both the AC side and the DC side of the rectifier. These non-characteristic harmonics cannot be calculated with the classic equations (15.1) and (15.2), since the key assumption of the classic model is that there is infinite inductance on the DC side of the rectifier and zero commutating reactance on the AC side. The former assumption means that at the end of the

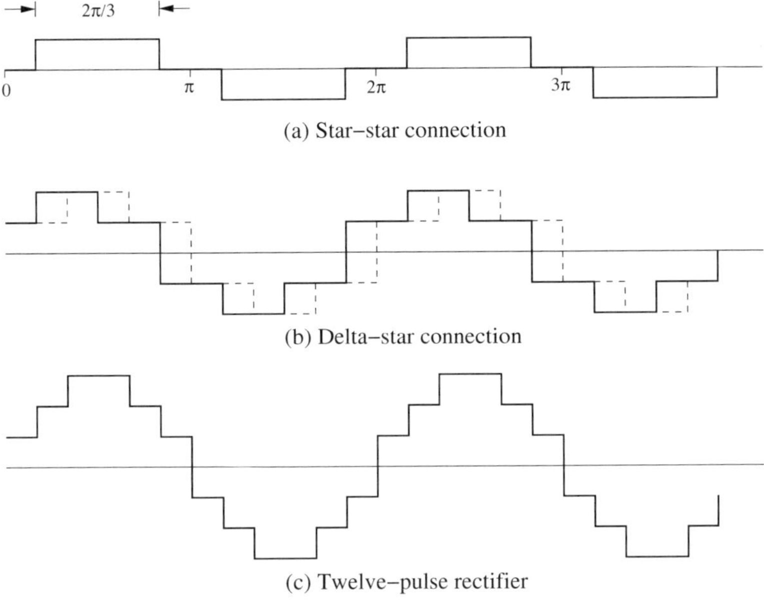

(a) Star–star connection

(b) Delta–star connection

(c) Twelve–pulse rectifier

Figure 15.1 AC current waveforms

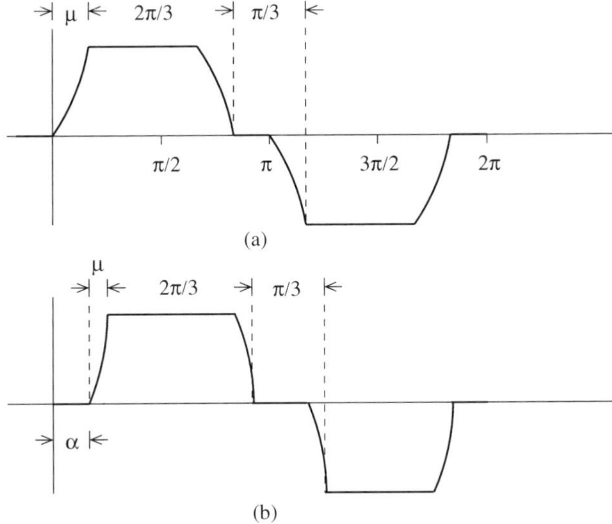

(a)

(b)

Figure 15.2 Effect of the phase control and overlap over the AC current waveform

commutation period the current waveform will be perfectly flat topped, as shown in Figure 15.2. This is an assumption that quite often does not conform to realistic converter design and operation.

Traditionally, the calculation of non-characteristic harmonics has been carried out using general time domain simulations, and very cumbersome, and yet contrived, methods based on analytical expressions [4].

In this chapter, the theory presented in [5] is used to model six-pulse rectifiers in order to calculate non-characteristic harmonics. This method uses switching functions for each phase of the converter which incorporates the conduction states of the thyristors in each phase. The analysis is carried out entirely in the harmonic domain where a harmonic admittance matrix of the rectifier is obtained.

15.2 Rectifier Operation Using Switching Functions

Very comprehensive models have been reported for the HVDC converter in the open literature over the last two decades. Both time [6,7] and frequency domain [8,9] formulations exist. The theory of converter modulation provides a simple alternative for the calculation of converter harmonics [10]. It uses switching functions to relate the voltages and currents that exist on both sides of the converter, the AC side and the DC side.

The six-pulse thyristor bridge, also known as the Graetz bridge, is shown in Figure 15.3(a). It consists of an arrangement of six thyristor valves with the firing sequence indicated by the numbers on the side of the thyristors. In this case, the valves are connected on the AC side to the secondary side of a star-connected, three-phase winding, whereas on the DC side they are connected to a load. To keep the explanation simple at this stage, it will be assumed that no overlap exist. Figure 15.3(b) shows the input AC voltage and the six switching functions, $s_1(t)$, $s_2(t)$, $s_3(t)$, $s_4(t)$, $s_5(t)$ and $s_6(t)$, which represent the conduction periods of the thyristors. A switching function takes a value of one when the thyristor is conducting and a value of zero when it ceases to conduct.

A basic analysis of the switching functions shown in this figure helps to understand their role in the calculation of harmonics. In this case, thyristor 1 is turned on at $\pi/6+\alpha$. It should be noted that thyristor 6 is already conducting and that thyristor 5 has ceased to conduct owing to natural commutation. It is said then that thyristor 1 commutes with thyristor 5. It is important to realise that during the interval $(\pi/6+\alpha) \leq \omega_0 t \leq (\pi/2+\alpha)$, thyristors 6 and 1 conduct and that the line-to-line voltage $v_{ab}(t)$ appears across the load. At $\omega_0 t = \pi/2+\alpha$, thyristor 2 is turned on and thyristor 6 turns off owing to natural commutation. During the interval $(\pi/2+\alpha) \leq \omega_0 t \leq (5\pi/6+\alpha)$, thyristors 1 and 2 conduct and the line-to-line voltage $v_{ac}(t)$ appears across the load. The following firing sequence would complete one conduction period: 1–2, 2–3, 3–4, 4–5, 5–6, 6–1. Assuming steady-state conditions, this basic switching mechanism continues for as long as the converter is in operation.

It should be noted from closer examination of Figure 15.3(b) that $s_1(t)$ is responsible for the positive contribution of $v_a(t)$ to $v_{dc}(t)$ and that $s_4(t)$ is responsible for the negative contribution of $v_a(t)$ to $v_{dc}(t)$. Similarly, $s_3(t)$ and $s_6(t)$ are responsible for the positive and negative contributions of $v_b(t)$ to $v_{dc}(t)$, respectively. Also, $s_5(t)$ and $s_2(t)$ are responsible for the positive and negative contributions of $v_c(t)$ to $v_{dc}(t)$, respectively. In this situation, the relationship between the AC and DC voltages is given by the following expression:

$$v_{dc}(t) = [s_1(t) - s_4(t)]\, v_a(t) + [s_3(t) - s_6(t)]\, v_b(t) + [s_5(t) - s_2(t)]\, v_c(t) \qquad (15.7)$$

The relationships between the AC and DC converter currents may be taken to be the dual relations of those in the voltages,

$$\begin{aligned} i_a(t) &= [s_1(t) - s_4(t)]\, i_{dc}(t) \\ i_b(t) &= [s_3(t) - s_6(t)]\, i_{dc}(t) \\ i_c(t) &= [s_5(t) - s_2(t)]\, i_{dc}(t) \end{aligned} \qquad (15.8)$$

Equations (15.7) and (15.8) describe very well the periodic steady-state operation of the three-phase, six-pulse rectifier in cases where no overlap is assumed to exist.

15.2.1 *Effect of the AC side inductance*

The effect of a finite AC inductance on the current commutation, as opposed to the zero inductance assumed in the previous section, may be included in the switching functions. However, the resulting switching functions will be far more complex than cases where the overlap is neglected, since the AC current during commutation is a function of system impedance, firing angle and DC current. It has been suggested that a simplifying approach that would reduce complexity while still yielding a satisfactory answer would be to use a linear rising function during the commutation period [10]. This is the approach that it is presented in this section.

With reference to Figure 15.4 and assuming that thyristors 5 and 6 have been previously conducting, at $\omega_0 t = \alpha$ the current begins to commutate from thyristor 5 to thyristor 1. The instant when $v_a(t)$ becomes more positive than $v_c(t)$ is chosen as the time origin $\omega_0 t = 0$. During the current-commutation interval μ, thyristors 1 and 5 conduct simultaneously and the phase voltages $v_a(t)$ and $v_c(t)$ are shorted together through the AC side inductance in each phase. The current $i_a(t)$ builds up from 0 to $i_{dc}(t)$ whereas $i_c(t)$ decreases from $i_{dc}(t)$ to 0, at which instant the current commutation from thyristor 5 to thyristor 1 is completed. Under such conditions the voltage waveform, during the commutation interval μ, is given by

$$v_P(t) = \frac{v_a(t) + v_c(t)}{2}$$

In order to incorporate such effects in the switching functions, the relevant voltages are averaged during the commutation interval. For the currents, the switching functions are represented by stepped functions.

The relationship between the voltages on the AC and DC sides of the rectifier is given by the following expression:

$$v_{dc}(t) = [s_{v_1}(t) - s_{v_4}(t)]\, v_a(t) + [s_{v_3}(t) - s_{v_6}(t)]\, v_b(t) + [s_{v_5}(t) - s_{v_2}(t)]\, v_c(t) \quad (15.9)$$

For the currents the relationship will be as follows:

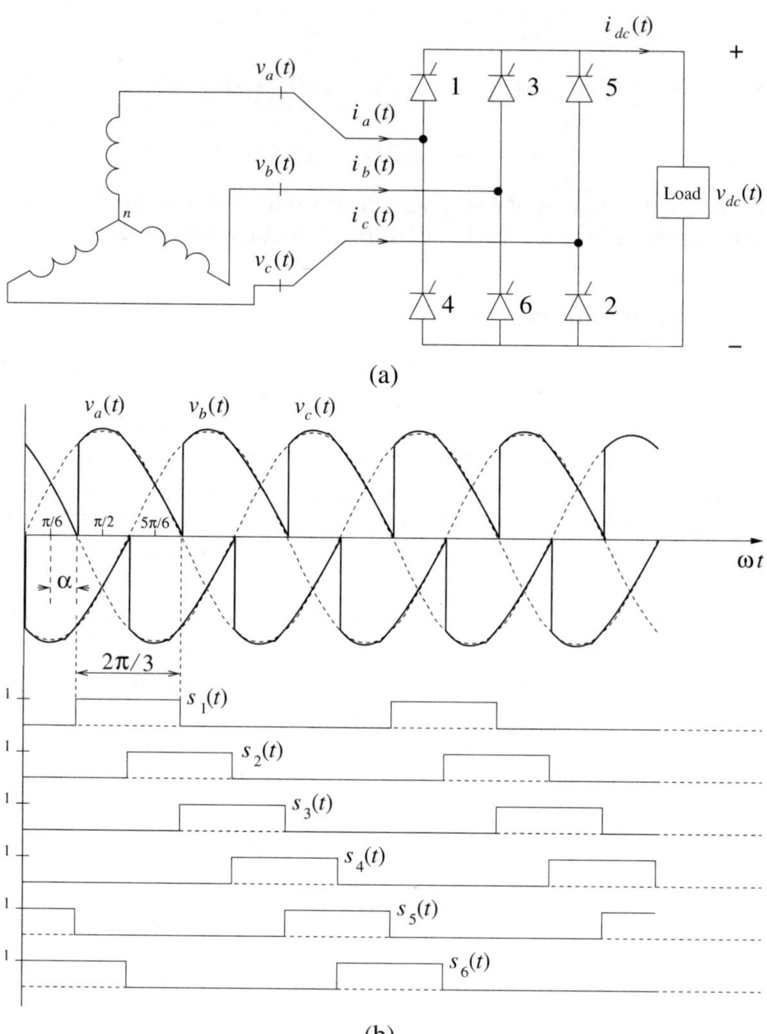

Figure 15.3 Three-phase, six-pulse rectifier

$$\begin{aligned} i_a(t) &= [s_{c_1}(t) - s_{c_4}(t)]\, i_{dc}(t) \\ i_b(t) &= [s_{c_3}(t) - s_{c_6}(t)]\, i_{dc}(t) \\ i_c(t) &= [s_{c_5}(t) - s_{c_2}(t)]\, i_{dc}(t) \end{aligned} \tag{15.10}$$

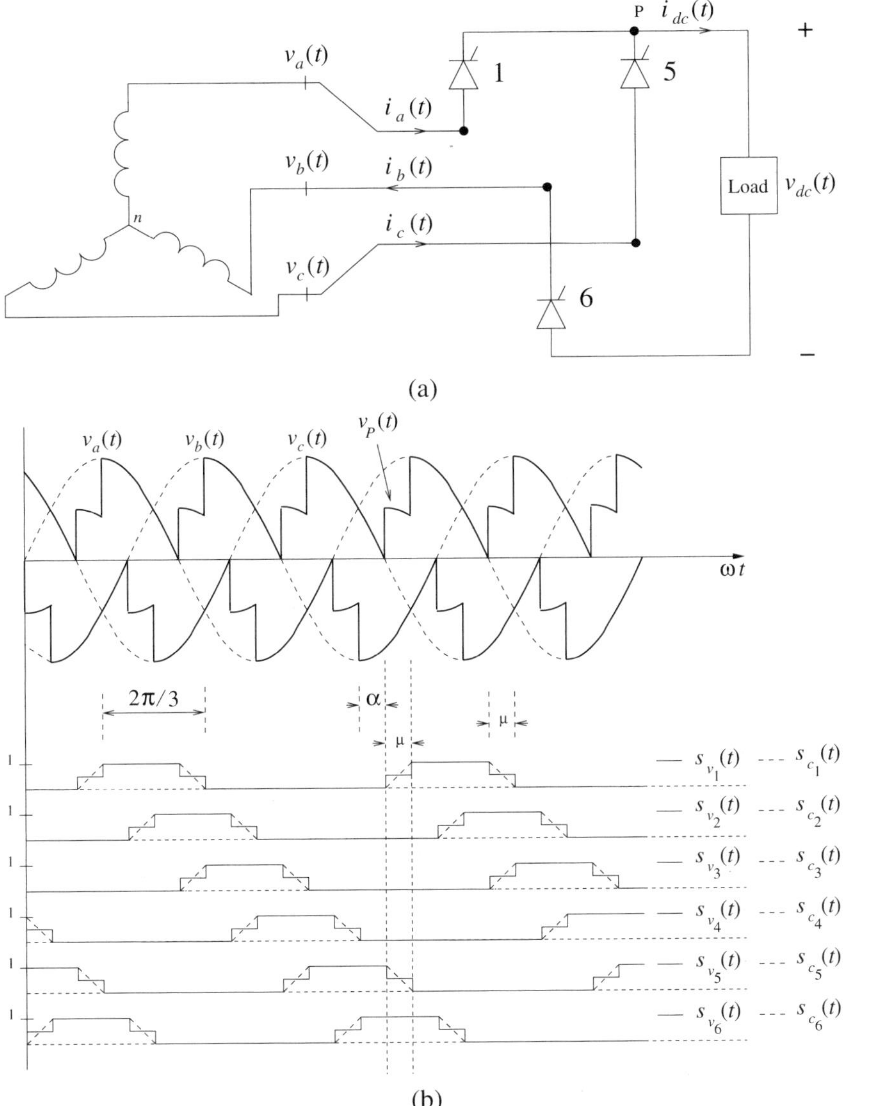

Figure 15.4 Commutation effect

15.2.2 Thyristor switching functions

The harmonic contents of the six switching functions of Figure 15.3(b) associated with the ideal operation of the six-pulse rectifier are given by

$$\begin{aligned} S_0 &= \frac{1}{3} \\ S_n &= \frac{1}{n\pi}\sin\left(\frac{n\pi}{3}\right)\mathrm{e}^{-\mathrm{j}n(\phi_\mathrm{r}+\phi_\mathrm{p}+\pi/2+\alpha)} \end{aligned} \tag{15.11}$$

where:

ϕ_r is the shift angle of the AC voltage supply with respect to the reference $v_a(t) = V_a \sin \omega_0 t$.

ϕ_p is the sequence phase angle required to fire the thyristors, i.e. 0 for thyristor 1; $\pi/3$ for thyristor 2; $2\pi/3$ for thyristor 3; π for thyristor 4; $4\pi/3$ for thyristor 5; and $5\pi/3$ for thyristor 6.

α is the firing angle.

A more realistic operation of the converter should include the representation of the commutation overlap angle. From Figure 15.4(b) the harmonics of the six switching functions for the voltages are

$$\begin{aligned} S_{v_0} &= \frac{1}{3} \\ S_{v_n} &= \frac{1}{2n\pi}\left[\sin n\left(\frac{\pi}{3}-\frac{\mu}{2}\right)+\sin n\left(\frac{\pi}{3}+\frac{\mu}{2}\right)\right]\mathrm{e}^{-\mathrm{j}n(\phi_\mathrm{r}+\phi_\mathrm{p}+\pi/2+\alpha)} \end{aligned} \tag{15.12}$$

Similarly, the six switching functions for the currents are

$$\begin{aligned} S_{c_0} &= \frac{1}{3} \\ S_{c_n} &= \frac{1}{\pi\mu n^2}\left[\cos n\left(\frac{\pi}{3}-\frac{\mu}{2}\right)-\cos n\left(\frac{\pi}{3}+\frac{\mu}{2}\right)\right]\mathrm{e}^{-\mathrm{j}n(\phi_\mathrm{r}+\phi_\mathrm{p}+\pi/2+\alpha)} \end{aligned} \tag{15.13}$$

It should be pointed out that when the converter is working under unbalanced supply conditions the commutation overlap for each phase will differ. This means that the switching function for each phase will have a different waveshape and therefore would need to be expressed by different equations. In this case, the calculation of the harmonics becomes more elaborated. In order to simplify the analysis, the average commutation overlap due to the positive sequence voltage may be used [10].

15.3 Model in the Harmonic Domain

The harmonic domain model of the three-phase, six-pulse rectifier presented below is based on the relationships between the harmonic voltages and currents in both the AC and DC sides of the rectifier. The equations which take into account the overlap angle, i.e. (15.9) and (15.10), are represented in the harmonic domain by

$$\mathbf{V}_{dc} = (\mathbf{S}_{v_1} - \mathbf{S}_{v_4})\,\mathbf{V}_a + (\mathbf{S}_{v_3} - \mathbf{S}_{v_6})\,\mathbf{V}_b + (\mathbf{S}_{v_5} - \mathbf{S}_{v_2})\,\mathbf{V}_c \tag{15.14}$$

and

$$\begin{aligned} \mathbf{I}_a &= (\mathbf{S}_{c_1} - \mathbf{S}_{c_4})\,\mathbf{I}_{dc} \\ \mathbf{I}_b &= (\mathbf{S}_{c_3} - \mathbf{S}_{c_6})\,\mathbf{I}_{dc} \\ \mathbf{I}_c &= (\mathbf{S}_{c_5} - \mathbf{S}_{c_2})\,\mathbf{I}_{dc} \end{aligned} \tag{15.15}$$

Equations (15.14) and (15.15) contain the harmonic coefficients of the voltages and currents existing in both sides of the converter. The harmonic coefficients of the AC and DC voltages and currents are related by the harmonic content of the switching functions. This is a general representation that caters for the incorporation of network and operational imbalances, pre-existing harmonic distortion in the AC supply and unbalanced converter operation.

It should be noted that the vector products in (15.14) and (15.15) represent mutual convolutions, where the switching functions become matrices with Toeplitz structures and entries obtained from (15.12) and (15.13).

It is not difficult to appreciate from (15.14) that if the AC voltages contain imbalances and non-characteristic harmonics their effects will be reflected in the harmonic content of $\mathbf{V}_{dc}$.

For the converter circuit to have practical meaning the admittance of the DC loads requires explicit representation,

$$\mathbf{I}_{dc} = \mathbf{Y}\mathbf{V}_{dc} \tag{15.16}$$

The harmonic coefficients of $\mathbf{I}_{dc}$ are a function of the voltage $\mathbf{V}_{dc}$ and the load admittance $\mathbf{Y}$, which clearly indicates that the DC component and the ripple of $i_{dc}(t)$ are represented explicitly in this relation. It should also be noted that the load in (15.16) may take any composition as long as it is static.

Using equations (15.14)–(15.16), the equivalent three-phase, six-pulse rectifier circuit, as seen from the AC side, is obtained:

$$\mathbf{I}_{abc} = \mathbf{Y}_{\text{rect}}\mathbf{V}_{abc} \tag{15.17}$$

where matrix $\mathbf{Y}_{\text{rect}}$ represents the three-phase, six-pulse rectifier equivalent admittance matrix. The current and voltage vectors are given by

$$\mathbf{I}_{abc} = \begin{bmatrix} \mathbf{I}_a \\ \mathbf{I}_b \\ \mathbf{I}_c \end{bmatrix}, \quad \mathbf{V}_{abc} = \begin{bmatrix} \mathbf{V}_a \\ \mathbf{V}_b \\ \mathbf{V}_c \end{bmatrix} \tag{15.18}$$

and the admittance matrix by

$$\mathbf{Y}_{\text{rect}} = \mathbf{S}_c\mathbf{Y}_l\mathbf{U}_u\mathbf{S}_v \tag{15.19}$$

where

$$\mathbf{S}_c = \begin{bmatrix} \mathbf{S}_{c_1} - \mathbf{S}_{c_4} & 0 & 0 \\ 0 & \mathbf{S}_{c_3} - \mathbf{S}_{c_6} & 0 \\ 0 & 0 & \mathbf{S}_{c_5} - \mathbf{S}_{c_2} \end{bmatrix} \tag{15.20}$$

$$\mathbf{Y}_l = \begin{bmatrix} \mathbf{Y} & 0 & 0 \\ 0 & \mathbf{Y} & 0 \\ 0 & 0 & \mathbf{Y} \end{bmatrix} \tag{15.21}$$

$$\mathbf{U}_u = \begin{bmatrix} \mathbf{U} & \mathbf{U} & \mathbf{U} \\ \mathbf{U} & \mathbf{U} & \mathbf{U} \\ \mathbf{U} & \mathbf{U} & \mathbf{U} \end{bmatrix} \tag{15.22}$$

and

$$\mathbf{S}_v = \begin{bmatrix} \mathbf{S}_{v_1} - \mathbf{S}_{v_4} & 0 & 0 \\ 0 & \mathbf{S}_{v_3} - \mathbf{S}_{v_6} & 0 \\ 0 & 0 & \mathbf{S}_{v_5} - \mathbf{S}_{v_2} \end{bmatrix} \tag{15.23}$$

where $\mathbf{U}$ is the identity matrix.

It should be mentioned that if no commutation angle is considered, then the switching functions $s_v(t)$ and $s_c(t)$ are both equal to $s(t)$, which in the harmonic domain becomes $\mathbf{S}_c = \mathbf{S}_v = \mathbf{S}$, with entries obtained from (15.11).

The three-phase, six-pulse rectifier model in the form of a harmonic transfer admittance matrix, $\mathbf{Y}_{\text{rect}}$, is both flexible and comprehensive. It caters for imbalances in the thyristors' firing angles, commutation effect, and a variety of static loads.

The following MATLAB™ function is used to calculate $\mathbf{Y}_{\text{rect}}$ when no commutation angle is considered:

```
function Yrect=calc_rect_S(R,X,phase,alpha,h)
 O=zeros(2*h+1,2*h+1);
 U=form_Zm(1,0,h);
 Y=inv(form_Zm(R,X,h));
 [s1,s2,s3,s4,s5,s6]=calc_TSF(phase,alpha,h);
 S14=calc_Fm(s1-s4,h);
 S36=calc_Fm(s3-s6,h);
 S52=calc_Fm(s5-s2,h);
 S=[S14 O O;O S36 O;O O S52];S=sparse(S);
 Yl=[Y O O;O Y O;O O Y];Yl=sparse(Yl);
 Uu=[U U U;U U U;U U U];Uu=sparse(Uu);
 Yrect=S*Yl*Uu*S;
```

The following MATLAB™ function calculates $\mathbf{Y}_{\text{rect}}$ when commutation angle is considered:

```
function Yrect=calc_rect_VC(R,X,phase,alpha,miu,h)
 O=zeros(2*h+1,2*h+1);
 U=form_Zm(1,0,h);
 Y=inv(form_Zm(R,X,h));
 [sv1,sv2,sv3,sv4,sv5,sv6,sc1,sc2,sc3,sc4,sc5,sc6]=
                              calc_TSFVC(phase,alpha,miu,h);
 Sv14=calc_Fm(sv1-sv4,h); Sv36=calc_Fm(sv3-sv6,h);
 Sv52=calc_Fm(sv5-sv2,h); Sc14=calc_Fm(sc1-sc4,h);
 Sc36=calc_Fm(sc3-sc6,h); Sc52=calc_Fm(sc5-sc2,h);
 Sv=[Sv14 O O;O Sv36 O;O O Sv52];Sv=sparse(Sv);
 Sc=[Sc14 O O;O Sc36 O;O O Sc52];Sc=sparse(Sc);
```

```
Yl=[Y O O;O Y O;O O Y];Yl=sparse(Yl);
Uu=[U U U;U U U;U U U];Uu=sparse(Uu);
Yrect=Sc*Yl*Uu*Sv;
```

The following MATLAB™ function is used to obtain the harmonic switching functions with no commutation angle; equation (15.11) is used:

```
function [s1,s2,s3,s4,s5,s6]=calc_TSF(phase,alpha,h)
% phase : phase angle with respect to sin(wt) in rads.
% alpha : firing angle in rads.
% h : number of harmonics
s1=zeros(1,2*h+1); s1(h+1)=1/3;
s2=s1; s3=s1; s4=s1; s5=s1; s6=s1;
for n=-h:h
  if n~=0
    agl=phase+pi/2+alpha;
    nh1=n+h+1;
    s1(nh1)=1/(pi*n)*sin(n*pi/3)*exp(-n*i*(agl));
    s2(nh1)=1/(pi*n)*sin(n*pi/3)*exp(-n*i*(agl+pi/3));
    s3(nh1)=1/(pi*n)*sin(n*pi/3)*exp(-n*i*(agl+2*pi/3));
    s4(nh1)=1/(pi*n)*sin(n*pi/3)*exp(-n*i*(agl+pi));
    s5(nh1)=1/(pi*n)*sin(n*pi/3)*exp(-n*i*(agl+4*pi/3));
    s6(nh1)=1/(pi*n)*sin(n*pi/3)*exp(-n*i*(agl+5*pi/3));
  end
end
```

The following MATLAB™ function is used to obtain the harmonic switching functions with commutation angle; equations (15.12) and (15.13) are used:

```
function [sv1,sv2,sv3,sv4,sv5,sv6,sc1,sc2,sc3,sc4,sc5,sc6]=
                                    calc_TSFVC(phase,alpha,miu,h)
% phase : phase angle with respect to sin(wt) in rads.
% alpha : firing angle in rads.
% miu : commutation overlap angle in rads.
a=pi/3-miu/2; b=pi/3+miu/2;
sv1=zeros(1,2*h+1); sv1(h+1)=1/3;
sv2=sv1;sv3=sv1;sv4=sv1;sv5=sv1;sv6=sv1;
sc1=sv1;sc2=sv1;sc3=sv1;sc4=sv1;sc5=sv1;sc6=sv1;
for n=-h:h
 if n~=0
  agl=phase+pi/2+alpha;
  nh1=n+h+1;
  sv1(nh1)=1/(2*n*pi)*(sin(n*a)+sin(n*b))*exp(-n*i*(agl));
  sv2(nh1)=1/(2*n*pi)*(sin(n*a)+sin(n*b))*exp(-n*i*(agl+pi/3));
  sv3(nh1)=1/(2*n*pi)*(sin(n*a)+sin(n*b))*exp(-n*i*(agl+2*pi/3));
  sv4(nh1)=1/(2*n*pi)*(sin(n*a)+sin(n*b))*exp(-n*i*(agl+pi));
  sv5(nh1)=1/(2*n*pi)*(sin(n*a)+sin(n*b))*exp(-n*i*(agl+4*pi/3));
  sv6(nh1)=1/(2*n*pi)*(sin(n*a)+sin(n*b))*exp(-n*i*(agl+5*pi/3));
  sc1(nh1)=1/(pi*n^2*(b-a))*(cos(n*a)-cos(n*b))*exp(-n*i*(agl));
  sc2(nh1)=1/(pi*n^2*(b-a))*(cos(n*a)-cos(n*b))*exp(-n*i*(agl+pi/3));
  sc3(nh1)=1/(pi*n^2*(b-a))*(cos(n*a)-cos(n*b))*exp(-n*i*(agl+2*pi/3));
  sc4(nh1)=1/(pi*n^2*(b-a))*(cos(n*a)-cos(n*b))*exp(-n*i*(agl+pi));
  sc5(nh1)=1/(pi*n^2*(b-a))*(cos(n*a)-cos(n*b))*exp(-n*i*(agl+4*pi/3));
  sc6(nh1)=1/(pi*n^2*(b-a))*(cos(n*a)-cos(n*b))*exp(-n*i*(agl+5*pi/3));
 end
end
```

Example 15-1: A six-pulse, three-phase rectifier feeds a load $z = R + \mathrm{j}X$. In its turn, the rectifier is fed from a sinusoidal, three-phase balanced voltage source of 1 p.u. magnitude. Simulations were carried out for two different representations of the rectifier: (i) With no commutation angle. (ii) With commutation angle.

(i) Neglecting the commutation angle, with a firing angle $\alpha = 20°$ and $h = 100$, three different loading conditions are considered, as shown in Table 15.1. Results are presented for the most relevant converter parameters.

Table 15.1 Neglecting commutation

	$R = 0.5, X = \infty$	$Z = 0.5 + \mathrm{j}100$	$Z = 0.5$
V_{dc}	1.5542	1.5542	1.5542
I_{dc}	3.1085	3.1085	3.1085
$I_{1_{\mathrm{rms}}}$	2.4237	2.4237	2.4293
I_{rms}	2.5381	2.5342	2.5525
DPF	0.9397	0.9397	0.9505
PF	0.8973	0.8987	0.9047
THD_I	31.08%	30.53%	32.25%
P_1	1.6104	1.6104	1.6328
Q_1	0.5862	0.5862	0.5336
S	1.7947	1.7919	1.8049
P	1.6104	1.6104	1.6328
Q_{H}	-	0.5862	0.5336
D_{H}	-	0.5234	0.5540

The first loading condition corresponds to the idealised operation, when no ripple exists on the DC side because of the infinite reactance of the load. These calculations are carried out by hand using the classic equations given below [1]:

$$V_{dc} = \frac{3\sqrt{3}V_{\max}}{\pi}\cos\alpha \qquad I_{dc} = \frac{1}{R_{dc}}V_{dc}$$

$$I_{1_{\mathrm{rms}}} = \frac{\sqrt{6}}{\pi}I_{dc} \qquad I_{\mathrm{rms}} = \sqrt{\frac{2}{3}}I_{dc}$$

$$\mathrm{DPF} = \cos\alpha \qquad \mathrm{PF} = \frac{3}{\pi}\cos\alpha$$

$$P_1 = \frac{V_{\max}}{\sqrt{2}}I_{1_{\mathrm{rms}}}\cos\alpha \qquad P = \frac{V_{\max}}{\sqrt{2}}I_{\mathrm{rms}}\frac{3}{\pi}\cos\alpha$$

$$Q_1 = \frac{V_{\max}}{\sqrt{2}}I_{1_{\mathrm{rms}}}\sin\alpha \qquad S = \frac{V_{\max}}{\sqrt{2}}I_{\mathrm{rms}}$$

$$\mathrm{THD}_I = 31.08\%$$

The results show that the computer solutions compare well with the analytical ones when the DC reactance of the load is large, i.e. 100 p.u. The results also show that the DC components of V_{dc} and I_{dc} are a function only of the resistance but not of the reactance, which is an expected result. The effect of the current and voltage ripple is most apparent in the current THD.

(ii) This case assumes a commutation angle $\mu = 10°$, a firing angle $\alpha = 20°$, and uses the more complete model.

This case cannot be solved using the analytical expressions given in case (i) because they make no allowances to accommodate commutation effects. The relevant converter parameters for this case are given in Table 15.2. Three different loading cases for the reactance are considered in this example, i.e. $X = 0$, 0.1 and 100. The resistance is kept at $R = 0.5$ in all three cases. It is seen from the results that the load reactance does not influence the DC variables. However, it has a more significant impact on the current THD and, as expected, in reactive power consumption.

Table 15.2 Including commutation

	$Z = 0.5 + \mathrm{j}100$	$Z = 0.5 + \mathrm{j}0.1$	$Z = 0.5$
V_{dc}	1.5483	1.5483	1.5483
I_{dc}	3.0966	3.0966	3.0966
$I_{1_{\mathrm{rms}}}$	2.4114	2.4174	2.4170
I_{rms}	2.4930	2.5033	2.5114
DPF	0.9397	0.9403	0.9462
PF	0.9089	0.9081	0.9107
THD_I	26.24%	26.89%	28.23%
P_1	1.6023	1.6073	1.6172
Q_1	0.5832	0.5817	0.5528
S	1.7628	1.7701	1.7759
P	1.6023	1.6073	1.6172
Q_{H}	0.5832	0.5817	0.5528
D_{H}	0.4475	0.4596	0.4824

The main waveforms of rectifier operation are shown in Figure 15.5 for a load of $Z = 0.5 + \mathrm{j}0.1$, with $\alpha = 20°$ and $\mu = 10°$. The DC parameters are shown in part (a) of the figure, together with the phase and line AC voltages, and the commutation voltage. Part (b) shows the AC line currents where the effect of the finite inductance on the DC side is clearly seen. These currents also show the effect of commutation.

It is well known that one effect of commutation is to make the converter waveforms more sinusoidal, thus reducing harmonic distortion. This can be appreciated by comparing the relevant THD_I values in Tables 15.1 and 15.2. The results presented in the latter table include commutation effect, and the THD_I values are shown to be consistently lower.

For this example the following steps were carried out in MATLAB™:

```
>> clear all
>> h=100;
>> phase=0;
>> alpha=20*pi/180;
>> miu=10*pi/180;
>> R=0.5;
```

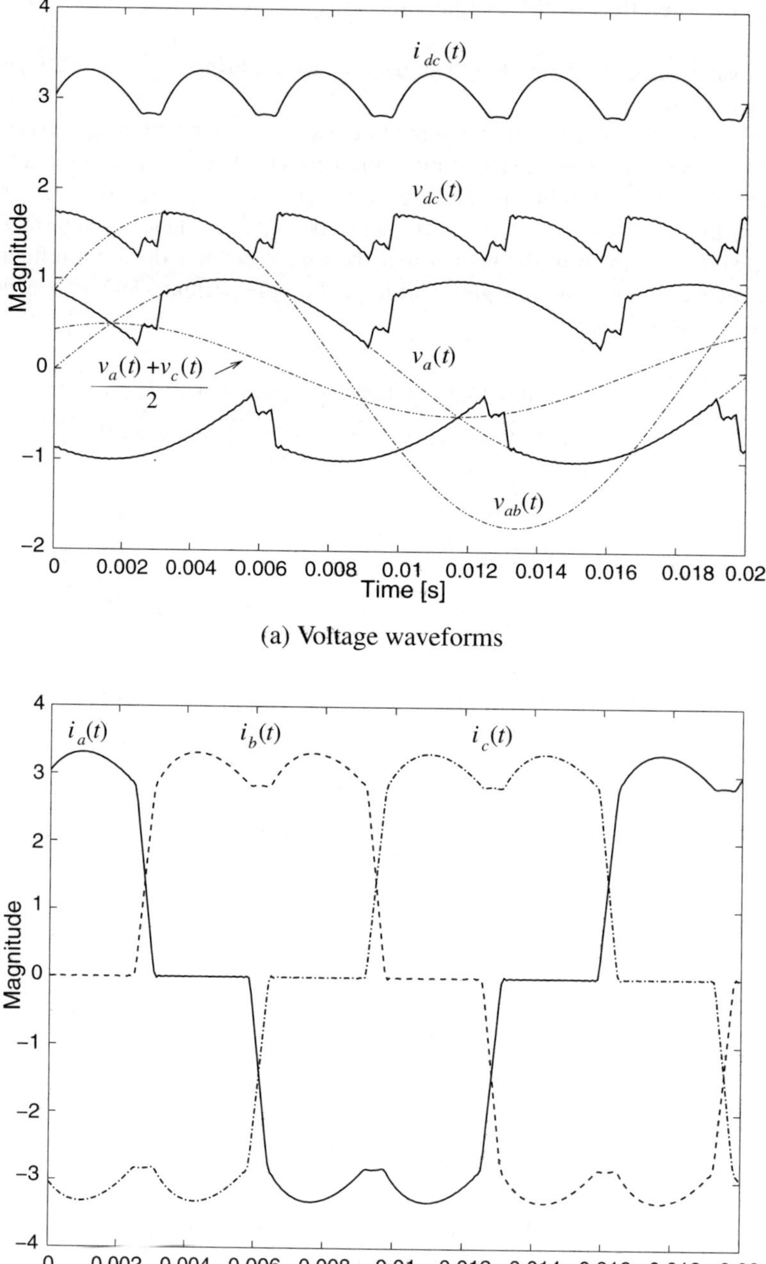

(b) Current waveforms

Figure 15.5 Time domain representation of rectifier operation including overlap

```
>> X=0.1;
>> [Va,Vb,Vc]=source_h(1,1,1,0,-120,120,h);
>> Vabc=[Va;Vb;Vc];
>> Yrect=calc_rect_VC(R,X,phase,alpha,miu,h)
>> Iabc=Yrect*Vabc;
>> Ia=Iabc(1:2*h+1);
>> [S,P,Q,D,Vrms,Irms,PF,VTHD,ITHD]=spqd(Va,Ia,h);
>> ia=plot_f(Ia,h,1);
>> plot(ia);
```

Example 15-2: A six-pulse, three-phase rectifier was assumed to exist in the power system shown in Figure 13.16. For the purpose of this example the rectifier replaces the TCR at Tiwai. A firing angle of $\alpha = 30°$, $\mu = 15°$ and $h = 15$ were used. The rectifier supplies an R–L series load of $Z = 2 + j0.5$ p.u. Figure 15.6 shows the three-phase AC voltage and current waveforms where the effects of commutation and a finite reactance on the DC side are clearly shown. Table 15.3 gives the harmonics as a percentage of the fundamental frequency. These results contain non-characteristic harmonics due to transmission system imbalances.

Table 15.3 Harmonic voltages and currents in the Tiwai node

h	$\%V_{a_1}$	$\%V_{b_1}$	$\%V_{c_1}$	$\%I_{a_1}$	$\%I_{b_1}$	$\%I_{c_1}$
3	0.0803	0.0731	0.0809	0.3458	0.3029	0.3374
5	6.0282	5.5978	7.9151	23.4431	23.5477	23.9537
7	0.5552	0.5542	0.5008	7.1556	7.1464	7.1656
9	0.0013	0.0022	0.0014	0.0815	0.1073	0.0834
11	0.1297	0.1142	0.1334	5.5328	5.5555	5.5463
13	0.2247	0.2132	0.2110	3.5893	3.5941	3.5843
15	0.0016	0.0018	0.0020	0.0140	0.0154	0.0163

15.4 Harmonics in Electrical Power Networks

The use of power electronic converters is quite extensive at all voltage levels. In distribution networks harmonics are normally due to large concentrations of single-phase rectifiers. In industrial installations, power electronics equipment is used extensively, mainly in motor drives, giving rise to considerable harmonic distortion. In this section, a series of harmonic measurements are presented which show that harmonics at the distribution and transmission levels come mainly from loads which use some kind of power electronic converter.

15.4.1 Harmonics produced by single-phase converters

A large number of domestic appliances use single-phase rectifiers. These devices generate all the odd harmonics, with the third harmonic being dominant. For example, Figure 15.7 shows the current waveform of a PC and its associated printer; this current is typical of this kind of equipment and has a THD of 110%, PF of 0.66 and DPF of 0.99.

The concentration of electronic appliances, such as TVs, fridges, PCs, energy saving lamps, etc., presents a very rich source of harmonics. For example, Figure 15.8 shows the current

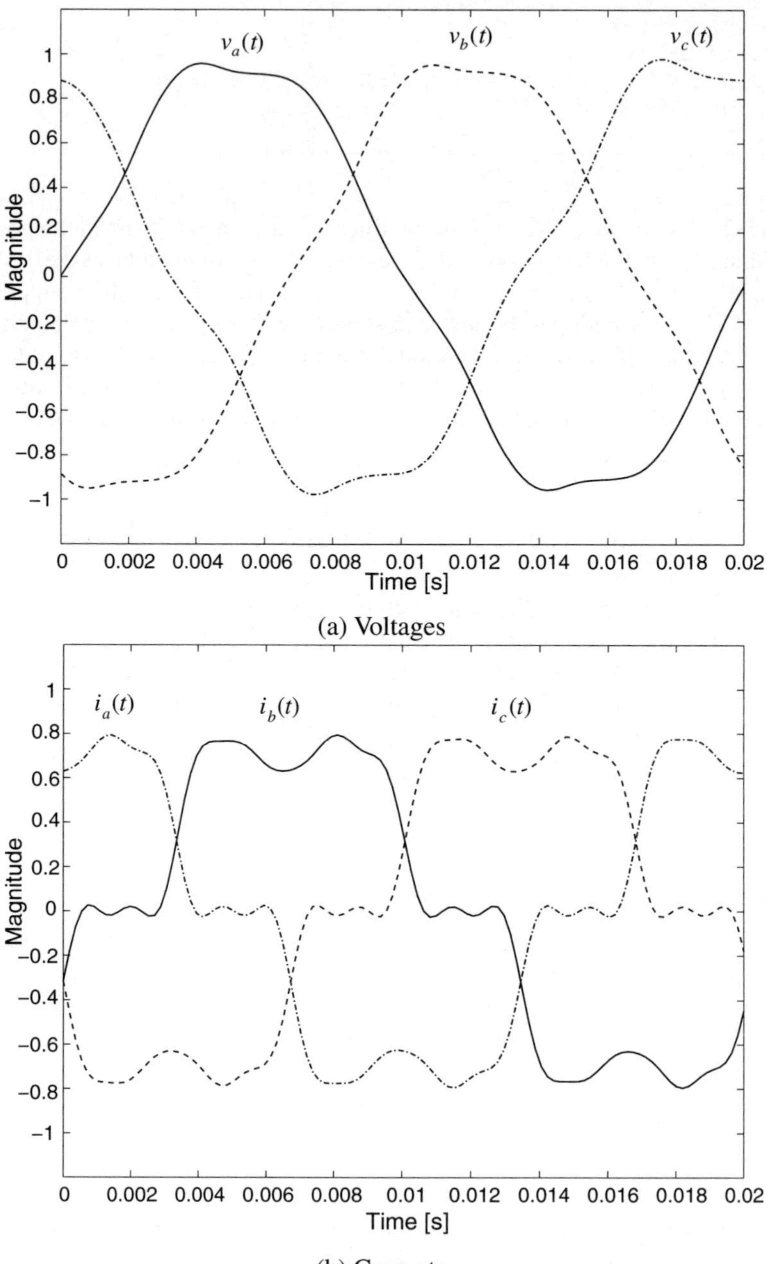

(a) Voltages

(b) Currents

Figure 15.6 Three-phase voltages and currents at Tiwai

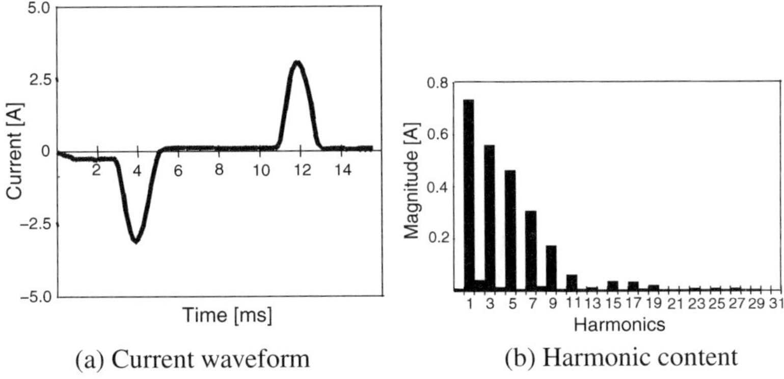

(a) Current waveform (b) Harmonic content

Figure 15.7 Current in a PC circuit

waveform measured at the main feeder of a residential circuit. The current THD is 17.5%, PF is 0.87, DPF is 0.89 and the load consumes 2.09 kW.

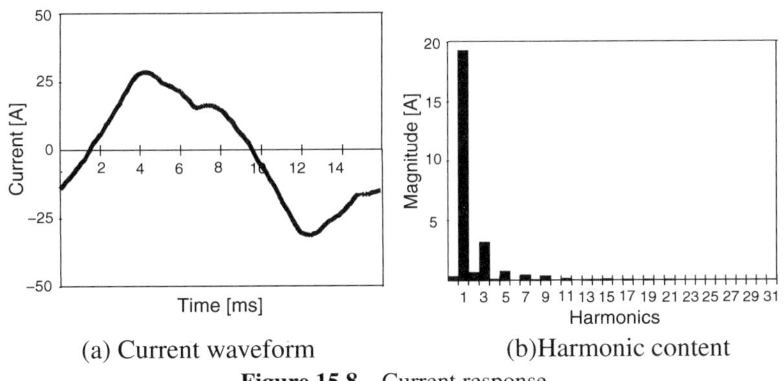

(a) Current waveform (b)Harmonic content

Figure 15.8 Current response

15.4.2 Harmonics produced by three-phase converters

Two of the main applications of power electronic converters in industry are motor speed drives and induction furnaces. Because high-frequency eddy currents are very effective in providing heating, induction furnaces are used to manufacture a variety of ferrous and non-ferrous grades. An induction furnace consists of a crucible with a coil wound around it, through which high-frequency AC currents are passed. A metallic charge is deposited in the crucible and heated magnetically using induced eddy currents. The high-frequency AC currents required to produce the induced eddy currents in the charge are produced from a constant frequency AC source using a large rectifier.

Figure 15.9 shows typical current and voltage waveforms found in induction furnaces. Table 15.4 gives the main parameters and harmonic information for the furnace. From the harmonic content it may be seen that the power rectifier used by the induction furnace is a six-pulse rectifier.

It is quite common to find induction furnaces in industrial installations, and motor drives

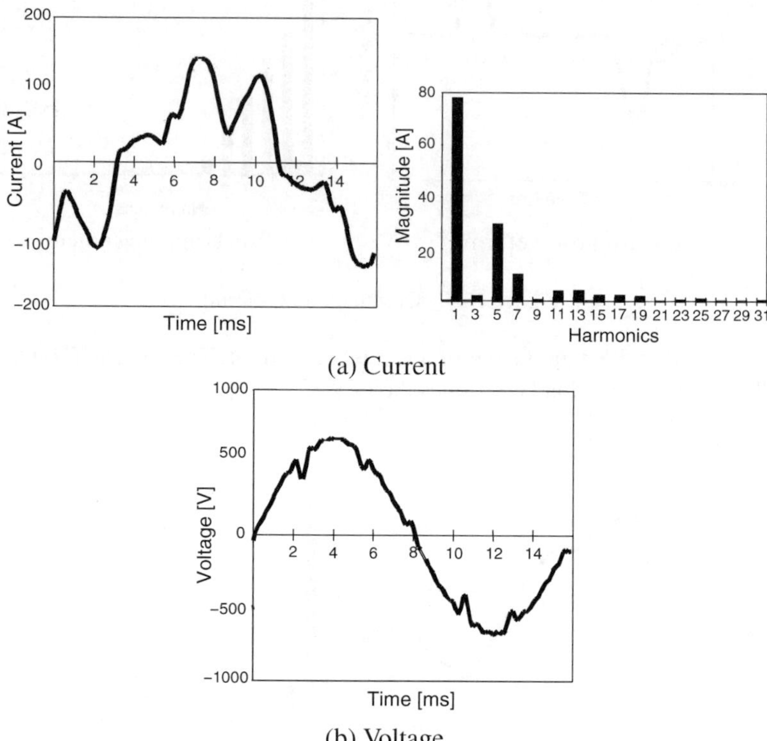

Figure 15.9 Induction furnace, phase a voltage and current

Table 15.4 Induction furnace characteristics

			Voltage	Current
kVA	40.7	RMS	481.2	84.62
kW	1.2	Peak	647.2	146.38
kVAR	37.5	DC offset	−0.2	−0.24
kVAD	15.8	THD	8.92	41.27
PF	0.03			
DPF	0.05			

feeding directly from the main transformer and the PF compensated by capacitor banks. Figure 15.10 shows the waveform and harmonic content of the current in a capacitor bank used to compensate PF in an induction furnace. It should be noted that some harmonic currents are larger than what one would expect, possibly due to near resonance problems between the capacitor bank and the transformer. Table 15.5 shows the measured values in phase a of the transformer. The results show that the DPF is one, but a high current distortion exists producing a PF equal to 0.91.

Table 15.5 Capacitor bank, 60 kVAR, 480 V

			Voltage	Current
kVA	36.1	RMS	471.5	76.47
kW	33.0	Peak	671.8	173.97
kVAR	0.4	DC offset	−0.3	−0.26
kVAD	14.63	THD	3.06	43.46
PF	0.91			
DPF	1.00			

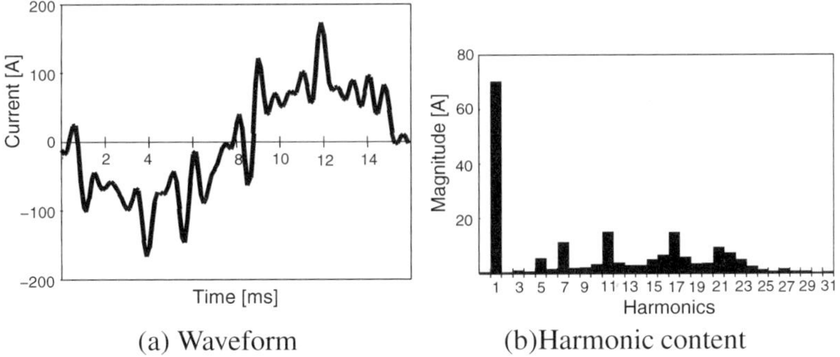

(a) Waveform (b)Harmonic content

Figure 15.10 Current in a capacitor bank

15.4.3 Harmonics in distribution systems

The harmonics produced by power electronics applications are increasing all the time, becoming an important problem to the utility company since these harmonics travel freely from the low-voltage distribution system to the transmission system. A 115/13.8 kV distribution system is shown in Figure 15.11. The distribution system [11] feeds a city via a ring system with six substations with transformers of 20 MVA 115/13.8 kV; 80% of the total power comes from a transformer bank of 100 MVA 230/115 kV located in the main power station.

Several measurements were taken to assess harmonic voltage distortion. Figure 15.12 shows the voltage THD at 115 kV in the six substations, whereas Figure 15.13 shows the voltage THD at 13.8 kV. The highest harmonics measured were in decreasing order of magnitude, the 5th, 7th, 3rd, 11th and 13th. These harmonics are typical of three-phase, six-

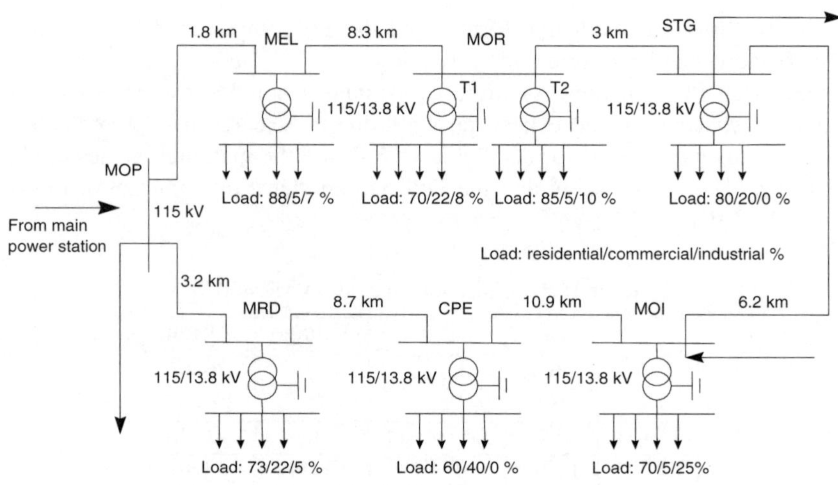

Figure 15.11 Distribution system at 115/13.8 kV

pulse converter systems commonly found in commercial centres and in industry. The third harmonic is typically generated by single-phase rectification systems commonly found in residential loads and office centres. These harmonics are also generated by TVs, PCs, energy saving lamps and photocopiers. The results show that during the evening the THD increases; this effect may be due to the fact that heavy-industry equipment, such as motor drives, do not operate at night. On the other hand, the use of electronic devices used in public, commercial and residential premises is prevalent at those times.

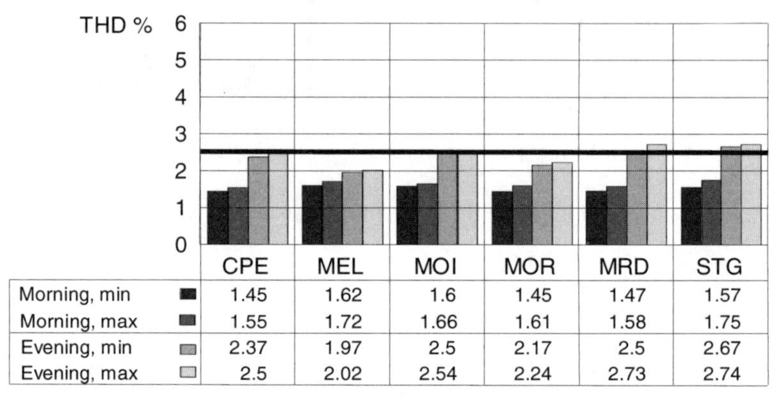

	CPE	MEL	MOI	MOR	MRD	STG
Morning, min	1.45	1.62	1.6	1.45	1.47	1.57
Morning, max	1.55	1.72	1.66	1.61	1.58	1.75
Evening, min	2.37	1.97	2.5	2.17	2.5	2.67
Evening, max	2.5	2.02	2.54	2.24	2.73	2.74

Figure 15.12 Minimum and maximum THD voltage at 115 kV

15.5 Summary

The thrust of this chapter has been on the derivation of harmonic domain models suitable for the study of conventional three-phase, six-pulse harmonic converters. Modulation theory based on switching functions and discrete convolutions are the cornerstones of these models.

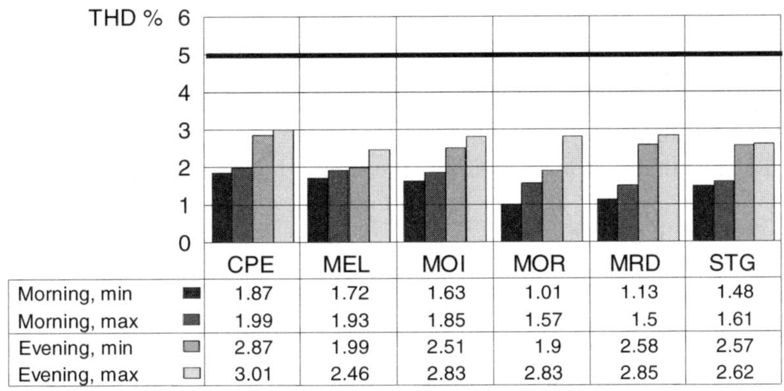

	CPE	MEL	MOI	MOR	MRD	STG
Morning, min	1.87	1.72	1.63	1.01	1.13	1.48
Morning, max	1.99	1.93	1.85	1.57	1.5	1.61
Evening, min	2.87	1.99	2.51	1.9	2.58	2.57
Evening, max	3.01	2.46	2.83	2.83	2.85	2.62

Figure 15.13 Minimum and maximum THD voltage at 13.8 kV

The harmonic converter models presented in this chapter build up incrementally, starting with the simplified case where no commutation effects are included. This simplicity is used to illustrate the simple relationship that exists between the AC and DC voltages and switching functions. The relevant MATLAB™ code is given for users to reproduce the numeric examples presented in the chapter and indeed to try out their own examples. This is followed by a fuller converter model, which uses averaged commutation effects in the switching functions, leading to a reasonably realistic representation while keeping simplicity within bounds. The characteristic and non-characteristic harmonics are implicitly included in the model, together with DC ripple, AC and converter imbalances, and pre-existing harmonics. The model comes in the form of a harmonic admittance where the cross-couplings between harmonics are explicitly shown. Also, the harmonic interaction and frequency conversion effects that exist between the AC and the DC sides of the converter are implicitly accounted for. MATLAB ™ code is also provided for this more comprehensive converter model.

At the end of the chapter, harmonic measurements are presented for completeness. They were carried out at different voltage levels, and relate to physical installations. In all cases, their harmonic spectrum shows the presence of characteristic harmonics.

15.6 Bibliography

1. N. Mohan, T.M. Undeland, W.P. Robbins, *Power Electronics: Converters, Applications, and Design*, John Wiley & Sons, New York, 1989.
2. E.W. Kimbark, *Direct Current Transmission*, Wiley-Interscience, New York, 1971.
3. J. Arrillaga, *High Voltage Direct Current Transmission, IEE Power Engineering Series* 6, Peter Peregrinus, Stevenage, 1983.
4. D.E. Rice, "A Detailed Analysis of Six-Pulse Converter Harmonic Currents", *IEEE Transactions on Industry Applications*, Vol. 30, No. 2, March/April 1994, pp. 294–304.
5. M. Sakui, H. Fujita, M. Shioya, "A Method for Calculating Harmonic Current of a Three-Phase Bridge Uncontrolled Rectifier with DC Filter", *IEEE Transactions on Industry Electronics*, Vol. 36, No. 3, 1989, pp. 434–440.
6. Manitoba HVDC Research Centre, PSCAD/EMTDC: Electromagnetic Transients Program Including DC Systems, 1994.

7. Analogy Inc., SaberDesigner™, 1998.
8. R. Yacamini, J.C. de Oliveira, "Harmonics in Multiple Convertor Systems: A Generalised Approach", *IEE Proceedings*, Part B, Vol. 127, No. 2, March 1980, pp. 96–106.
9. J. Arrillaga, B.C. Smith, N.R. Watson, A.R. Wood, *Power System Harmonic Analysis*, John Wiley & Sons, Chichester, 1997.
10. L. Hu, R. Yacamini, "Calculation of Harmonic Interface in HVDC Systems with Unbalance", *Proceedings of the International Conference on AC and DC Power Transmission*, 1991, pp. 390–394.
11. J.L. Campos, M. Madrigal, C. Barnaba, "Resultados de la Medición de Voltajes y Corrientes Armónicas en las Subestaciones de Distribución 115/13.8 kV del Area Morelia de la Comisión Federal de Electricidad", *Proceedings of IEEE-RVP*, Acapulco, Mexico, July 1998.
12. M.E. Amoli, T. Florence, "Voltage and Current Harmonic Content of a Utility System: A Summary of 1120 Test Measurements", *IEEE Transactions on Power Delivery*, Vol. 5, No. 3, July 1990, pp. 1552–1557.

16

Static Compensator with PWM Converters

16.1 Introduction

The static compensator (STATCOM) is a custom power device comprising a voltage source inverter, a DC capacitor and a coupling transformer. The inverter uses gate-turn-off thyristor (GTO) or insulated gate bipolar transistor (IGBT) switches, which are turned on and off by pulse-width modulation (PWM) techniques. The STATCOM is connected in parallel with the AC system and is used to exchange active and reactive power with the AC system.

Over the last few years, an increasing amount of research work addressing the STATCOM has been published in the open literature mainly on design, operation and control issues. A fair amount of work has been reported on modelling and analysis techniques oriented towards time domain simulations. An accurate representation of PWM harmonics is of paramount importance because the harmonic spectrum generated by PWM control is shifted towards the high frequencies, a fact that may lead to resonance problems if one of the many alternate resonant points of the transmission system impedance becomes excited.

A harmonic voltage source may be used to represent the STATCOM at harmonic frequencies but this will include no explicit representation of the capacitor. However, the capacitor plays a key role in STATCOM operation and a great deal in modelling flexibility is gained by having direct access to the capacitor parameters, such as capacitor size, and DC voltage and current. In this chapter, a comprehensive harmonic domain model of a three-phase, six-pulse PWM STATCOM is presented. The model takes proper account of the DC capacitor effect and comes in the form of a three-phase Thévenin equivalent expressed in the harmonic domain, where switching functions are used to represent with ease the PWM control firing sequences.

The harmonic impedance matrix of the Thévenin equivalent shows high cross-couplings between phases and between harmonics, an effect which is strongly influenced by the STATCOM capacitor size. Results are presented which show that the PWM STATCOM observes quite different harmonic voltage responses when it is made to operate as a reactive power source and when it is made to operate as a sink. It should be mentioned that this effect cannot be observed with steady-state models that use a voltage source to represent the STATCOM.

The STATCOM may be seen as the basic building block, with which to assemble a wide range of power electronic controllers aimed at solving a variety of power quality problems in low-voltage distribution networks and power control in high-voltage transmission networks.

The most popular controllers that use the STATCOM are: the dynamic voltage restorer (DVR), the unified power flow controller (UPFC), shunt and series active filters, and power factor correctors. The STATCOM used in high-voltage transmission is designed to switch at the fundamental frequency, whereas in low-voltage distribution the STATCOM is designed to switch at PWM frequencies. The thrust of this chapter is on PWM STATCOM modelling and simulation at harmonic frequencies with emphasis on reactive power compensation. The role of the STATCOM as active filter is emphasised towards the end of the chapter, where an outline on active filters is presented.

16.2 STATCOM Operation

Following the commissioning of industry-size prototype installations in various parts of the world, and after a suitable appraisal period has elapsed, it is quite safe to say that the custom power technology is here to stay. Hence, system-level studies of interactions between custom power equipment and other plant components in the network become an issue of great importance. In particular, a thorough examination of the load must be carried out at the point where the custom power plant will be installed. As suggested in [8], this is not a straightforward task and a suitable combination of measurements and simulation studies is required in order to carry out effective solutions that will pinpoint trouble spots.

Bearing this in mind, this chapter presents a comprehensive and flexible harmonic domain model for the three-phase STATCOM. Figure 16.1 shows the schematic representation of a STATCOM connected to an AC equivalent circuit, where the reactive power exchange between the STATCOM and the AC power system is controlled by adjusting the voltage magnitude difference across the coupling transformer, i.e. $|V_i| - |V_s|$, where V_i is the STATCOM's output voltage and $|V_s|$ is the voltage at the common point of connection. In principle, the exchange of active power can be controlled by adjusting the phase angle difference of the PWM converter output voltage and the voltage at the point of common coupling, i.e. $\delta_i - \delta_s = \delta_{is}$. The classic active and reactive power expressions provide a very useful first approach to appreciate the effectiveness of the control strategies based on the voltage magnitude and angle differences,

$$P \approx \frac{|V_i||V_s|}{X_\mathrm{e}} \sin \delta_{is} \tag{16.1}$$

$$Q \approx \frac{|V_i|^2}{X_\mathrm{e}} - \frac{|V_i||V_s|}{X_\mathrm{e}} \cos \delta_{is} \tag{16.2}$$

where X_e is the equivalent reactance of the coupling transformer.

It follows from the approximate expressions (16.1) and (16.2) that the STATCOM neither injects nor absorbs active and reactive power from the system when $|V_i| - |V_s| = 0$ and $\delta_{is} = 0$. Conversely, the STATCOM injects reactive power when $|V_i| > |V_s|$ and it absorbs reactive power when $|V_i| < |V_s|$.

16.3 STATCOM Based on PWM Converters

The STATCOM comprises a voltage source inverter, a DC capacitor and a coupling transformer. The inverter is assumed to use GTO or IGBT switches, which can be turned on and off at a rate considerable higher than power frequencies. PWM techniques are used

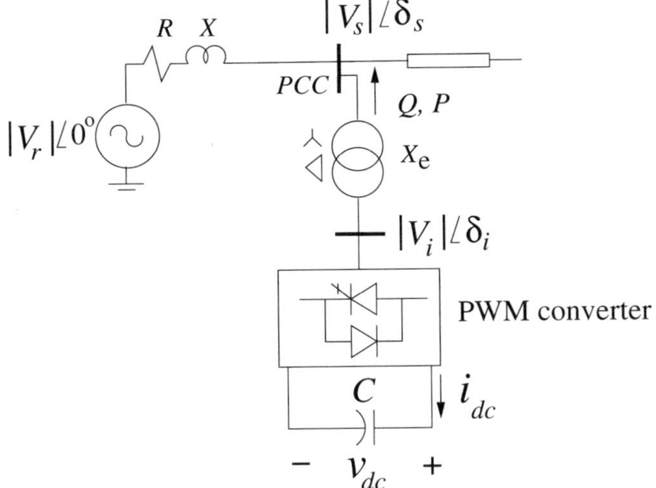

Figure 16.1 STATCOM equivalent connected at point of common coupling (PCC)

to control the inverter output harmonics. The switches are represented by switching functions governed by the PWM control mechanism.

In actual operation the DC side capacitor charges and discharges to a greater or lesser degree during the course of each cycle. However, for the purpose of steady-state operation the capacitor voltage may be assumed to take an average constant value. In this situation the capacitor is well represented by an equivalent capacitance C having a DC voltage $v_{\text{cap}}(0^+)$.

16.4 Analytical Model

The equation that describes the capacitor voltage of the STATCOM in Figure 16.2 is

$$v_{\text{cap}}(t) = \frac{1}{C}\int_0^t i_{dc}(t)\text{d}t + v_{\text{cap}}(0^+) \tag{16.3}$$

and the voltage on the DC side is

$$v_{dc}(t) = v_{\text{cap}}(t) \tag{16.4}$$

In this circuit the line voltages on the AC side and on the DC side are related to each other by the following expressions:

$$\begin{aligned} v_{ab}(t) &= s_{ab}(t)v_{dc}(t) \\ v_{bc}(t) &= s_{bc}(t)v_{dc}(t) \\ v_{ca}(t) &= s_{ca}(t)v_{dc}(t) \end{aligned} \tag{16.5}$$

where $s_{ab}(t)$, $s_{bc}(t)$ and $s_{ca}(t)$ are the PWM switching functions that govern the switches 1–6, 3–2 and 5–4, respectively. Assuming lossless switches, the instantaneous power on the AC and DC sides of the STATCOM would match with each other,

$$v_{ab}(t)i_{ab}(t) + v_{bc}(t)i_{bc}(t) + v_{ca}(t)i_{ca}(t) = v_{dc}(t)i_{dc}(t) \tag{16.6}$$

Substitution of (16.5) in (16.6) leads to a very useful expression that relates the AC side and DC side currents as a function of the switching functions:

$$i_{dc}(t) = s_{ab}(t)i_{ab}(t) + s_{bc}(t)i_{bc}(t) + s_{ca}(t)i_{ca}(t) \tag{16.7}$$

It can be seen from Figure 16.2 that the phase voltages $v_a(t)$, $v_b(t)$ and $v_c(t)$ on the primary side of the transformer are proportional to the line voltages $v_{ab}(t)$, $v_{bc}(t)$ and $v_{ca}(t)$ on the secondary side. Also, the line currents $i_a(t)$, $i_b(t)$ and $i_c(t)$ on the primary side of the transformer are proportional to the phase currents $i_{ab}(t)$, $i_{bc}(t)$ and $i_{ca}(t)$ on the secondary side. The AC line currents of the inverter are given by $i_{\mathrm{inv}_a}(t) = i_{ab}(t) - i_{ca}(t)$, $i_{\mathrm{inv}_b}(t) = i_{bc}(t) - i_{ab}(t)$ and $i_{\mathrm{inv}_c}(t) = i_{ca}(t) - i_{bc}(t)$.

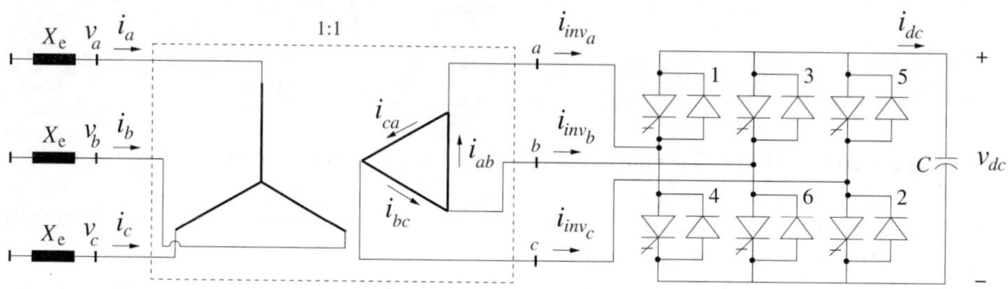

Figure 16.2 Three-phase STATCOM

Equation (16.3) may be expressed in the harmonic domain by

$$\mathbf{V}_{\mathrm{cap}} = \frac{1}{C}\mathbf{D}^{-1}(\mathrm{j}h\omega_0)\mathbf{I}_{dc} + \mathbf{E}_{\mathrm{dc}} \tag{16.8}$$

where $\mathbf{V}_{\mathrm{cap}}$ is assembled with the harmonic content of $v_{\mathrm{cap}}(t)$. Likewise, vector $\mathbf{I}_{dc}$ includes the harmonic content of $i_{dc}(t)$, and $\mathbf{E}_{\mathrm{dc}}$ contains only a DC term, i.e.,

$$\mathbf{E}_{\mathrm{dc}} = \begin{bmatrix} 0 & \cdots & 0 & v_{\mathrm{cap}}(0^+) & 0 & \cdots & 0 \end{bmatrix}^{\mathrm{T}}$$

Equation (16.8) can be written as

$$\mathbf{V}_{\mathrm{cap}} = \mathbf{Z}_{\mathrm{cap}}\mathbf{I}_{dc} + \mathbf{E}_{\mathrm{dc}} \tag{16.9}$$

where $\mathbf{Z}_{\mathrm{cap}} = \frac{1}{C}\mathbf{D}^{-1}(\mathrm{j}h\omega_0)$ represents the equivalent impedance of the capacitor on the DC side.

From (16.4) and (16.5), the voltage on the DC side and its relationship to the AC phase voltages may be expressed in the harmonic domain as

$$\mathbf{V}_{dc} = \mathbf{V}_{\mathrm{cap}} \tag{16.10}$$

and

$$\begin{aligned} \mathbf{V}_a &= \mathbf{S}_{ab}\mathbf{V}_{dc} \\ \mathbf{V}_b &= \mathbf{S}_{bc}\mathbf{V}_{dc} \\ \mathbf{V}_c &= \mathbf{S}_{ca}\mathbf{V}_{dc} \end{aligned} \tag{16.11}$$

Similarly, the relationship between the line currents and the direct current, in the harmonic domain, is given by

$$\mathbf{I}_{dc} = \mathbf{S}_{ab}\mathbf{I}_a + \mathbf{S}_{bc}\mathbf{I}_b + \mathbf{S}_{ca}\mathbf{I}_c \tag{16.12}$$

By a straightforward manipulation of equations (16.9)–(16.12) the Thévenin equivalent of the three-phase, six-pulse PWM converter is arrived at,

$$\mathbf{V}_{abc} = \mathbf{E}_{\mathrm{Th}} + \mathbf{Z}_{\mathrm{Th}}\mathbf{I}_{abc} \tag{16.13}$$

where $\mathbf{V}_{abc}$ and $\mathbf{I}_{abc}$ are harmonic vectors of the phase voltages and line currents, respectively.

The equivalent harmonic impedance, i.e. Thévenin impedance, as seen from the AC side of the PWM converter is

$$\mathbf{Z}_{\mathrm{Th}} = \mathbf{S}_1\mathbf{Z}_{\mathrm{CAP}}\mathbf{S}_2 \tag{16.14}$$

and the Thévenin equivalent voltage is

$$\mathbf{E}_{\mathrm{Th}} = \mathbf{S}_1\mathbf{E}_{\mathrm{DC}} \tag{16.15}$$

where

$$\mathbf{S}_1 = \begin{bmatrix} \mathbf{S}_{ab} & 0 & 0 \\ 0 & \mathbf{S}_{bc} & 0 \\ 0 & 0 & \mathbf{S}_{ca} \end{bmatrix}$$

$$\mathbf{S}_2 = \begin{bmatrix} \mathbf{S}_{ab} & \mathbf{S}_{bc} & \mathbf{S}_{ca} \\ \mathbf{S}_{ab} & \mathbf{S}_{bc} & \mathbf{S}_{ca} \\ \mathbf{S}_{ab} & \mathbf{S}_{bc} & \mathbf{S}_{ca} \end{bmatrix}$$

$$\mathbf{Z}_{\mathrm{CAP}} = \begin{bmatrix} \mathbf{Z}_{\mathrm{cap}} & 0 & 0 \\ 0 & \mathbf{Z}_{\mathrm{cap}} & 0 \\ 0 & 0 & \mathbf{Z}_{\mathrm{cap}} \end{bmatrix}$$

and

$$\mathbf{E}_{\mathrm{DC}} = \begin{bmatrix} \mathbf{E}_{\mathrm{dc}} \\ \mathbf{E}_{\mathrm{dc}} \\ \mathbf{E}_{\mathrm{dc}} \end{bmatrix}; \quad \mathbf{I}_{abc} = \begin{bmatrix} \mathbf{I}_a \\ \mathbf{I}_b \\ \mathbf{I}_c \end{bmatrix}; \quad \mathbf{V}_{abc} = \begin{bmatrix} \mathbf{V}_a \\ \mathbf{V}_b \\ \mathbf{V}_c \end{bmatrix}$$

The Thévenin equivalent voltage is a constant three-phase harmonic voltage source, which includes the effect of the PWM switching functions over the DC voltage $v_{\mathrm{cap}}(0^+)$. It should be noted that a simpler model corresponds to the case when the capacitor is not explicitly represented in the model. In this situation $\mathbf{Z}_{\mathrm{Th}} = 0$, and the capacitor effect is represented only as a DC source, i.e. $v_{\mathrm{cap}}(0^+) = E_{\mathrm{dc}}$.

The complete model of the STATCOM shown in Figure 16.3 is obtained by including the reactance of the coupling transformer in series with $\mathbf{Z}_{\mathrm{Th}}$.

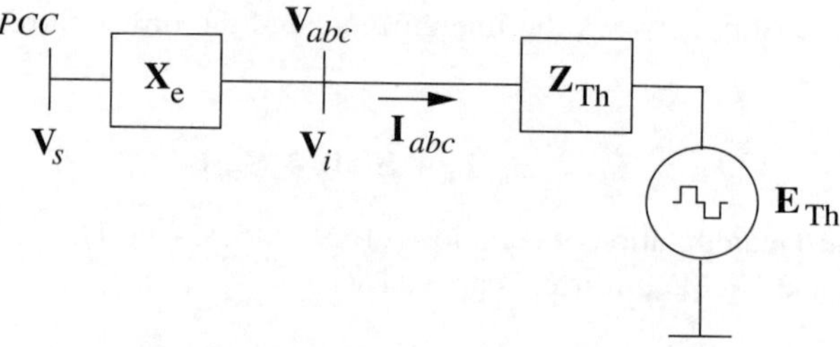

Figure 16.3 STATCOM equivalent circuit

16.5 PWM Converter Switching Functions

PWM is a technique which enables effective control of the harmonic content and harmonic magnitude generated by IGBT and GTO converters, by controlling the turn-on and turn-off of the power electronic switches in the STATCOM. The PWM technique bases its principle of operation on carrier signal comparisons.

For example, a PWM with unipolar voltage switching is represented in Figure 16.4. In this figure v_r is the carrier signal with frequency f_r, $v_{sa} = -v_{sb}$ is the modulation signal with frequency f_s , e.g. 50 Hz, and m is a modulation factor which corresponds to the maximum value of v_{sa}. The functions $s_a(t)$ and $s_b(t)$ are obtained by comparing v_{sa} and v_{sb} with v_r. Both switched waveforms take a value of one when $v_{sa} > |v_r|$ and $v_{sb} > |v_r|$, respectively, and zero otherwise. If the conduction of GTO number 1, in Figure 16.2, is governed by the pulses of $s_a(t)$, and the GTO number 6 by $s_b(t)$, then the voltage $v_{ab}(t)$ is proportional to the PWM switching function $s_{ab}(t)$.

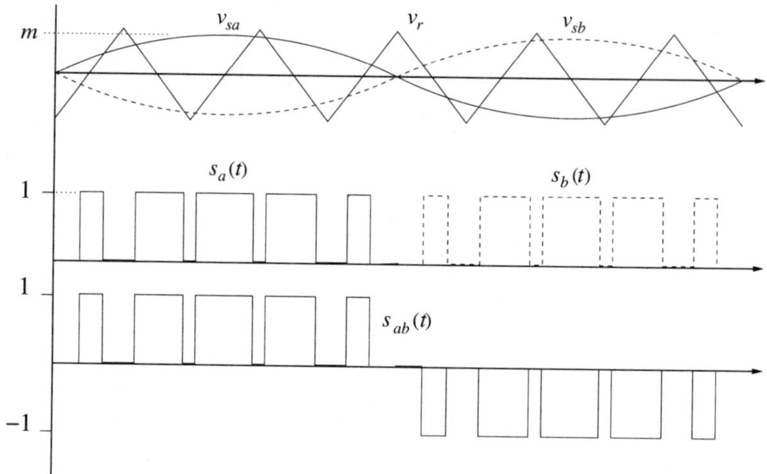

Figure 16.4 PWM converter with unipolar voltage switching

16.5.1 Selective harmonic elimination

The general switching function of Figure 16.5 is used to explain the selective harmonic elimination technique. It has the harmonic content given by the following Fourier series:

$$s(\omega t) = \sum_{n=-h}^{h} S_n e^{jn\omega_0 t}$$

where

$$S_n = -\frac{2j}{n\pi}\sum_{k=1}^{M}(-1)^{k+1}\cos n\beta_k \tag{16.16}$$

for $n = 1,\ 3,\ 5,\ 7,\ \ldots.$ and $S_{-n} = S_n^*$.

A harmonic elimination technique is used in order to calculate the M angles, β's. The required M non-linear equations can be obtained by fixing the magnitude of M harmonics to zero, i.e. $S_{n_1} = 0,\ S_{n_2} = 0, \ldots,\ S_{n_M} = 0$,

$$S_{n_i} = -\frac{2j}{n_i\pi}\sum_{k=1}^{M}(-1)^{k+1}\cos n_i\beta_k = 0 \tag{16.17}$$

for $i = 1, 2, \ldots, M$, where $n_1, n_2, \ldots, n_M$ are the harmonics to be eliminated.

Solving (16.17) for the M angles, using Newton's method, we have

$$\Delta\beta_k = -\left.\frac{\partial S_{n_i}}{\partial\beta_k}\right|^{-1} S_{n_i} \tag{16.18}$$

where

$$\frac{\partial S_{n_i}}{\partial\beta_k} = \frac{2j}{\pi}(-1)^{k+1}\sin n_i\beta_k \tag{16.19}$$

for $k = 1, 2, \ldots, M$, and $\Delta\beta_k = \beta_k^{(\mathrm{k}+1)} - \beta_k^{(\mathrm{k})}$, where k is an iteration counter.

The following equation is solved by iterations:

$$\begin{bmatrix} \Delta\beta_1 \\ \Delta\beta_2 \\ \vdots \\ \Delta\beta_M \end{bmatrix} = -\frac{\pi}{2j}\begin{bmatrix} \sin n_1\beta_1 & -\sin n_1\beta_2 & \cdots & (-1)^{M+1}\sin n_1\beta_M \\ \sin n_2\beta_1 & -\sin n_2\beta_2 & \cdots & (-1)^{M+1}\sin n_M\beta_M \\ \vdots & \vdots & \ddots & \vdots \\ \sin n_M\beta_1 & -\sin n_M\beta_2 & \cdots & (-1)^{M+1}\sin n_M\beta_M \end{bmatrix}^{-1}\begin{bmatrix} S_{n_1} \\ S_{n_2} \\ \vdots \\ S_{n_M} \end{bmatrix} \tag{16.20}$$

The selected angles must satisfy the following condition: $\beta_1 < \beta_2 < \cdots < \beta_M$.

For example, to eliminate the third harmonic, the calculated angle is $\beta_1 = 30°$. To eliminate the 5th, 7th, 11th, 13th and 17th harmonics, the calculated angles are $\beta_1 = 11.35°$, $\beta_2 = 17.27°$, $\beta_3 = 23.81°$, $\beta_4 = 34.88°$ and $\beta_5 = 37.27°$.

In three-phase PWM converter applications, the switching functions for the other two phases are calculated by shifting the switching function by 120° and −120°, respectively.

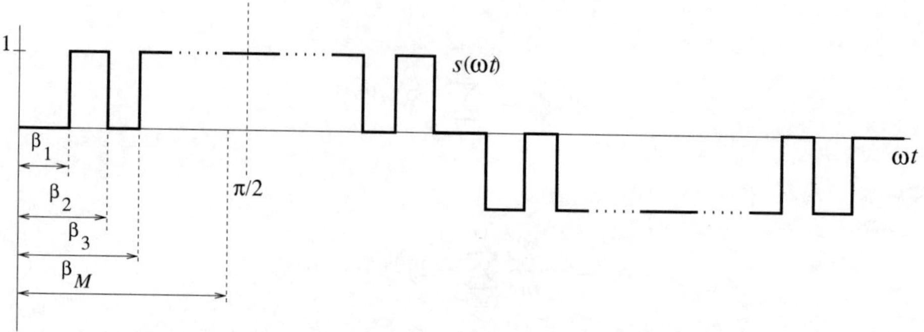

Figure 16.5 Switching function for selective elimination

The following MATLAB™ function may be used to obtain the switching functions of the PWM control with selective harmonic elimination:

```
function [Sab,Sbc,Sca,beta_M]=calc_PWM_select(M,h,harm_M,beta_M0,phase)
% S's : harmonic vectors 2h+1 of switching functions
% beta_M : is a vector which contains the M switching angles
% M : is the number of harmonics to be eliminated
% h : number of harmonics to be calculated
% harm_M : is a vector which contains the M harmonics to be eliminated
% beta_M0: init. conditions in rad: 0<beta1<beta2<....<betaM<pi/2
% phase : shift angle, in radians, of phase a
J      = zeros(M,M);
Sni    = zeros(M,1);
beta_M = zeros(M,1);
Sn     = zeros(h,1);
% initial condition for beta_M
beta_M=beta_M0;
error=1;
while error>1e-6
 for k=1:M
  for j=1:M
   J(k,j)= 2*i/pi*(-1)^(j+1)*sin(harm_M(k)*beta_M(j));
  end
 end
 for k=1:M
  Sni(k)=0;
  for j=1:M
   Sni(k)=Sni(k)+(-1)^(j+1)*cos(harm_M(k)*beta_M(j));
  end
  Sni(k)= -Sni(k)*2*i/(harm_M(k)*pi);
 end
 Dbeta_M  = -inv(J)*Sni;
 beta_M   = beta_M+Dbeta_M;
 error=norm(Dbeta_M);
end
for n=-h:h
 k=n+h+1;
 if n~=0
 Sn(k)=0;
 for j=1:M
  Sn(k)=Sn(k)+(-1)^(j+1)*cos(n*beta_M(j));
```

```
    end
    Sn(k)=-i*2*Sn(k)/(n*pi);
    end
  end
  for k=h+3:2:(2*h+1)
    Sn(k)=0;
    Sn(2*h+2-k)=0;
  end
  for n=-h:h
    Sab(n+h+1)=Sn(n+h+1)*exp(-i*n*(phase));
    Sbc(n+h+1)=Sn(n+h+1)*exp(-i*n*(phase+2*pi/3));
    Sca(n+h+1)=Sn(n+h+1)*exp(-i*n*(phase-2*pi/3));
  end
```

16.5.2 Multi-module PWM technique

Figure 16.6 shows a multi-pulse STATCOM configuration, which enables harmonic elimination at the electromagnetic interface level, without resorting to increased switching frequencies in the PWM control scheme. Here the object is to reduce the harmonic content of the output voltage by cascading the n_i PWM voltage converters using a suitable magnetic circuit arrangement. In general, harmonic cancellation is achieved by phase shifting the harmonic components to be cancelled in such a way that when the output voltages of individual units are added together, the targeted harmonic components are cancelled out. This can be achieved by using phase-shifting transformers. One simple way to represent this effect in harmonic domain calculations is to use a phase-shifted carrier, a technique which is explained below.

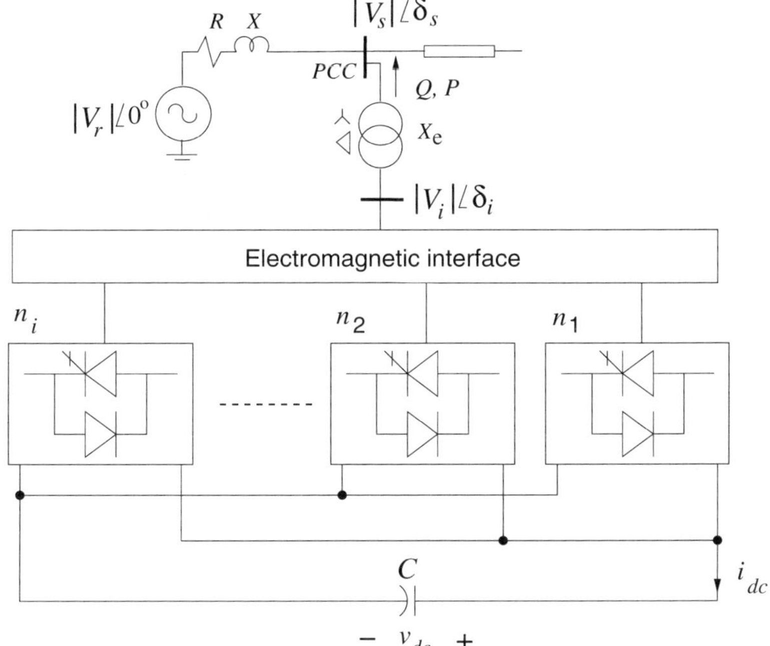

Figure 16.6 Multi-pulse STATCOM with n_i PWM converters

The single-phase PWM converter unit, n_i, uses unipolar voltage switching, as shown in Figure 16.4. The number of pulses per half cycle of the switching function $s_{n_i}(t)$ is given by the frequency modulation ratio m_{f},

$$m_{\mathrm{f}} = \frac{f_r}{f_s} \tag{16.21}$$

where the harmonics generated by the voltage converter i are given by

$$h = 2jm_{\mathrm{f}} \pm k \qquad j, k = 1, 2, 3, \ldots \tag{16.22}$$

which are the harmonics of $s_{n_i}(t)$.

Harmonic cancellation at the secondary side of the STATCOM transformer is possible by shifting the carrier signal of each individual PWM control scheme, say converter i, by δ_{r_i} rad, and using equal modulation signals, v_s, for all the PWM converters. The shifted angles for the n_i carrier signals are given by

$$\delta_{r_i} = 2\pi\,(i-1)\,/n_i \tag{16.23}$$

Using n_i PWM converters in parallel, the harmonics generated by the STATCOM are given by the following sequence,

$$h = 2jn_i m_{\mathrm{f}} \pm k \qquad j, k = 1, 2, 3, \ldots \tag{16.24}$$

and the magnitudes of the harmonics are given by

$$S_{\mathrm{STATCOM}_h} = \frac{S_{n1_h} + S_{n2_h} + \cdots + S_{ni_h}}{n_i} \tag{16.25}$$

where the magnitude of the fundamental frequency component has a value of m.

Figure 16.7 shows the waveforms and harmonic content of the switching function, $s_{ab}(t)$, for the multi-pulse STATCOM, with $f_r = 250$ Hz, $f_s = 50$ Hz and $m = 0.8$. Cases are presented for one PWM converter, two PWM converters and three PWM converters. The results were obtained using the following MATLAB™ steps:

```
>> [Sab,Sbc,Sca,Sabt]=calc_PWM_multipulse(0,3,50,0.8,5,100);
>> plot(Sabt)
```

The MATLAB™ function used to obtain the three switching functions of the STATCOM using the multi-module PWM technique is:

```
function [Sab,Sbc,Sca,Sabt]=calc_PWM_multipulse(Fis,nc,f,m,hr,h)
% Fis    : is the phase angle in rad of the modulation signal fs.
% nc     : is the number of PWM converters
% Fir    : is the phase angle in rad of the carrier signal fr.
% f      : is the frequency in Hz of the modulation signal fs.
% m      : is the modulation index, i.e. fs=m*sin(wt+Fis)
% hr     : is the harmonic frequency of the carrier signal, e.g for
%          250 Hz hr=5 for f=50 Hz.
% h      : is the number of harmonics to use in the vector UpwmF
% Sab,Sbc,Sca: harmonic vectors of the obtained switching functions
%           [-h .... -1 0 1 .... h] transposed
% Sabt   : is the switching function in the time domain for Sab
 Fir=0;
 [Sab,Sabt]=calc_PWM_unipolar(Fis,Fir,f,m,hr,h);
```

```
for k=2:nc
 Fir=2*pi*(k-1)/nc;
 [Sk,Skt]=calc_PWM_unipolar(Fis,Fir,f,m,hr,h);
 Sab=Sab+Sk;
 Sabt=Sabt+Skt;
end
Sab=Sab/nc;
Sabt=Sabt/nc;
for k=1:h
 Sab(h+1-k)=Sab(h+1+k)';
end
for n=-h:h
 Sbc(n+h+1)=Sab(n+h+1)*exp(-i*n*(2*pi/3));
 Sca(n+h+1)=Sab(n+h+1)*exp(-i*n*(-2*pi/3));
end
```

```
function [Spwm,Spwmt]=calc_PWM_unipolar(Fis,Fir,f,m,hr,h)
% Fis   : is the phase angle in rad of the modulation signal fs.
% Fir   : is the phase angle in rad of the carrier signal fr.
% f     : is the frequency in Hz of the modulation signal fs.
% m     : is the modulation index, i.e. fs=m*sin(wt+Fis)
% hr    : is the harmonic frequency of the carrier signal, e.g
%         for 250 Hz hr=5 for f=50 Hz.
% h     : is the number of harmonics to use in the vector Spwm
% Spwm  : is a harmonic vector of the obtained switching function
%         [-h .... -1 0 1 .... h]'
% Spwmt : is the switching function in the time domain
p_trian=63;  % odd points per triangle in the carrier signal
tril=triang(p_trian);
tril=[0;tril];
fr=tril;
for k=1:(2*hr-1)
 fr=[fr;(-1)^(k)*tril];
end
points=size(fr);
points=points(1);
t=0:1/f/(points-1):1/f;
fs=m*sin(2*pi*f.*t+Fis);
points_rad=p_trian/(2*pi);
points_shifted=round(points_rad*abs(Fir));
if sign(Fir)==1
 fr=[fr(points_shifted:points);fr(1:points_shifted-1)];
end
if sign(Fir)==-1
 fr=[fr(points-points_shifted:points);fr(1:points-points_shifted-1)];
end
for k=1:points
 if abs(fs(k))<abs(fr(k))
  Upwm(k)=0;
 else
  Upwm(k)=sign(fs(k));
 end
end
UpwmF=fft(Upwm)/points;
UpwmF=fftshift(UpwmF);
x=round(points/2)-points/2;
```

```
if x==0.5
 centre=round(points/2);
else
 centre=round(points/2)+1;
end
Spwm=UpwmF(centre-h:centre+h);
Spwmt=Upwm;
```

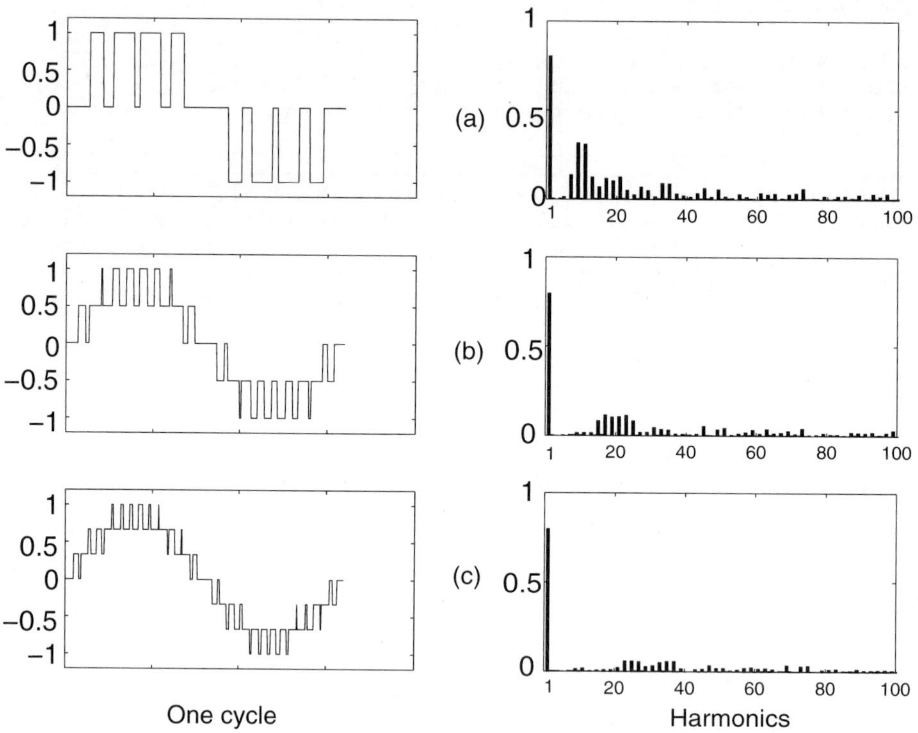

Figure 16.7 Switching functions of a STATCOM with: (a) one PWM converter, (b) two PWM converters, (c) three PWM converters

16.6 Implementation of the Thévenin Equivalent of the STATCOM

The Thévenin equivalent of the STATCOM presented in Section 16.4, given by equation (16.13), incorporates the switching functions $s_{ab}(t)$, $s_{bc}(t)$ and $s_{ca}(t)$. These switching functions may be obtained by using either the harmonic elimination technique or the multi-module technique. In fact any PWM techniques may be used as long as the switching function is amenable to a harmonic vector representation. In this section the harmonic selection technique was implemented in MATLAB™ code to obtain the Thévenin equivalent of the STATCOM.

```
function [Eth,Zth,beta_M]=calc_STATCOM(M,h,harm_M,beta_M0,phase,E,C,w)
% Eth : Is the three-phase constant voltage given by the inverter
% Zth : Is the imp. equiv. of the PWM including the capacitor
% beta_M : Is a vector which contains the angles b (0<b1<b2<...bM)
% M : Number of harmonics to be eliminated in the PWM
% h : Number of harmonics to be considered
% harm_M  : Vector which contains the M harmonics to be eliminated
% beta_M0 : Vector which contains the guess values of b (in radians)
% phase   : Shift angle of E with respect to terminals
% E       : DC capacitor voltage
% C       : Capacitor in the DC side, farads
[Sab,Sbc,Sca,beta_M]=calc_PWM_select(M,h,harm_M,beta_M0,phase);
Sab =calc_Fm(Sab,h); Sbc =calc_Fm(Sbc,h); Sca =calc_Fm(Sca,h);
Xc=1/(w*C); Zcap=form_Zm(0,-Xc,h);
O=zeros(2*h+1,2*h+1);
Edc=zeros(2*h+1,1);
Edc(h+1)=E;
S1=[ Sab O O; O Sbc O; O O Sca ];
S1=sparse(S1);
S2=[Sab Sbc Sca; Sab Sbc Sca; Sab Sbc Sca];
S2=sparse(S2);
ZCAP=[ Zcap O O; O Zcap O; O O Zcap ];
ZCAP=sparse(ZCAP);
Zth=S1*ZCAP*S2;
EDC=[Edc;Edc;Edc];
Eth=S1*EDC;
```

16.6.1 Analysis of the equivalent model

The three-phase circuit of Figure 16.1 was used to test the new model under different reactive power conditions, with the constraint $\delta_{is} = 0$, since no active power exchange exists. The solution is reached by iteration, where the source E_{Th} is shifted by means of switching functions until $\delta_{is} < \epsilon$ is satisfied.

The system data are as follows: $|V| = 1$ p.u., $R = 0.0033\,\Omega$, $X = 0.1571\,\Omega$, $R_{\mathrm{e}} = 0.01\,\Omega$ and $X_{\mathrm{e}} = 0.4712\,\Omega$ at 50 Hz.

Harmonic Thévenin impedance

Figure 16.8 shows the structure of $\mathbf{Z}_{\mathrm{Th}}$ for the case when $C = 1000\,\mu$F and the conduction angle of the switching functions is 120°. Fifteen harmonics are considered in the analysis. The matrix structure shows that the STATCOM impedance has a very strong inter-coupling between phases and between harmonics. In this harmonic domain application $\mathbf{Z}_{\mathrm{Th}}$ is a skew Hermitian matrix, i.e. $\mathbf{Z}_{\mathrm{Th}} = -\mathbf{Z}^{*}_{\mathrm{Th}}$, where $*$ represents the transposed complex conjugate operation.

STATCOM response

Two cases are considered below, corresponding to two different operating conditions. They represent cases when the STATCOM absorbs and supplies reactive power, respectively. To achieve such operating conditions, the following DC voltage source values are used:

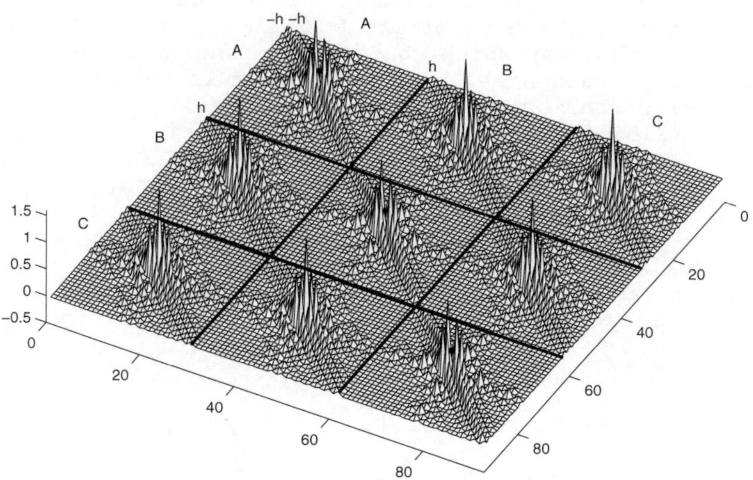

Figure 16.8 PWM converter equivalent impedance, $\mathbf{Z}_{\mathrm{Th}}$, for $C = 1000\ \mu F$ and conduction angles of $120°$

$E_{\mathrm{dc}} = 0.7$ and 1.2, respectively. In this case the PWM control was set to eliminate only the third harmonic.

Absorbing reactive power: In this test case, a value of E_{dc} equal to 0.7 p.u. enables the STATCOM to absorb reactive power equal to 0.2836 p.u. Table 16.1 shows the harmonic voltages of $v_s(t)$, $v_i(t)$, and current $i_s(t)$ for two STATCOM representations, namely Thévenin equivalent and voltage source. Only the results for phase a are presented since balanced conditions were assumed to exist in this example.

As discussed in Section 16.4, in the latter case the effect of the capacitor is not explicitly represented, a fact that introduces modelling constraints and approximations. The numeric values of $v_s(t)$, $v_i(t)$, and current $i_s(t)$, given in Table 16.1 for both STATCOM representations, give a clear indication of the errors expected when the simpler model is used. A sufficiently large number of harmonics are included, i.e. up to the 49th, and THD values are also given. The voltage and current waveforms corresponding to this example are shown in Figure 16.9.

Supplying reactive power: Increasing the value of E_{dc} to 1.2 p.u. enables the STATCOM to supply reactive power equal to 0.7048 p.u. Table 16.2 shows similar information to the case when the STATCOM absorbs reactive power. The corresponding voltage and current waveforms are shown in Figure 16.10.

By comparing Tables 16.1 and 16.2, it can seen that the STATCOM operating mode affects very significantly the harmonic magnitudes of voltage $v_s(t)$. When the capacitor effect is neglected the harmonic pattern is rather predictable but this is clearly not the case when the capacitor effect is explicitly taken into account. It is shown in Figure 16.9(a) that when the STATCOM absorbs reactive power, the voltage waveforms at the terminals of the PWM converter are formed of rounded sections, with the cusps pointing upside down, but when

Table 16.1 Harmonic content for the case when the STATCOM absorbs reactive power

h	$i_s(t)$ with Z_{Th}	$v_s(t)$ with Z_{Th}	$v_i(t)$ with Z_{Th}	$i_s(t)$ without Z_{Th}	$v_s(t)$ without Z_{Th}	$v_i(t)$ without Z_{Th}
1	100	100	100	100	100	100
5	8.1105	2.4970	12.2514	13.5361	4.0927	20.0000
7	10.9782	4.7317	23.2162	6.9062	2.9234	14.2857
11	2.9648	2.0081	9.8527	2.7967	1.8603	9.0909
13	2.7366	2.1905	10.7477	2.0024	1.5741	7.6923
17	1.3121	1.3734	6.7385	1.1709	1.2037	5.8824
19	1.2299	1.4388	7.0597	0.9374	1.0770	5.2632
23	0.7329	1.0378	5.0922	0.6397	0.8897	4.3478
25	0.6965	1.0722	5.2607	0.5414	0.8185	4.0000
29	0.4662	0.8325	4.0845	0.4024	0.7056	3.4483
31	0.4475	0.8541	4.1907	0.3521	0.6601	3.2258
35	0.3217	0.6932	3.4013	0.2762	0.5847	2.8571
37	0.3111	0.7087	3.4771	0.2472	0.5531	2.7027
41	0.2336	0.5897	2.8935	0.2013	0.4991	2.4390
43	0.2275	0.6025	2.9560	0.1830	0.4759	2.3256
47	0.1673	0.4841	2.3754	0.1532	0.4354	2.1277
49	0.1656	0.4997	2.4516	0.1409	0.4176	2.0408
THD	14.4089	6.8730	33.7224	15.6922	6.1422	30.0153

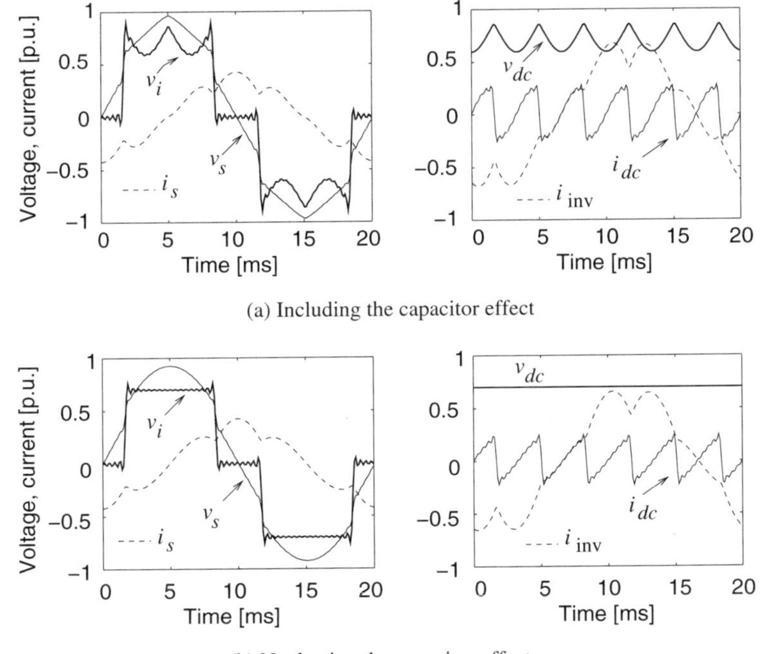

Figure 16.9 Waveform results for the case when the STATCOM absorbs reactive power

Table 16.2 Harmonic content for the case when the STATCOM supplies reactive power

h	I_s with Z_{Th}	V_s with Z_{Th}	V_i with Z_{Th}	I_s without Z_{Th}	V_s without Z_{Th}	V_i without Z_{Th}
1	100	100	100	100	100	100
5	24.3781	9.3490	30.3995	16.3802	6.1214	20.000
7	1.3824	0.7422	2.4134	8.3573	4.3724	14.2857
11	3.0817	2.6000	2.6000	3.3844	2.7824	9.0909
13	1.0068	1.0038	3.2641	2.4231	2.3544	7.6923
17	1.1440	1.4916	4.8502	1.4170	1.8004	5.8824
19	0.5557	0.8098	2.6331	1.1344	1.6109	5.2632
23	0.5902	1.0412	3.3857	0.7741	1.3307	4.3478
25	0.3456	0.6627	2.1549	0.6552	1.2243	4.0000
29	0.3596	0.7998	2.6006	0.4869	1.0554	3.4483
31	0.2351	0.5590	1.8175	0.4261	0.9873	3.2258
35	0.2428	0.6519	2.1198	0.3343	0.8745	2.8571
37	0.1710	0.4852	1.5776	0.2991	0.8272	2.7027
41	0.1776	0.5585	1.8159	0.2436	0.7465	2.4390
43	0.1322	0.4361	1.4182	0.2215	0.7118	2.3256
47	0.1500	0.5409	1.7587	0.1854	0.6512	2.1277
49	0.1197	0.4498	1.4626	0.1706	0.6246	2.0408
THD	24.6811	10.1359	32.9581	18.9893	9.1867	30.0153

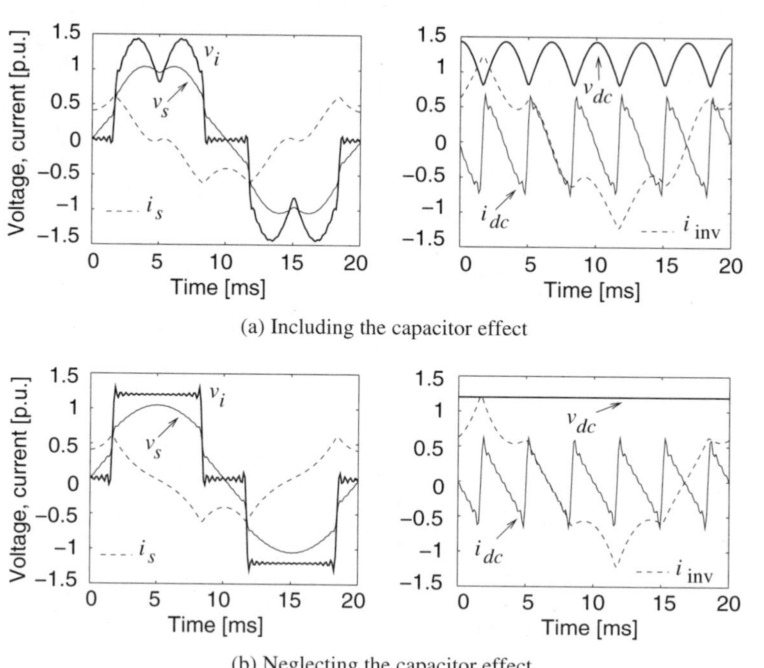

(a) Including the capacitor effect

(b) Neglecting the capacitor effect

Figure 16.10 Waveform results for the case when the STATCOM supplies reactive power

it generates reactive power, the cusps are more prominent and point upwards as shown in Figure 16.10(a). The voltage at the point of common coupling looks more sinusoidal but it still contains a large amount of harmonic distortion. The converter current is non-sinusoidal. In contrast, if the DC capacitor is not explicitly taken into account, and a DC voltage source is used instead, i.e. $\mathbf{Z}_{\mathrm{Th}} = 0$, the voltage at the terminals of the converter is given by a train of square pulses and the voltage at the point of common coupling will be largely sinusoidal, as shown in Figures 16.9(b) and 16.10(b).

16.7 Active Filters

Active filters are the new trend in harmonic filtering technology. They make use of state-of-the-art power electronic switches and advanced control techniques. The basic principle of operation of an active filter is to inject a suitable non-sinusoidal voltage and current into the system in order to achieve a clean voltage and current waveforms at the point of filtering.

The most promising active filter technology uses the PWM voltage source inverter as its core component. In theory, the main difference between the PWM STATCOM and the active filter lies in the control blocks used by both applications. The STATCOM's primary concern is to provide reactive power support and adaptive voltage regulation whereas that of the active filter is to provide adaptive filtering action. Nevertheless, the great progress made in the development of advanced digital signal processors and sophisticated algorithms is enabling both applications to converge. At the research level at least, prototypes are being developed where a PWM STATCOM has the ability to provide reactive power support, adaptive voltage regulation and harmonic filtering.

The basic principles of operation of active filters are outlined below together with their general classification [9]. Only filters that are based on PWM voltage source inverters are addressed here since they are a more popular alternative than current source inverters. Depending on control strategy and configuration of the PWM converter, many different kinds of active filters may be realised.

16.7.1 Basic principles of operation

Figure 16.11 illustrates the case when a general non-sinusoidal voltage source composed of the fundamental and harmonic frequencies, i.e. $V_{\mathrm{f}} + V_{\mathrm{h}}$, supplies a non-linear load. The load current may be represented as the vector sum of the active part of the current at fundamental frequency, I_{r}, the reactive part of the current at fundamental frequency, I_{i}, and the harmonic currents I_{h}. An ideal active filter would remove all harmonics from the line current and from the voltage supply, i.e. I_{h} and V_{h}, respectively. Also, an ideal active filter would supply reactive current compensation, I_{i}, without affecting the flow of active load current I_{r}. In principle, such a filter could be realised by ideal harmonic current and voltage generators and by an ideal reactive power compensator. Under such a condition, the non-linear load is supplied from a perfect sinusoidal voltage, V_{t}, and draws a perfect sinusoidal current, I_{r}, from the voltage source.

16.7.2 Shunt active filter

Provided a comprehensive control function is included in the shunt active filter, the device may be used to replace the function of a conventional voltage and VAR compensator, but

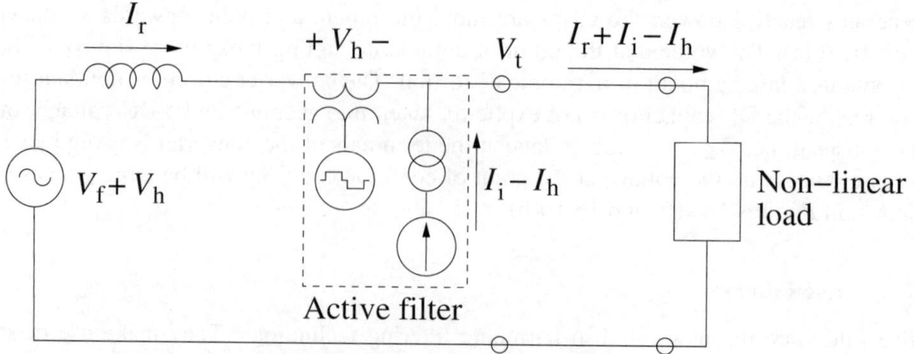

Figure 16.11 Basic concept of active filters

with the added advantage that it can perform harmonic current compensation. Shunt active filters are becoming quite popular owing to their light construction and, above all, their expandable design. They are controlled with ease and are mostly used in low- and medium-power applications.

Figure 16.12 shows the schematic representation of a shunt active filter and the external network, where the filter is connected between an equivalent voltage source and a non-linear load. At this high level of representation it is not possible to distinguish between an active filter and a PWM STATCOM. As explained above, the difference between the two applications may lie in the control scheme.

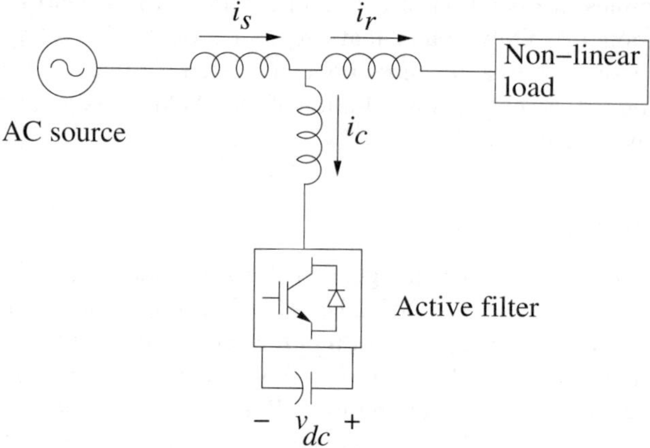

Figure 16.12 Shunt active filter

Figure 16.13 shows the basic configuration and control scheme of shunt active filters. The filter consists of a PWM voltage source inverter and a control scheme. It is connected in shunt with the distribution system in order to compensate the load harmonic currents and reactive power. It should be mentioned that the same configuration but connected in series with the distribution system can be used for voltage distortion compensation.

The three-phase currents from the load and the active filter are transformed into pq orthogonal components [10]. For instance, the current i_{abc_L} is decomposed into components i_p and i_q as shown below:

$$i_p = i_{\bar{p}} + i_{\tilde{p}} \tag{16.26}$$

$$i_q = i_{\bar{q}} + i_{\tilde{q}} \tag{16.27}$$

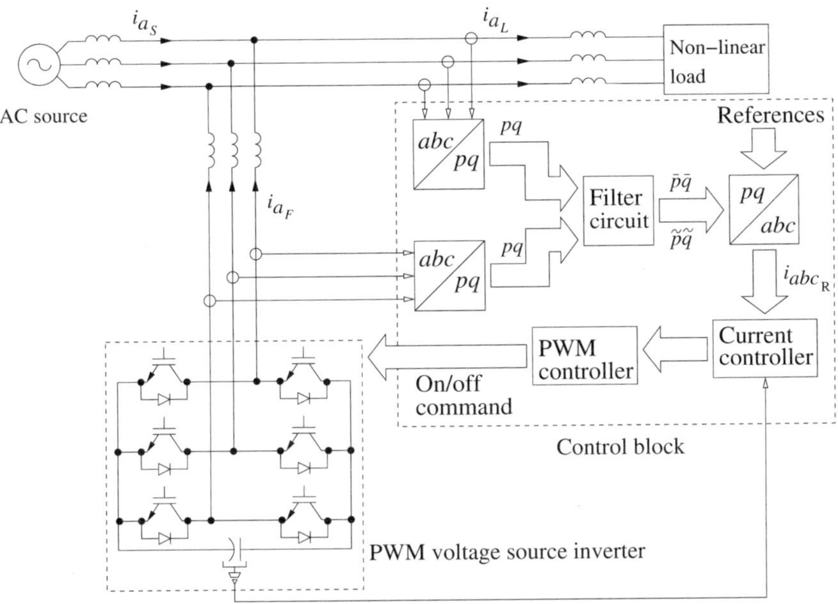

Figure 16.13 Basic configuration and control scheme of shunt active filters

where:

$i_{\bar{p}}$ is the DC component of i_p associated with the fundamental frequency active component of the load current.

$i_{\bar{q}}$ is the DC component of i_q associated with the fundamental frequency reactive component of the load current.

$i_{\tilde{p}}$ is the AC component of i_p associated with the harmonic active components of the load current.

$i_{\tilde{q}}$ is the AC component of i_q associated with the harmonic reactive components of the load current.

Also, the components of i_p and i_q may be related to sequence quantities as shown in Table 16.3, where the AC components are further separated into two terms,

$$i_{\tilde{p}} = i_{\tilde{p}_{2\omega_0}} + i_{\tilde{p}_{h\omega_0}}$$

$$i_{\tilde{q}} = i_{\tilde{q}_{2\omega_0}} + i_{\tilde{q}_{h\omega_0}}$$

Table 16.3 Parameters in pq coordinates

pq parameters	Physical interpretation in abc coordinates
$i_{\overline{p}}$ and $i_{\overline{q}}$	Positive sequence components of the load
$i_{\underset{p}{\sim}}$ and $i_{\underset{q}{\sim}}$ with $\omega = 2\omega_0$	Negative sequence components of the load
$i_{\underset{p}{\sim}}$ and $i_{\underset{q}{\sim}}$ with $\omega = h\omega_0 \pm \omega_0$	Harmonic components of the load

The output signal of the abc/pq conversion circuit is applied to a filter circuit whose output gives the DC and AC components of i_p and i_q. The filter circuit consists of a combination of high-pass filters and a control algorithm, where suitable control parameters are selected according to the frequency band required for compensation. The conditioned and the reference parameters are compared, and transformed back to abc components in a conversion circuit. The currents, $i_{abc_{\mathrm{R}}}$, are sent to the current controller. The signal from the DC capacitor voltage is also used by the current controller circuit. The PWM control circuit is responsible for sending the switch-on/off signals to the GTO valves.

The active filter performance is strongly influenced by the characteristics of the high-pass filters since the control commands are carried through such filters. The high-pass characteristics of the filters are application dependent and they are optimised for each application.

Depending on requirements and control strategy and configuration, active filters can be used to eliminate harmonic currents and voltages, to regulate terminal voltage, to suppress voltage flicker, and to improve voltage balance in three-phase systems.

The flexibility of the control block is influenced by the choice of electronic technology: analogue control, digital control embedded in microcontrollers or digital signal processor (DSP) controllers. With the rapid evolution of microelectronics, the implementation of complex on-line algorithms for the control of active filters has made it possible to provide fast correction actions almost on a cycle-by-cycle basis, even when non-linear loads experience important disturbances.

Of the four components demanded by the non-linear load, the source must always provide the $i_{\overline{p}}$ component, and the active filter must provide some or all of the remaining components depending on the application of the active filter. Relevant components of i_p and i_q are chosen as reference in the control block circuit, and they vary from application to application. Table 16.4 shows the basic parameters required for each one of the applications listed in the table. If required, a combination of the basic sets may also be used.

Table 16.4 Control parameters and application

Type of control	Application	Control parameters
A	Harmonic current compensation	$i_{\underset{p}{\sim}}$ and $i_{\underset{q}{\sim}}$
B	Reactive power compensation	$i_{\overline{p}}$ and $i_{\overline{q}}$
C	Flicker compensation	$i_{\underset{p}{\sim}}$, $i_{\underset{q}{\sim}}$ and $i_{\overline{q}}$
D	Unbalanced voltage control	$i_{\underset{p}{\sim}}$, $i_{\underset{q}{\sim}}$ and $i_{\overline{q}}$

16.7.3 *Series active filter*

This filter is used to eliminate voltage harmonics, phase imbalances and to provide voltage regulation at the point of connection [9]. This configuration is becoming quite popular in distribution systems applications where it is termed the dynamic voltage restorer (DVR). The DVR may be seen as a PWM STATCOM connected in series with the AC system, as shown in Figure 16.14. In distribution systems, the DVR has been used to control the voltage applied to a load by injecting a voltage of variable amplitude and phase angle, with its main application being voltage sag (dip) support.

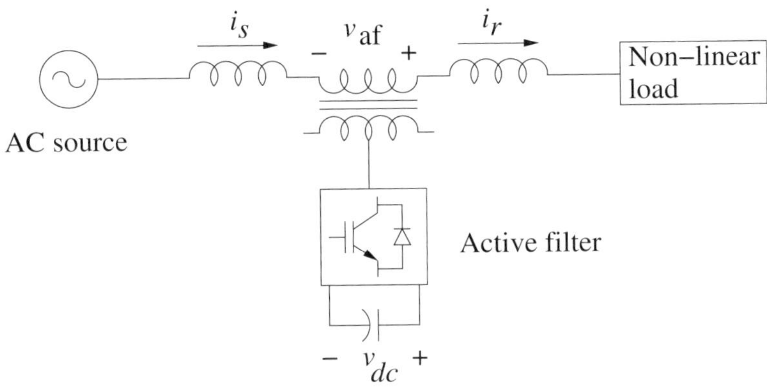

Figure 16.14 Series active filter

16.7.4 *Hybrid filter*

This filter combines active and passive shunt filters. Many combinations are possible in hybrid active filter design, with one possible arrangement shown in Figure 16.15 where the active filter is connected in series with the system.

Active filters are still characterised by their relatively high cost, compared to the cost of passive filters. This factor has somehow limited the widespread incorporation of active filters in industry except in special cases. Hybrid active filters provide a viable alternative to the use of active filters only, since the active unit may be sized to only a fraction of the total compensating power, thus limiting the overall cost.

16.7.5 *Universal active filter*

This filter is also known as the unified power quality conditioner. It may be considered an ideal power quality compensator in the sense that it eliminates voltage and current harmonics, in addition to providing compensation for voltage sags, power factor and nodal voltage magnitude, enabling the supply of clean power to sensitive loads. The filter consists of two PWM voltage source inverters connected back to back and sharing a capacitor. One of the voltage source inverters is connected in shunt with the network whereas the other is connected in series. Figure 16.16 shows the arrangement of the universal active filter in diagrammatic form.

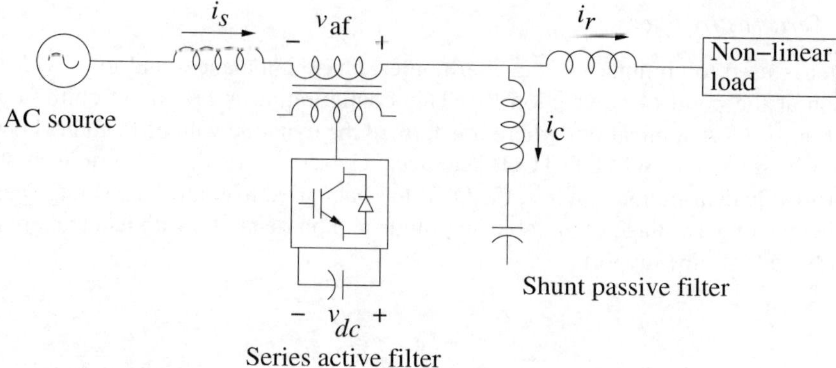

Figure 16.15 Hybrid filter

A similar configuration is finding favour in high-voltage transmission but with the emphasis on active and reactive power flow control, as opposed to power quality. The device is termed the unified power flow controller. It should be mentioned that the unified power flow controller is designed to switch at the fundamental frequency because of switching loss considerations.

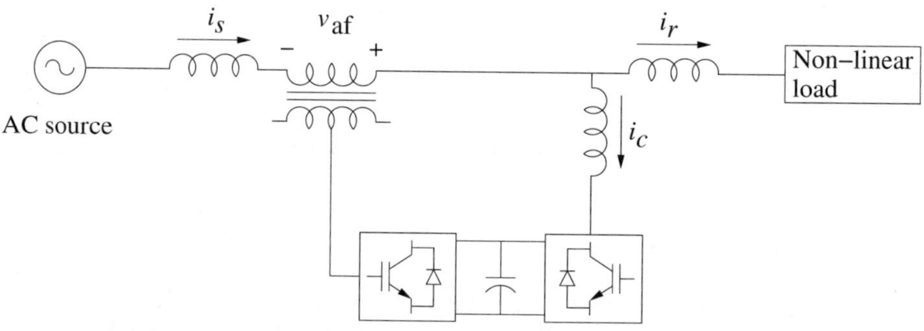

Figure 16.16 Universal Active Filter

16.8 Summary

In this chapter a flexibly harmonic domain model of a STATCOM intended for custom power studies has been introduced. It represents a three-phase, six-pulse PWM converter where the capacitor effect is represented explicitly. The analysis and results presented demonstrate that the three-phase equivalent impedance of the STATCOM shows a very strong cross-coupling between phases and between harmonics.

It should be noted that the currents on the DC and AC sides of the converter do not differ by much in cases when the capacitor is explicitly represented and when it is not, but the voltage waveforms are very different. An interesting observation emanating from the simulation studies concerns the very significant influence that the size of the STATCOM capacitor has on governing the STATCOM harmonic interactions with the network. It is clear from these results that the STATCOM capacitor must be represented explicitly for more realistic harmonic studies.

An overview of active filters has been presented, emphasising the similarities that exist between the PWM STATCOM and active filters.

16.9 Bibliography

1. B.M. Han, G.G. Karady, J.K. Park, "Interaction Analysis Model for Transmission Static Compensator with EMTP", *IEEE Transactions on Power Delivery*, Vol. 13, No. 4, October 1998, pp. 1297–1302.
2. J.E. Hill, W.T. Norris, "Exact Analysis of a Multipulse Shunt Converter Compensator or Statcon. Part 1: Performance", *IEE Proceedings General Transmission and Distrib*ution, Vol. 144, No. 2, March 1997, pp. 213–218.
3. N.R. Raju, S.S. Venkata, V.V. Sastry, "The Use of Decoupled Converters to Optimize the Power Electronics of Shunt and Series AC System Controllers", *IEEE Transactions on Power Delivery*, Vol. 12, No. 2, April 1997, pp. 895–900.
4. H.S. Patel, R.G. Hoft, "Generalized Techniques of Harmonic Elimination and Voltage Control in Thyristor Inverter: Part I-Harmonic Elimination", *IEEE Transactions on Industry Applications*, Vol. IA-9, Vol. 9, No. 3, May/June 1973, pp. 310–317.
5. J.G. Mayordomo, M. Lopez, A. Asensi, L.F. Beites, J.M. Rodriguez, "A General Treatment of Traction PWM Converters for Load Flow and Harmonic Penetration Studies", *Proceedings of the IEEE-ICHQP 1998*, Athens, Greece, 14–16 October, pp. 685-692.
6. G. Joos, L. Moran, P. Ziogas, "Performance Analysis of a PWM Inverter VAR Compensator", *IEEE Transactions on Power Electronics*, Vol. 6, No. 3, July 1991, pp. 380–391.
7. D. Shen, W. Liu, Z. Wang, "Study on the Operation Performance of STATCOM under Unbalanced and Distorted System Voltage", *Proceedings of 2000 IEEE Power Engineering Society Winter Meeting*, January 2000, Singapore, paper 0-7803-5938-0/00.
8. S. Nilsson, "Special Application Considerations for Custom Power Systems", 1999 *IEEE-PES Winter Meeting Proceedings*, pp. 1127–1131.
9. B. Singh, K. Al-Haddad, A. Chamdra, "A Review of Active Filters for Power Quality Improvement", *IEEE Transactions on Industry Electronics*, Vol. 46, No. 5, October 1999, pp. 960–971.
10. H. Akagi, Y. Kanazawa, A. Nabae, "Instantaneous Reactive Power Compensators Comprising Switching Devices without Energy Storage Components", *IEEE Transactions on Industry Applications*, Vol IA-20, No. 3, May/June 1984, pp. 625–630.
11. M. Takeda, T. Aritsuka, "Practical Applications of Active Filters for Power Conditioning in Distribution Networks", *Proceedings of IEEE-ICHQP 1998,* Athens, Greece, 14–16 October, pp. 304–309.
12. A. Nava-Segura, M. Carmona, "Instantaneous Active and Reactive Current Compensation for AC/DC Three-Phase Harmonic Generating Loads", *Proceedings of IEEE-ICHQP-1998*, Athens, Greece, 14–16 October, pp. 521–526.
13. M. Takeda, K. Ikeda, Y. Tominaga, "Harmonic Current Compensation with Active Filter", *IEEE-IAS Annual Meeting*, 1997, pp. 808–815.
14. L. Moran, P.D. Ziodas, G. Joos, "Analysis and Design of a Novel 3-ϕ Solid-State Power Factor Compensator and Harmonic Suppressor System", *IEEE Transactions on Industry Applications*, Vol. 25, No. 4, July/August 1989, pp. 609–619.
15. F.Z. Peng. H. Akagi, A. Nabae, "A New Approach to Harmonic Compensation in Power Systems - A Combined System of Shunt Passive and Series Active Filter", *IEEE*

Transactions on Industry Applications, Vol. 26, November/December 1990, pp. 983–990.

16. L. Xu, O. Anaya, V.G. Agelidis, E. Acha, "Development of Prototype Custom Power Devices for Power Quality Enhancement", *Proceedings of IEEE-ICHQP 2000*, Orlando, USA, pp. 775–783.
17. M. Madrigal, O. Anaya, E. Acha, J.G. Mayordomo, R. Asensi, "Single-Phase PWM Converters Array for Three-Phase Reactive Power Compensation. Part I: Time Domain Studies", *Proceedings of IEEE-ICHQP 2000*, Orlando, USA, pp. 451–547.
18. M. Madrigal, E. Acha, J.G. Mayordomo, R. Asensi, A. Hernandez, "Single-Phase PWM Converters Array for Three-Phase Reactive Power Compensation. Part II: Frequency Domain Studies", *Proceedings of IEEE-ICHQP 2000*, Orlando, USA, pp. 645–651.

Index